STEAM-H: Science, Technology, Engineering, Agriculture, Mathematics & Health

STEAM-H: Science, Technology, Engineering, Agriculture, Mathematics & Health

The goal of this interdisciplinary series is to highlight the wealth of recent advances in the pure and applied sciences made by researchers collaborating between fields of study where mathematics is a core focus. As we continue to make fundamental advances in various scientific disciplines, the most powerful applications will increasingly be revealed by an interdisciplinary approach. This series will serve as a catalyst for these researchers, to develop novel applications of, and approaches to, the mathematical sciences. As such, we expect this series to become a national and international reference in STEAM-H education and research. Because the series will be interdisciplinary by design, a major focus will be on scientists and mathematicians developing novel methodologies and research techniques that have benefits beyond a single research community.

This approach seeks to connect researchers from across the globe, united in the common language of the mathematical sciences. Thus, volumes in this series will be suitable for both students and researchers in a variety of interdisciplinary fields, such as: mathematics as it applies to engineering; physical chemistry and material sciences; environmental, health, behavioral and life sciences; nanotechnology and robotics; computer science, signal/image processing, and machine learning; as well as finance, economics, operations research, and game theory.

Originating from the STEAM-H Lecture Series at Virginia State University, all talks are given by invited researchers only, recognized as experts in their respective fields. The weekly yearlong seminar is a dynamic forum for the best faculty to present contributions reflecting the most recent advances in scientific knowledge, delivered in a standardized, self-contained and pedagogically-oriented manner, to a multidisciplinary audience of faculty and students, with the objective of fostering student interest and participation in the STEAM-H disciplines as well as genuine interdisciplinary collaborative research. Plenary talks are followed by a one-hour discussion post-presentation on research collaboration and visits, mentorships, internships, and grant proposals.

More information about this series at http://www.springer.com/series/15560

Bourama Toni
Editor

New Trends and Advanced Methods in Interdisciplinary Mathematical Sciences

Springer

Editor
Bourama Toni
Department of Mathematics & Economics
Virginia State University
Petersburg, VA, USA

ISSN 2520-193X ISSN 2520-1948 (electronic)
STEAM-H: Science, Technology, Engineering, Agriculture, Mathematics & Health
ISBN 978-3-319-55611-6 ISBN 978-3-319-55612-3 (eBook)
DOI 10.1007/978-3-319-55612-3

Library of Congress Control Number: 2017943497

Printed on acid-free paper

This Springer imprint is published by Springer Nature
The registered company is Springer International Publishing AG
The registered company address is: Gewerbestrasse 11, 6330 Cham, Switzerland

Preface

The multidisciplinary STEAM-H series (Science, Technology, Engineering, Agriculture, Mathematics, and Health) brings together leading researchers to present their own work in the perspective to advance their specific fields and in a way to generate a genuine interdisciplinary interaction transcending disciplinary boundaries. All chapters therein were carefully edited and peer-reviewed; they are reasonably self-contained and pedagogically exposed for a multidisciplinary readership.

Contributions are invited only and reflect the most recent advances delivered in a high-standard, self-contained way. The goals of the series are as follows:

1. To enhance multidisciplinary understanding between the disciplines by showing how some new advances in a particular discipline can be of interest to the other discipline or how different disciplines contribute to a better understanding of a relevant issue at the interface of mathematics and the sciences
2. To promote the spirit of inquiry so characteristic of mathematics for the advances of the natural, physical, and behavioral sciences by featuring leading experts and outstanding presenters
3. To encourage diversity in the readers' background and expertise while structurally fostering genuine interdisciplinary interactions and networking

Current disciplinary boundaries do not encourage effective interactions between scientists; researchers from different fields usually occupy different buildings on university campuses, publish in journals specific to their field, and attend different scientific meetings. Existing scientific meetings usually fall into either small gatherings specializing on specific questions, targeting specific and small group of scientists already aware of each other's work and potentially collaborating, or large meetings covering a wide field and targeting a diverse group of scientists but usually not allowing specific interactions to develop due to their large size and a crowded program. Traditional departmental seminars are becoming so technical as to be largely inaccessible to anyone who did not coauthor the research being presented. Here contributors focus on how to make their work intelligible and accessible to a diverse audience, which in the process enforces mastery of their own field of expertise.

This volume, as the previous ones, strongly advocates multidisciplinarity with the goal to generate new interdisciplinary approaches, instruments, and models including new knowledge, transcending scientific boundaries to adopt a more holistic approach. For instance, it should be acknowledged, following Nobel laureate and president of the UK's Royal Society of Chemistry, Professor Harry Kroto, "that the traditional chemistry, physics, biology departmentalised university infrastructures—which are now clearly out-of-date and a serious hindrance to progress—must be replaced by new ones which actively foster the synergy inherent in multidisciplinarity." The National Institutes of Health and the Howard Hughes Medical Institute have strongly recommended that undergraduate biology education should incorporate mathematics, physics, chemistry, computer science, and engineering until "interdisciplinary thinking and work become second nature." Young physicists and chemists are encouraged to think about the opportunities waiting for them at the interface with the life sciences. Mathematics is playing an ever more important role in the physical and life sciences, engineering, and technology, blurring the boundaries between scientific disciplines.

The series is to be a reference of choice for established interdisciplinary scientists and mathematicians and a source of inspiration for a broad spectrum of researchers and research students, graduates, and postdoctoral fellows; the shared emphasis of these carefully selected and refereed contributed chapters is on important methods, research directions, and applications of analysis including within and beyond mathematics. As such the volume promotes mathematical sciences, physical and life sciences, engineering, and technology education, as well as interdisciplinary, industrial, and academic genuine cooperation.

Toward such goals the following chapters are featured in the current volume.

Chapters 1, 2, 3, and 4 by Raymond Fletcher extensively study perfect polygons, geometric triple systems, and geometric ramifications in the ternary hypercommutative variety.

Chapter 5 by Larry Keen investigates the psychoneuroimmunological influences of cannabinoids, offering an opportunity for mathematical modeling techniques in this area of neuropsychology.

In Chapter 6 by Andrei Khrennikov et al., the authors propose a methodology for image segmentation with the aid of p-adic metrics.

Chapter 7 by Klaudia Oleschko et al. shows that primes are everywhere but nowhere.

In Chapter 8 by Martha Ortiz-Posadas, the author presents a logical combinatorial approach for pattern recognition in medicine.

Chapter 9 by Rana Parshad and Brian Ewald is on the uniqueness of invariant measures for the stochastic infinite Darcy-Prandtl number model.

Chapter 10 by K.C. Patidar et al. is concerned with pricing barrier options using integral transform.

In Chapter 11 Matti Pitkänen discusses at length the philosophy of adelic physics.

Chapter 12 by Bourama Toni proposes a game-theoretical analysis of cultural integration and assimilation in America as a way of understanding the evolution of American social dynamics.

The book as a whole certainly enhances the overall objective of the series, that is, to foster the readership interest and enthusiasm in the STEAM-H disciplines (science, technology, engineering, agriculture, mathematics, and health), stimulate graduate and undergraduate research, and generate collaboration among researchers on a genuine interdisciplinary basis.

The STEAM-H series is hosted at Virginia State University, Petersburg, Virginia, USA, an area that is socially, economically, intellectually very dynamic and home to some of the most important research centers in the USA, including NASA Langley Research Center, manufacturing companies (Rolls-Royce, Canon, Chromalloy, Sandvik, Siemens, Sulzer Metco, NN Shipbuilding, Aerojet), and their academic consortium (CCAM), the University of Virginia, Virginia Tech, the Virginia Logistics Research Center (CCAL), Virginia Nanotechnology Center, The Aerospace Corporation, C3I Research and Development Center, Defense Advanced Research Projects Agency, Naval Surface Warfare Center, National Accelerator Facility, and the Homeland Security Institute.

The STEAM-H series, by now well established with a high impact through its intensive seminars and books published by Springer, a world-renowned publisher, is expected to become a national and international reference in interdisciplinary education and research.

Petersburg, VA, USA Bourama Toni

Acknowledgements

We would like to express our sincere appreciation to all the contributors and to all the anonymous referees for their professionalism. They all made this volume a reality for the greater benefit of the community of science, technology, engineering, agriculture, mathematics, and health.

Contents

Contributors

Clive Callender College of Medicine, Howard University Hospital, Washington, DC, USA

Alfonso Campbell Department of Psychology, Howard University, Washington, DC, USA

Brian Ewald Department of Mathematics, Florida State University, Tallahassee, FL, USA

Raymond R. Fletcher III Department of Mathematics and Economics, Virginia State University, Petersburg, VA, USA

Larry Keen II Department of Psychology, Virginia State University, Petersburg, VA, USA

Andrei Khrennikov International Center for Mathematical Modelling, Physics, Engineering, Economics, and Cognitive Science, Linnaeus University, Växjö, Sweden

Nikolay Kotovich Institute of System Analysis of Russian Academy of Science, Moscow, Russia

Edgard Ngounda Department of Mathematics and Applied Mathematics, University of the Western Cape, Bellville, South Africa

Klaudia Oleschko Centro de Geociencias, Universidad Nacional Autonoma de Mexico (UNAM), Queretaro, Mexico

Beatriz F. Oleshko Universidad Autonoma de México (UNAM), Centro de Ciencias de la Complejidad (C3), Cd. Mx., México

Martha R. Ortiz-Posadas Electrical Engineering Department, Universidad Autónoma Metropolitana-Iztapalapa, Ciudad de México, Mexico

Jean-Francois Parrot Instituto de Geografía, Universidad Autonoma de México (UNAM), Cd. Universitaria, Delegación Coyoacán, Ciudad de México, México

Rana D. Parshad Department of Mathematics, Clarkson University, Potsdam, NY, USA

Kailash C. Patidar Department of Mathematics and Applied Mathematics, University of the Western Cape, Bellville, South Africa

Deidre Pereira Department of Clinical and Health Psychology, University of Florida, Gainesville, FL, USA

Edson Pindza Department of Mathematics and Applied Mathematics, University of the Western Cape, Bellville, South Africa

Matti Pitkänen Karkinkatu 3 I 3, Karkkila, Finland

Bourama Toni Department of Mathematics & Economics, Virginia State University, Petersburg, VA, USA

Arlener D. Turner Department of Behavioral Sciences, Rush University Medical Center, Chicago, IL, USA

Perfect Polygons and Geometric Triple Systems

Raymond R. Fletcher III

Abstract A *perfect n-gon* is an abstraction of a regular n-gon when regarded in the real projective plane. The vertices of a regular n-gon P lie on n parallel classes of lines. The lines in any parallel class meet at a point at infinity. We call these points the *perspective points* of P. The vertices of P lie on a circle and the perspective points of P lie on the line at infinity in the projective plane, so we can say that the combined set of vertices and perspective points lie on a (reducible) cubic curve consisting of a line and a circle. In our Main Theorem we show that the combined set of vertices and perspective points of any perfect polygon lie on a cubic curve which may be irreducible. In case the cubic is irreducible, a well-known algebra which we call a *geometric triple system* can be defined on its points. We show that perfect polygons can be obtained as translates of these algebras.

Keywords Polygon • Cubic curve • Conic • Triple system • Flex

1 Introduction

Let P be a set of $n \geq 5$ points in the real projective plane, no three of which are collinear, labeled with the integers mod n (Z_n). For each $k \in Z_n$ let W_k denote the set of lines $\{[x,y] : x \neq y, x+y = k(\text{mod } n)\}$. If for each k the lines in W_k are concurrent, then we call P a *perfect n-gon.* We denote by X_k the point of concurrence of the lines in W_k. We insist that $n \geq 5$ so that each set W_k contains at least two lines and thus each perspective point is defined. The n labeled points are the *vertices* of P and the set $\{X_k : k \in Z_n\}$ are the *perspective points* of P. We also require that no point is simultaneously a vertex and a perspective point. A regular polygon with its vertices labeled consecutively with elements from Z_n is an example of a perfect n-gon. In this case the vertices of P are equally spaced around a circle and the perspective points all lie on the line at infinity in the projective plane. If the vertices of P lie

R.R. Fletcher III (✉)
Department of Mathematics and Economics, Virginia State University, Petersburg, VA 23806, USA
e-mail: rfletcher210@gmail.com

B. Toni (ed.), *New Trends and Advanced Methods in Interdisciplinary Mathematical Sciences*, STEAM-H: Science, Technology, Engineering, Agriculture, Mathematics & Health, DOI 10.1007/978-3-319-55612-3_1

on a circle or any conic and the perspective points lie on a line, we call P a *cyclic perfect polygon*. We will demonstrate, as our Main Theorem, that the vertices and perspective points of any perfect polygon P lie on a cubic curve α which we call the *cubic envelope of P*. If α is reducible, then P may have all its vertices and perspective points on three lines, or P is cyclic and all perspective points must be collinear.

The vertices of a perfect polygon will be referred to by their labels (from Z_n), and the notation [A,B], where A,B $\in Z_n$ will be used to represent the line containing points A,B. The notation $[A_1, A_2, \ldots, A_n]$ will be used to indicate a line containing points $A_1, A_2, \ldots, A_n$, or as an abbreviation for the phrase: "the points $A_1, A_2, \ldots, A_n$ are collinear." We will make reference to the following two theorems from Euclidean geometry. These can be found in many places, e.g., [1, 2].

E1: (Pascal's Theorem) Let (A,B,C,D,E) be a hexagon inscribed on a conic. Then the points [A,B]∩ [D,E], [B,C]∩[E,F], [C,D]∩[F,A] are collinear.

E2: (Converse of Pascal's Theorem) If A,B,C,D,E are six points in the projective plane such that the points [A,B]∩ [D,E], [B,C]∩[E,F], [C,D]∩[F,A] are collinear, then the six points lie on the same conic.

Some preliminary results regarding cubic curves are also needed. A cubic curve has the general form:

(1) $f(x,y) = ax^3 + bx^2y + cxy^2 + dy^3 + ex^2 + fxy + gy^2 + hx + jy + k = 0.$

The cubic f(x,y) is *reducible* if it factors as the product of two nonconstant real polynomials, and we call it *irreducible* if no such factorization exists. When a cubic is reducible it consists either of three lines or a line and a conic. A curve is *nonsingular* if it has tangents at all of its points. The following property of nonsingular irreducible cubics can be found in [3].

P1: Let j be a line which intersects a nonsingular irreducible cubic curve α at least twice counting multiplicities. Then j intersects α exactly three times counting multiplicities.

Property P1 allows us to define a binary operation * on the nonsingular points of an irreducible cubic curve α by setting x*y equal to the unique third point on [x,y] and on α. The expression x*x denotes the point besides x which lies on α and on the tangent to α at x. If a tangent to α intersects α at a point x with multiplicity three, then x is called a *flex* of α. In this case x *x = x, i.e., x is idempotent in the algebra (α,*). Three identities are satisfied by *:

A1: x*y = y*x
A2: x*(x*y) = y
A3: (x*y)*(z*w) = (x*z)*(y*w).

The first two identities are clear, but A3 is a remarkable, (and well known), property of nonsingular irreducible cubic curves. An algebra satisfying A1, A2, A3 is called a *thirdpoint groupoid* in [4] and a *binary hypercommutative algebra (BHA)* in [5]. We call A3 the *hypercommutative axiom* and illustrate it on a typical irreducible cubic curve in Fig. 1. The *cancellation property*:

A4: x*y = x*z → y = z

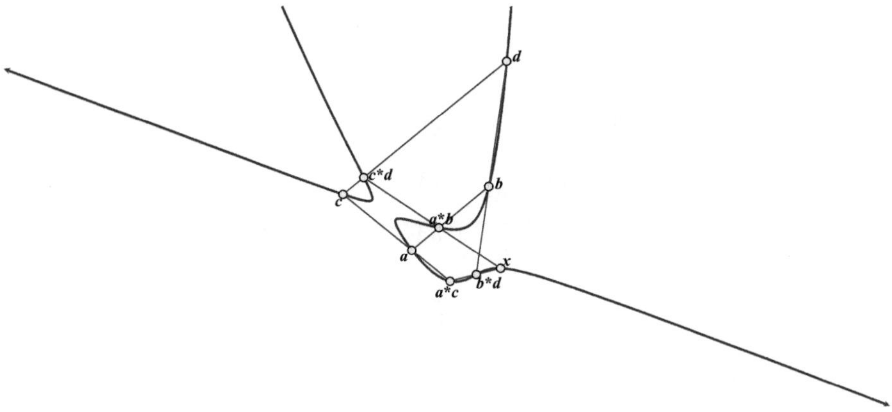

Fig. 1 The hypercommutative axiom: (a*b)*(c*d)=(a*c)*(b*d)=x

is a consequence of A2. The following properties, also found in [3], are concerned with the determination of conics and cubics.

P2: Five points, no four collinear, lie on a unique curve of degree 2.

P3: Let P_1, P_2, ..., P_8 be eight points, no four of which are collinear, and no seven of which lie on a conic. Then there exists a cubic curve which contains the eight points.

P4: Let P_1, P_2, ..., P_8 be eight points, no four of which are collinear, and no seven of which lie on a conic. Then all pairs of cubics containing P_1, P_2, ..., P_8 intersect in the same nine points, listed by multiplicity, i.e., *there is a point P_9 such that any two cubics containing P_1, P_2, ..., P_8 also contain P_9.*

P5: Let P_1, P_2, ..., P_8 be eight points, no four of which are collinear, and no seven of which lie on a conic, and let P_9 be the point stipulated in P4. If Q is any point other than P_1, P_2, ..., P_8, P_9, then the nine points P_1, P_2, ..., P_8, Q lie on a unique cubic.

2 The Special Perfect Hexagon

We shall consider first a special type of perfect polygon which contains two perspective points and a vertex on the same line. Suppose $[X_s, X_t, k]$ where $s \neq t$. Then we must have $[X_s, X_t, k, t\text{-}k, s\text{-}k]$. We are not allowed three collinear vertices, so we must have either $t\text{-}k = k$ or $s\text{-}k = k$. Without loss of generality, assume $s\text{-}k = k$, so we have $[X_{2k}, X_t, k, t\text{-}k]$, or replacing t-k by m, we have $[X_{2k}, X_{m+k}, k, m]$ with $m \neq k$. The point $2k - m$ must also lie on this line, in which case we must have $2k = 2m \pmod{n}$. If n is odd, then $2k = 2m \pmod{n}$ implies $m = k$. We conclude that no perfect polygon with an odd number of vertices can have two perspective points and a vertex on the same line. There does exist a perfect hexagon with such a line, as we show in our first theorem.

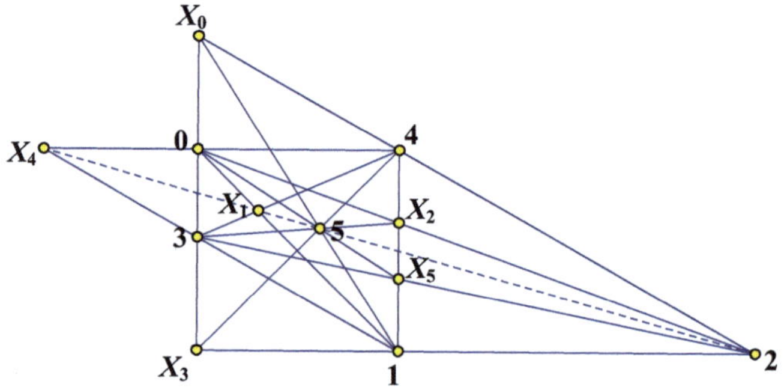

Fig. 2 The special perfect hexagon

Theorem 1 *A perfect hexagon P = (0,1,2,3,4,5) with [X_0, X_3, 0, 3] can be constructed as follows: (1) Put X_0 on a line with vertex 0 and X_3; (2) place the points 1,2,3,4,5 in the plane so that [1, 2]∩[4, 5] = X_3 and [1, 5]∩[2, 4] = X_0 as in* Fig. 2; *(3) Let X_1= [2, 5]∩[0,1] and let vertex 3 = [X_1,4]∩[X_3,0]. We also have [X_1, X_4, 2, 5] and [X_2, X_5, 1, 4].*

Proof Via a transformation we can assume that the points X_3, 0, 4, 1 form a square with coordinates X_3= (0,0); 0 = (0,1); 4 = (1,1), and 1 = (1,0). We can then set X_0= (0,t) where t is a real number different from -1,0,1,2,3. Thus the points X_3, 0, X_0 lie on the line x = 0. By intersecting lines [X_0, 4], [X_3, 1] we find vertex 2 = (t/(t-1), 0). Intersecting lines [X_3, 4], [X_0,1] gives vertex 5 = (t/(t+1), t/(t+1)), and then intersecting lines [2, 5], [0,1] we obtain X_1= ((2-t)/(3-t), (1/(3-t)). Intersecting [4,X_1] [0,X_3] gives vertex 3 = (0,t-1). Finally we obtain X_2= [0,2]∩ [3, 5] = (1,(t-1)/t); X_4= [0,4]∩[1, 3] = ((t-2)/(t-1), 1) and X_5= [0,5]∩[2, 3] = (1, (t-1)/t). With these coordinates we immediately see that X_2, X_5 lie on the line x = 1 along with vertices 1, 4. This is all that is needed to show that P is a perfect hexagon.

Line [2, 5] has equation 2y = (1-t)x + t. The coordinates of X_4 satisfy this equation so we also have [X_1, X_4, 2,5]. Note: if t = -1, then vertex 5 is the point at infinity on lines with slope -1, and if t = 3, then X_1 is the point at infinity on lines with slope -1. In either of these cases the construction given in the Theorem is easily proved. Also t = 2 is not possible since then the vertices 2,5,0 would be collinear. □

We call the perfect hexagon constructed in Theorem 1, the *special perfect hexagon.* There is no other perfect polygon like it as we show next.

Theorem 2 *The special perfect hexagon is the only perfect polygon which contains two perspective points and a vertex on the same line.*

Proof Suppose P is a perfect n-gon with two perspective points and a vertex on the same line. As per the discussion preceding Theorem 1, we can assume that n is even. If n = 6 we obtain the special perfect hexagon, so suppose n ≥ 8. Also from the discussion preceding Theorem we can suppose that the line with two perspective

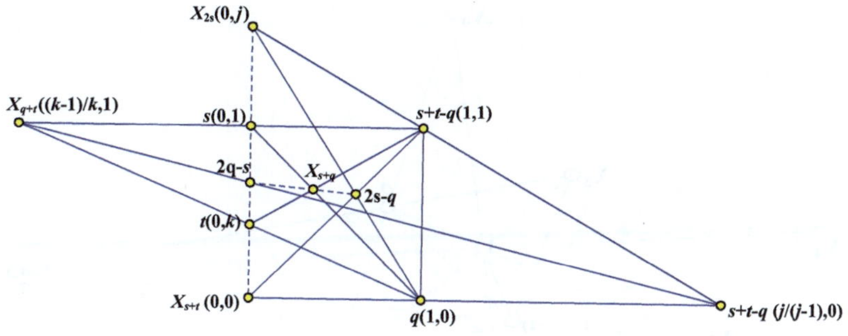

Fig. 3 Coordinatization used to show uniqueness of special perfect hexagon

points and a vertex contains the points s, t, X_{s+t},X_{2s} where s $\neq$ t but 2s = 2t. This actually requires that t = s + (n/2). Since n $\geq$ 8 we can find vertex q of P such that the points {s,t,q,s+t-q,s-t+q,2s-q,2q-s} are distinct. For example, if we set q = s+1, then this set becomes {s, s+(n/2), s+1, s-1+(n/2), s+1+(n/2), s−1, s+2} which is easily seen to be a set of seven distinct points (provided n is even and n $\geq$ 8). We transform, similar to Theorem 1, so that the points X_{s+t}, s, s-t, s-t+q, q become the vertices of a unit square, i.e., X_{s+t}= (0,0); s = (0,1); s-t+q = (1,1), and q = (1,0). We let X_{2s}= (0,j) and t = (0,k) so that the required line [s,t, X_{s+t}, X_{2s}] is the y-axis (see Fig. 3). We then obtain the following coordinates for points of interest:

s+t-q =[X_{s+t}, q]∩[X_{2s}, s-t+q] = (j/(j-1),0),
2s-q = [X_{s+t},, s-t+q]∩[X_{2s}, q] = (j/(j+1), j/(j+1)),
X_{s+q} = [s,q]∩[t, s-t+q] = ((k-1)/(k-2), -1/(k-2)),
X_{q+t}= [s, s-t+q]∩[q,t],
2q-s = [X_{q+t}, s+t-q]∩[X_{s+q}, 2s-q] = (0, jk/(j+k-1)).

The x-coordinate of the vertex 2q-s indicates that it lies on the y-axis along with vertices s,t. Since we chose q so that the vertices s,t, 2q-s are distinct, we find that P contains three collinear vertices. Since this is not allowed, we conclude that no perfect n-gon with n >6 exists which has two perspective points and a vertex on the same line. □

3 Some Constructions

In Fig. 4 we have constructed a perfect hexagon using the following procedure: (1) put vertices 0,1,2 in the plane so that they are noncollinear; (2) put X_1 on [0,1], X_2 on [0,2], and X_3 on [1, 2]; (3) put vertex 3 on [X_3, 0]; (4) let vertex 5 = [X_1,2]∩[X_2,3]; (5) let X_5= [0,5]∩[2, 3]; (6) let vertex 4 = [X_1,3]∩[X_5,1]; (7) let X_4 = [0,4]∩[1, 3] and X_0 = [1, 5]∩[2, 4]. Let P denote the resulting configuration of six vertices and six perspective points. To show P is a perfect hexagon it remains only to show

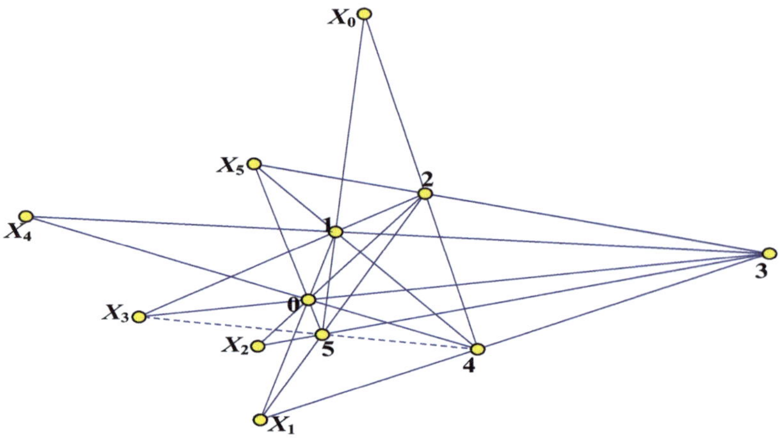

Fig. 4 Perfect hexagon

[4,5,X_3]. This is easily accomplished by applying the Theorem of Pappus to the points 1,0, X_1 on one line, and 3,2, X_5 on a second line. In Fig. 5 we construct a perfect 7-gon as follows: Steps (1),(2),(3) are the same as in the above construction; (4) let vertex 6 = [X_1,2]∩[X_2,3]; (5) put vertex 4 on [X_3,6]; (6) let vertex 5 = [X_2,4]∩[X_1,3]; (7) shift vertex 4 on [X_3,6] until the lines [2, 3], [1, 4], and [0,5] are concurrent, and let X_5 denote the point of concurrence; (8) Let X_4= [0,4]∩[1, 3]; X_0= [1, 6]∩[2, 5] and let X_6= [2, 4]∩[1, 5]. Let P denote the resulting configuration of vertices and perspective points and let us suppose that no four perspective points of P are collinear. To show P is a perfect 7-gon it remains to show [5,6,X_4], [3,4,X_0], and [0,6,X_6]. It seems quite difficult to do this using classical methods, i.e., the Theorems of Pappus and Desargues, so we shall instead show that the vertices of P all lie on an irreducible cubic curve and then use the axioms, particularly A3, which define a BHA.

Theorem 3 *The configuration P, as constructed above, is a perfect 7-gon.*

Proof Consider the eight points {6,0,1,2,3, X_1, X_2, X_3}. No four of these points are collinear, and if any one point is removed from this set, the remaining seven points contain three collinear points and thus cannot belong to the same conic. In accordance with properties P4 and P5, there exists a cubic curve α which contains the nine points S = {6,0,1,2,3, X_1, X_2, X_3, X_5}. We have stipulated that no four perspective points are collinear, and so α must be irreducible. Also, a line through a singular point of α cannot contain two more points of α, but each point of S lies on a line with two other points of S, so each point in S is nonsingular. We can thus define a BHA on the nonsingular points of α which includes all the points in S. We will show that the remaining vertices and perspective points of P lie on α, and that α is uniquely determined. We have $X_3* 6 = (3*0)*(2*X_1) = (3*2)*(0*X_1) = X_5*1$, and thus $X_3* 6 = X_5*1 = q$ is a point on α. But then q = [X_3,6]∩[X_5,1]

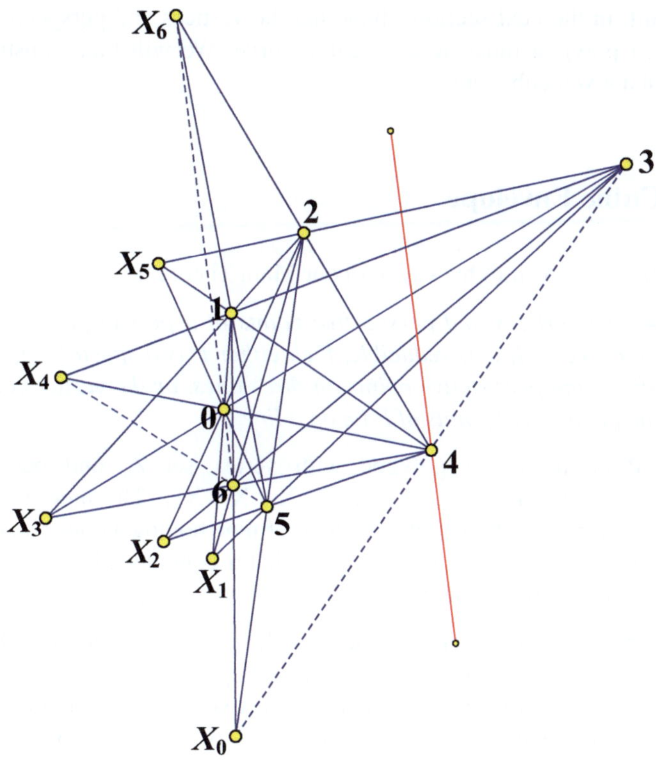

Fig. 5 Perfect 7-gon

= 4 lies on α. One of the two points X_5, 4 must be different from the point P_9 stipulated in property P4, so by property P5 we can conclude that α is unique. To show that vertex 5 of P lies on α, consider: $4 * X_2 = (X_5*1)*(2*0) = (X_5*2)*(1*0) = 3*X_1$ and thus $4 * X_2 = 3*X_1 = [4,X_2]\cap[3,X_1] = 5$ which must then lie on α. We have shown that every vertex of P lies on α. Consider: $1*6 = (X_5*4)*(3*X_2) = (X_5*3)*(4*X_2) = 2*5$ and thus $1*6 = 2*5 = [1, 6]\cap[2, 5] = X_0$ lies on α. Consider: $2*4 = (X_3*1)*(2*X_6) = (X_3*2)*(1*X_6) = 1*5$ and thus $2*4 = 1*5 = X_6$ lies on α. Consider: $3*4 = (X_3*0)*(1*X_5) = (X_3*1)*(0*X_5) = 2*5 = X_0$. This proves $[3,4,X_0]$. Consider: $0*6 = (X_2*2)*(4*X_3) = (X_2*4)*(2*X_3) = 5*1 = X_6$. This proves $[0,6,X_6]$. We have now shown that all the perspective points of P lie on α and we have shown two of the required collinearities. It remains only to show $[5,6,X_4]$. Consider: $5*6 = (X_1*3)*(1*X_0) = (X_1*1)*(3*X_0) = 0*4 = X_4$, and thus $[5,6,X_4]$. □

We call the construction of Theorem 3 a *one-point slide construction* since, once vertex 4, in step (7), is shifted to the right spot on line $[X_3,6]$, the positions of the remaining vertices and perspective points are all forced and the result is the desired perfect polygon. We claim that for every positive integer n > 6 a perfect n-gon can be constructed, as in Theorem 3, by a one-point slide. Instead of attempting to prove this, we shall, in the next section, show that the vertices and perspective points of every perfect polygon must lie on a cubic curve. We will then construct perfect polygons on a given cubic curve.

4 The Cubic Envelope

We begin this section straightaway with our Main Theorem.

Theorem 4 *The vertices and perspective points of a perfect polygon P lie on a unique cubic curve α. If α is reducible, then either P is a special perfect hexagon with all vertices and perspective points on three lines, or the vertices of P lie on a conic and the perspective points of P lie on a line.*

Proof Let P be a perfect polygon with vertex set Z_n and perspective set $\{X_k: k \in Z_n\}$, and let $S = \{-1,0,1,2,3, X_0, X_1, X_2, X_3\}$. The nine points in S lie on a cubic curve α since they yield a system of nine homogeneous equations in the nine coefficients of the cubic. We will show that α contains all the vertices as well as all the perspective points of P.

Case 1 Suppose α is a reducible cubic. A reducible cubic consists either of three lines or a line and a conic. We have shown that the special perfect hexagon is the only perfect polygon all of whose vertices and perspective points lie on three lines, so we may suppose that α consists of a line j and a conic β. Suppose j contains one of the vertices in S. Then j can contain at most one more vertex and one more perspective point. Removing any such line from S leaves a figure which contains at least one set of three collinear points, i.e., the remaining points in S cannot possibly lie together on the conic β. We conclude that all four perspective points $\{X_0, X_1, X_2, X_3\}$ must lie on j, and the five vertices in S lie on β. Note that by property *P2*, the conic β is uniquely determined. We now present an algorithm which will terminate when it is shown that all the perspective points of P lie on j and all the vertices of P lie on β. Let V denote the set of vertices of P so far proved to lie on β, and let L denote the set of perspective points of P so far shown to lie on j. The variable s is a counter. These variables are initialized in step (1).

(1) $V \leftarrow \{-1,0,1,2,3\}$; $L \leftarrow \{X_0, X_1, X_2, X_3\}$; $s \leftarrow 1$.
(2) If $n = 2s+3$ go to step (7), else we have $n \geq 2s+4$; $V = \{-s, \ldots ,s+2\}$; $L = \{X_{-s+1}, \ldots , X_{s+2}\}$.

Since $n \geq 2s+4$, the consecutive elements $\{-s-1, \ldots ,s+2\}$ are distinct and thus $H = (0,1,-1,s+2,-s-1,s+1)$ is a hexagon with six distinct points. The opposite sides of H meet in the collinear points $\{X_1,X_0,X_{s+1}\}$ and so by *E2*, the vertices of H lie on a conic ρ. But five points of H, namely $\{0,1,-1,s+1,s+2\}$ lie in V and so have been shown to lie on the conic β. Since five points determine a conic by *P2*, we must have ρ=β, and thus the vertex -s-1 lies on β. So we can augment V:

$$V \leftarrow V \cup \{-s-1\}.$$

(3) The hexagon F = (0,-s,s+1,1,-s-1,s+2) consists of six distinct points all of which lie in V and hence on the conic β. The opposite sides of F meet in the three points $\{X_{-s},X_1,X_{s+2}\}$. By Pascal's Theorem (*E1*), these three points are collinear. Since X_1 and X_{s+2} lie in L and hence on the line j, we conclude that the perspective point X_{-s} must also lie on j. So we can augment L:

$L \leftarrow L \cup \{X_{-s}\}$

(4) If n = 2s+4 go to step (7), else we have n ≥ 2s+5; V = {-s-1, ... , s+2}; L = $\{X_{-s}, \ldots , X_{s+2}\}$.

Since n ≥ 2s+5, the points {-s-1, ... , s+3} are distinct, and so the hexagon G = (s+3,-1,-2,0,s+2,-s-1) has six distinct points. The opposite sides of G meet in the collinear points $\{X_{s+2},X_1,X_2\}$, so the vertices of G lie on a conic which must be β since G contains the five points {-1,2,0,s+2,-s-1} which lie in V hence on β. We conclude that the vertex s+3 lies on β, so we can augment V:
$V \leftarrow V \cup \{s+3\}$.

(5) The hexagon K = (s+3,0,-1,s+2,1,-2) consists of six distinct points, all in V. By Pascal's Theorem we must have $[X_{s+3}, X_{-1}, X_{s+1}]$. Since X_{-1}, X_{s+1} lie in L hence on j, we deduce that X_{s+3} also lies on j and thus:

$L \leftarrow L \cup \{X_{s+3}\}$

(6) s ← s+1; go to step (2).
(7) At this point in the algorithm n = 2s+3 or 2s+4; V contains all the vertices of P and L = $\{X_{-s-1}, \ldots , X_{s+2}\}$ contains every perspective point of P except for X_{-s}. If s = 1, (n=5 or n=6), go to step (8). If s>1, then the possibly degenerate hexagon M = (-s,0,2,-s+1,-1,3) whose vertices all lie on β implies by Pascal's Theorem, that $[R, X_2, X_{-s+3}]$, where R = [-s,0]∩[-s+1,-1]. If s>2 this implies that R = X_{-s} lies on j. If s=2, then n = 7 or n = 8 and the set of points {-3,-2,-1,0,1,2,3} are distinct and all in V. So (-3,1,2,-2,0,3) is a hexagon on β which implies $[X_{-2}, X_3, X_0]$ and thus $X_{-s} = X_{-2}$ lies on j. We may thus increment L:

$L \leftarrow L \cup \{X_{-s}\}$, and halt.

(8) At this point we have n = 5 or n = 6. First suppose n = 6. It remains only to show that X_{-2} lies in L. The hexagon (-2,0,-1,1,-3,2) contains the six distinct vertices of P all of which have been shown, at this point, to lie on β. The opposite sides of this hexagon meet at the collinear points $\{X_{-2}, X_{-1}, X_0]$. Since X_{-1}, X_0 have been shown to lie on j, we conclude that X_{-2} lies on j as well:

$L \leftarrow L \cup \{X_{-2}\}$.

If n = 5 we have only to show that X_{-1} lies on j. The vertices of the degenerate hexagon (3,-1,2,1,1,0) which contains all five vertices of P lie on the conic β. Applying Pascal's Theorem, we obtain $[Q, X_1, X_3]$ where Q = [1]∩[3,-1]. (Here, [1] denotes the tangent to β at vertex 1.) Since X_1, X_3 lie on j, we have that Q lies on j also. Then Q = j ∩ [3,-1] = X_2, and thus [1,1,X_2]. The degenerate hexagon (0,-1,1,1,-2,2) on β implies [X_{-1}, X_0, X_2] and thus X_{-1} lies on j.

L ← L ∪ {X_{-1}}. The algorithm now halts and Case 1 is proved.

Case 2 α is an irreducible cubic curve. Let k ≥3 and assume inductively that the vertices {-1,0,1,2,... , k} and the perspective points {X_0, X_1, ... , X_k} lie on α. This is so when k = 3. The binary operation * introduced in Section 1 can be defined on the nonsingular points of α yielding a BHA. We will employ the axioms A1,A2,A3,A4 of a BHA to obtain the desired result. We complete the induction by showing that vertex k+1 and perspective point X_{k+1} lie on α.

Consider: (0*0)*(X_1*X_2) = (0*X_1)*(0*X_2) = 1*2 = X_3, and also X_0*(X_1*X_2) = (1* -1)*(X_1*X_2) = (1*X_1)*(-1*X_2) = 0*3 = X_3. Now by cancellation (A4) we can deduce that 0*0 = X_0. The geometric interpretation of this result is that the tangent to α at vertex 0 contains the perspective point X_0. Another useful preliminary result is 1*1 = X_2 which can be shown as follows. Consider: (1*1)*(X_0*X_3) = (1*X_0)*(1*X_3) = -1*2 = X_1, and (0*2)*(X_0*X_3) = (0*X_0)*(2*X_3) = 0*1 = X_1. We can then deduce by cancellation that 1*1 = 0*2 = X_2.

Consider: (X_0*2)*1 = (X_0*2)*(X_1*0) = (X_0*X_1)*(0*2) = (X_0*X_1)*X_2 = (X_0*X_1)*(3*-1) = (X_0*-1)*(X_1*3) = (X_1*3)*1, so we can conclude by cancellation that X_0*2 = X_1*3. If we let A = X_0*2 = X_1*3, then A lies on α and also A = [X_0,2]∩[X_1,3] = -2. Thus vertex -2 lies on α.

Consider: (1*k)*(X_{k-1}*X_k) = (1* X_{k-1})*(k* X_k) = (k-2)*0 = X_{k-2}. Note that when k = 3 we have used the preliminary result 1*1 = X_2, or 1*X_2= 1. Also ((k-1)*2)*(X_{k-1}*X_k) = ((k-1)* X_{k-1})*(2*X_k) = 0*(k-2) = X_{k-2}. Consequently 1*k = (k-1)*2. Consider: (X_k*-1)*(0*X_0) = (X_k*0)*(-1*X_0) = k*1 = (k-1)*2 = (X_{k-1}*0)*(-2*X_0) = (X_{k-1}*-2)*(0*X_0), so by cancellation X_k*-1 = X_{k-1}*-2. Then the point B = X_k*-1 = X_{k-1}*-2 lies on α, but also B = [X_k,-1]∩[X_{k-1},-2] = k+1, so vertex k+1 lies on α.

Finally consider: (k*1)*(X_k*X_1) = (k*X_k)*(1*X_1) = 0*0 = X_0, and ((k+1)*0)*(X_k*X_1) = ((k+1)*X_k)*(0*X_1) = -1*1 = X_0, so we can conclude that k*1 = (k+1)*0 is a point C on α. But C = [k,1]∩[k+1,0] = X_{k+1}. So X_{k+1} lies on α completing the induction. We observe, in this case, that once a tenth point of P has been shown to lie on α, then α is uniquely determined by property *P5*. □

If s ∈Z_n is a vertex of a perfect n-gon with cubic envelope α, we have defined the line [s,s] to be the tangent to α at s. In the course of the proof of Theorem 4 we showed that X_0 lies on [0,0] and X_2 lies on [1]. We will now show that the perspective point X_{2s} lies on the tangent [s,s] for every s ∈Z_n.

Theorem 5 *Let P be a perfect n-gon with vertex set Z_n, perspective set {X_k: k ∈Z_n}, and cubic envelope α. If s ∈Z_n, then the tangent line to α at vertex s contains the perspective point X_{2s}.*

Proof First suppose α is irreducible and let (α,*) denote the corresponding BHA. We have shown in Case 2 of Theorem 4, that 0*0 = X_0. We observe that any integer s can be added to the vertices of P and 2s to the subscripts of the perspective points of P and that with this relabeling P remains a perfect n-gon. It then follows that s*s = X_{2s} for any s in Z_n.

Now suppose α is reducible so the vertices of P lie on a conic β and the perspective points lie on a line j. Let k = [0,0] denote the tangent to β at vertex 0. In this case it suffices to show that k contains the perspective point X_0. Consider the degenerate hexagon (0,0,-2,-1,1,-3) which is inscribed in β. By Pascal's Theorem *(E1)* the points k ∩[-1,1], X_{-2}= [0,-2]∩[1,-3], and X_{-3}= [-2,-1]∩[0,-3] are collinear. This means that k ∩[-1,1] lies on line j, but j∩[-1,1] = X_0 and it follows that k contains X_0. □

5 Cyclic Perfect Polygons

As a consequence of Theorem 4, perfect polygons can be put into three Classes: (i) those whose cubic envelope consists of three lines; (ii) those whose cubic envelope consists of a line which contains all the perspective points and a conic which contains all the vertices, and (iii) those whose cubic envelope is irreducible. In Section 2 we have discussed perfect polygons, (the special perfect hexagons), which constitute Class (i). In this section we consider perfect polygons in Class (ii). Let P be such a perfect polygon with cubic envelope consisting of line j containing the perspective points of P, (we call j the *perspective axis of P),* and conic β containing the vertices of P. In accordance with [3], there exists an invertible transformation T which sends j to another line k and β to a circle, so to prove the existence of P, it suffices to show that P can be constructed with vertices on a given circle and perspective points on a given line. We call such a perfect polygon *circular.* To analyze perfect circular polygons the following basic geometric results are needed, all of which can be found in [1] or [2].

E3: (Theorem of Desargues) Two triangles are in perspective from a line iff they are in perspective from a point.

The *power of a point P with respect to a circle ω* is given by $d^2 - r^2$ where d is the length of the line segment joining P to the center of ω, and r is the radius of ω.

E4: If P lies outside of a circle ω with center O and l is a line through P which meets ω in two points A, B, then (PA)(PB) = $d^2 - r^2$ *where r is the radius of ω and d is the length of segment PO. If T is the point of contact of a tangent to ω from P, then* $(PT)^2 = d^2 - r^2$. *If P lies in the interior of ω and l is a line through P which meets ω at A, B, then (PA)(PB) =* $r^2 - d^2$.

E5: (Radical axis of two circles) Let ω,ν be two nonconcentric circles. Then the locus of points with equal powers with respect to ω,ν is a lone perpendicular to the line of centers of ω,ν. This line is called the radical axis of ω,ν. In case ω,ν meet in two points then the radical axis is the line containing these points.

E6: (Radical center) Let η,ν,ω be three circles whose centers form a triangle. Then the radical axes of these three circles taken in pairs concur at a point called the radical center of η,ν,ω.

Some basic properties from inversive geometry will also be useful. Let ω be a circle with center O and radius r and let P ≠ O be a point in the plane of ω. The inverse of P with respect to ω is the point P′ lying on [O,P] such that [O,P][O,P′] = r^2. If P lies in the exterior of ω, then the inverse of P with respect to ω can be constructed as follows. Let A,B be the contact points on ω of the tangents from P, then P′ = [A,B]∩[O,P]. If P lies in the interior of ω, then let [A,B] be the chord through P and perpendicular to [O,P], then P′ is the intersection point of the tangents to ω at A,B. If P lies on ω, then P′ = P and if P = O, then P′ is not defined. Note that inversion through ω is an involutory transformation of the inversive plane.

E7: If a line k meets a circle ω in two points A, B, then the inverse of k wrt ω is the circle through A, B and the center O of ω.

E8: All circles orthogonal to two given circles α, β cross the line of centers of α, β in the same two points.

E9: The inverse of a circle which does not contain the center of the circle of inversion is a circle.

E10: Inversion preserves angles.

In what follows we assume that Z_n is the vertex set of a perfect cyclic polygon P and $\{X_k : k \in Z_n\}$ is the perspective set of P. Any line [x,y] where x,y ∈ Z_n is a *diagonal* of P. If n is even, then the lines {[x, x+(n/2)]: x ∈ Z_n } are the *long diagonals* of P.

Theorem 6 *Let P be an even cyclic perfect n-gon. Then the long diagonals of P concur at a point Z which we call the center of P.*

Proof Let t ∈ Z_n and consider the triangles (t, t+1, t+2), (t+(n/2), t+1+(n/2), t+2+(n/2)). Since [t,t+1]∩[t+(n/2),t+1+(n/2)] = X_{2t+1}; [t+1, t+2]∩[t+1+(n/2), t+2+(n/2)] = X_{2t+3}; [t, t+2)]∩[t+(n/2), t+2+(n/2)] = X_{2t+2}, and X_{2t+1}, X_{2t+2}, X_{2t+3} lie on the perspective axis j of P, these triangles are in perspective from j (see Fig. 6). By Desargue's Theorem (E3), these triangles are in perspective from a point, i.e., the lines [t, t+(n/2)], [t+1, t+1+(n/2)], and [t+2,

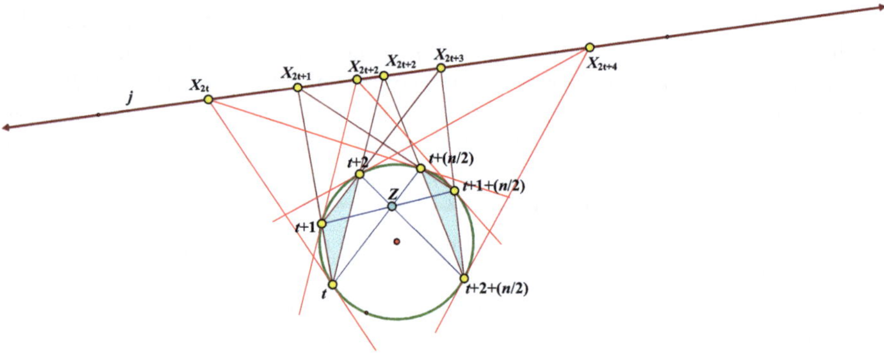

Fig. 6 The center of a cyclic perfect polygon

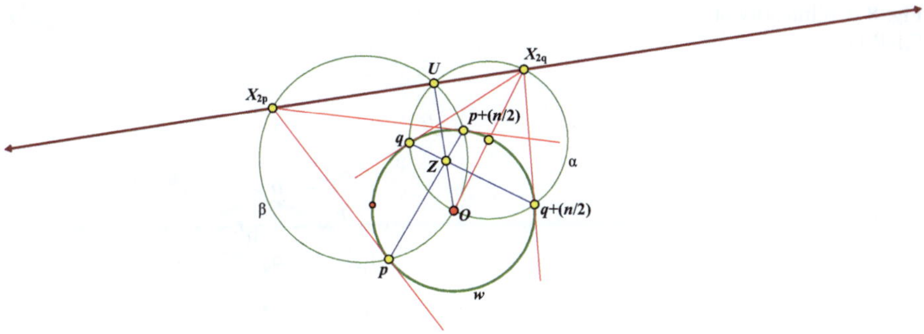

Fig. 7 Points U, Z are inverses wrt w

t+2+(n/2)] are concurrent. Since $t \in Z_n$ is arbitrary, we have shown that any three consecutive long diagonals of P are concurrent. It follows that there is a point Z which lies on each of the long diagonals of P. □

Theorem 7 *Let P be an even perfect circular polygon with center Z and perspective axis j. Let ω denote the circle which contains the vertices of P; let O denote the center of ω and let Q denote the foot of the perpendicular from O to j. Then Z is the inverse of Q wrt ω. (Thus OZ is perpendicular to j.)*

Proof Let X_{2p}, X_{2q} be any two distinct even perspective points of P (see Fig. 7). By Theorem 5, the tangent lines to ω from X_{2p} touch ω at the points p, p+(n/2), and similarly the tangent lines from X_{2q} touch ω at q, q+(n/2). Let α denote the circle (X_{2q}, q, q+(n/2)) and let β denote the circle (X_{2p}, p, p+(n/2)). The long diagonal (p, p+(n/2)) is the radical axis of the circle pair {ω, β}, and the long diagonal (q, q+(n/2)) is the radical axis of the circle pair {ω, α} and thus Z = [p, p+(n/2)] ∩[q, q+(n/2)] is the radical center of the circles α,β,ω by (E6). Both circles α, β pass through the center O of ω. Let U denote the second point of intersection of α, β. The line OU is the radical axis of {α,β} by (E4) and thus Z lies on the line OU. By (E7) the inverse of line (p, p+(n/2)) wrt ω is the circle β and the inverse of (q, q+(n/2)) wrt ω is the circle α. As a consequence, the inverse of Z wrt ω is the point U. Since U lies on β and segment [O, X_{2p}] is a diameter of β, it must be that $\angle OUX_{2p} = 90°$. Similarly, $\angle OUX_{2q} = 90°$, and so U lies on j and OU is perpendicular to j. Then U = Q and the theorem is proved. □

In order to achieve our final description of perfect circular n-gons, we need to show how a given perfect circular n-gon can be *doubled* to obtain a perfect circular 2n-gon. For this the following three Lemmas are required.

Lemma 8 *Let A,B,C,D be four points on a circle ω and let P = [A,D]∩[B,C]; Q = [A,B]∩[C,D]; R = [B,D]∩[A,C] and E,F the contact points of the tangent lines to ω from P. Then [E,R,F,Q].*

Proof Let J = [A,E]∩[B,F]; K = [A,F]∩[B,E]; L = [E,C]∩[F,D], and M = [E,D]∩[F,C], then the inscribed hexagons (E,A,F,F,B,E), (A,D,F,B,C,E),

Fig. 8 Collinearity of E,F,R,Q

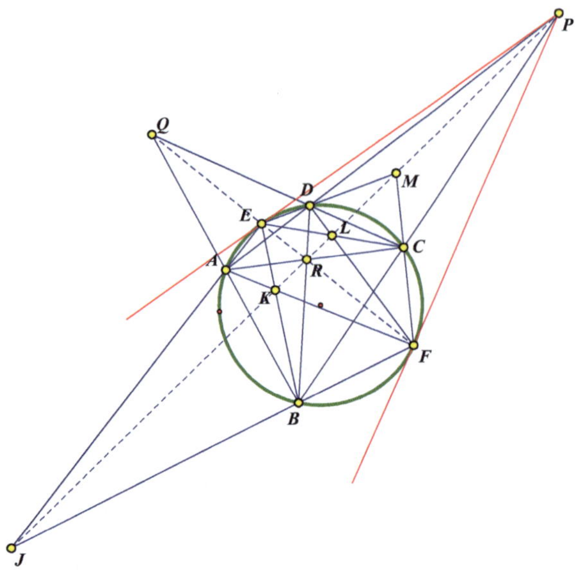

(A,D,E,B,C,F), (A,C,E,B,D,F) imply [J,K,P], [P,L,J], [P,M,K], and [R,L,K], respectively, by Pascal's Theorem (E1) (see Fig. 8). Combining these results we obtain [J,K,P,L,M,R] which we call line j. Triangles (F,B,C), (E,A,D) are in perspective from line j and thus lines [E,F], [A,B], [C,D] are concurrent and we have [E,F,Q]. The inscribed hexagon (E,F,D,A,C,E) yields [[E,F]∩[A,C], L,P], and thus [E,F]∩[A,C] lies on j. But [A,C]∩j = R, so we may conclude [E,F,R] and thus [E,F,R,Q]. □

Lemma 9 *Let A,B,C,D be four points on a circle ω and let P = [A,D]∩[B,C]. Let [E,F] be a chord of ω whose extension contains P; let Q = [E,E]∩[A,B] and R = [F,F]∩[C,D]. Then [P,Q,R].*

Proof Let k denote the line which contains the contact points of the tangents from P and let S = [A,B]∩[C,D]; T = [E,B]∩[C,F]; R = [F,B]∩[C,E], and V = [F,F]∩[E,E] (see Fig. 9). Applying Lemma 8 to the points {A,B,C,D} we obtain that S lies on k, and applying the same Lemma to {E,B,C,F} we obtain that T, R lie on k. The inscribed degenerate hexagon (E,E,B,F,F,C) yields [V,T,R] and thus V also lies on line k. Corresponding sides of triangles (E,B,Q), (F,C,R) meet at points T, S, V which all lie on k, so these triangles are in perspective from k and consequently the lines [E,F], [B,C], [Q,R] are concurrent. Since [B,C]∩[E,F] = P, we conclude that [P,Q,R]. □

Lemma 10 *Let A,B,C,D be four points in clockwise order on circle ω and let P = [A,D]∩[B,C]. Let k be a line through P which lies entirely in the exterior of ω and let R = k∩[D,C] and Q = k∩[A,B]. Let E be the contact point between A,B of the tangent from Q and let F denote the contact point between C, D of the tangent from R. Then [E,F,P].*

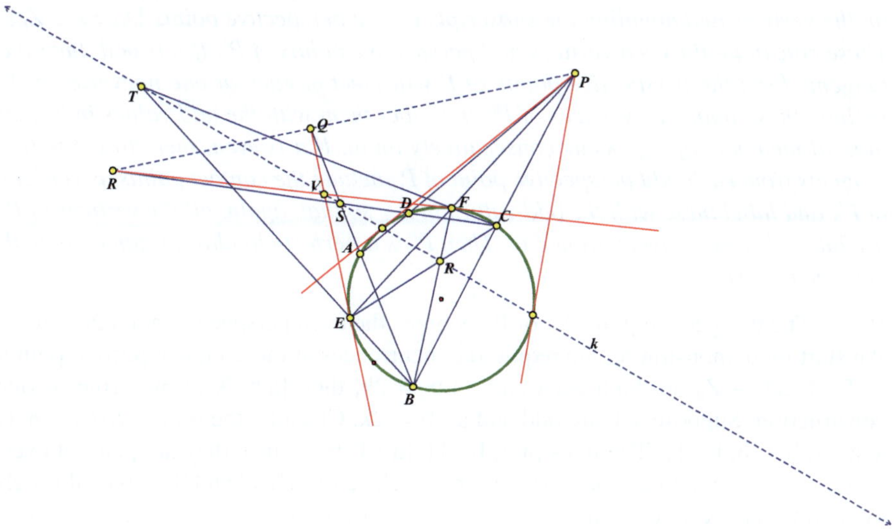

Fig. 9 Collinearity of P,Q,R

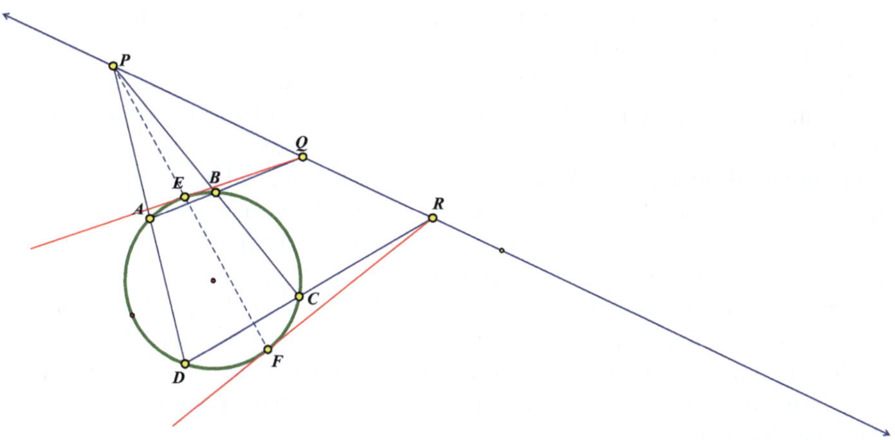

Fig. 10 Collinearity of P, E, F

Proof Let E′ = [F,P]∩ω. By Lemma 9 we have [R,P,Q′] where Q′ = [E′, E′]∩[A,B] (see Fig. 10). Since R,P,Q all lie on k, we have [R,P,Q] and thus [R,P,Q, Q′]. But then Q′ = [A,B]∩k = Q, and so E, E′ are both contact points of tangents from Q. Since F lies between C,D and E′ = [F,P]∩ω, it must be that E′ lies between A,B and thus E′ = E. But then [P, E′,F] implies [P,E,F]. □

Now we are ready to present our doubling theorem.

Theorem 11 *Let P be a perfect circular n-gon with circle ω, vertex set Z_n, and perspective axis j. We construct a perfect 2n-gon P′ on ω by first doubling the labels* *on the vertices and doubling the subscripts on the perspective points* $\{X_k : k \in Z_n\}$. *These constitute the even vertices and perspective points of P′. If n is odd, then the tangents from the perspective points of P will meet ω each at one nonvertex of P. Include these points as vertices of P′ and label them with the odd values in* Z_{2n} *so that all members of* Z_{2n} *occur consecutively on ω. If n is even, then draw the two tangents from each odd perspective point of P, include the contact points as vertices of P′, and label these with the odd values in* Z_{2n} *so that, again, all the vertices of P′ are labeled consecutively around ω. Then P′ is a perfect circular 2n-gon on ω with perspective axis j.*

Proof The perspective point X_k of P becomes the even perspective point X_{2k} of P′. We start by demonstrating the required concurrences at the even perspective points of P′. If a,b ∈ Z_n are both even and a+b = 2k, then [a,b, X_{2k}] by virtue of our construction. Suppose a, b are odd and a+b = 2k. Consider the set of vertices {a-1, a, a+1, b-1, b, b+1}. The lines [a-1, b+1], [a+1, b-1] since they are pairs of even vertices which sum to 2k contain the point X_{2k}. The lines [a,a] and [b,b] pass through the points X_{2a}, X_{2b}, respectively, by virtue of our construction. The line [a+1, a-1] joins two even vertices, so we know it contains X_{2a}, and similarly the line [b+1, b-1] contains X_{2b}. Since X_{2a}, X_{2b}, and X_{2k} lie on j, we may apply Lemma 10 to conclude [a,b, X_{2k}]. If n is even, then it may be that $X_{2a} = X_{2b}$. In this case [a,b, X_{2k}] follows directly from Lemma 8.

It remains to show for each odd m ∈ Z_{2n} that the lines {[a,b]: a+b = m } concur at a point X_m. Let p, q, r be three distinct odd members of Z_{2n} and let W = [m-p, p]∩ [m-q,q]. By Pascal's Theorem, (E1), the inscribed hexagon (m-q, q, -q, p, m-p, p-m) implies that the points W, X_0 = [q, -q] ∩ [m-p, p-m] and $X_{p\text{-}q}$ = [p-m, m-q] ∩ [p, -q)] are collinear. (Note that, by the first paragraph, we have shown the required concurrences at the even perspective points X_0, $X_{p\text{-}q}$.) The point W thus lies on the perspective axis j. Now let W′ = [m-p, p]∩[m-r, r], then similarly the points W′, X_0, $X_{p\text{-}r}$ are collinear and so W′ also lies on j. But W = j ∩ [m-p, p] = W′ and so the diagonals [p, m-p], [q, m-q], and [r, m-r] concur at the same point (W = W′) on j. Since p, q, r were chosen freely from among the odd values in Z_{2n}, this implies that all the diagonals {[k, m-k]: k is odd} meet at the same point on j (which we designate by X_m). □

With Theorem 11 in hand, we can now define the center of an odd perfect circular n-gon P by first doubling P to obtain an even perfect circular 2n-gon P′ and then taking the center of P′ to be the center of P. The following is our intended description of perfect circular n-gons.

Theorem 12 *Every perfect circular n-gon is the inverse of a regular n-gon.*

Proof Let P be a perfect cyclic n-gon with circular envelope ω, vertex set Z_n, perspective set X = {X_k: k ∈ Z_n}, and perspective axis j (see Fig. 11). For each k ∈ Z_n let ω_k denote the circle with center X_k which is orthogonal to ω. We regard j as a circle with in finite radius and the line of centers for circles ω, j is the line k through the center O of ω and perpendicular to j. The circles ω_k are orthogonal to

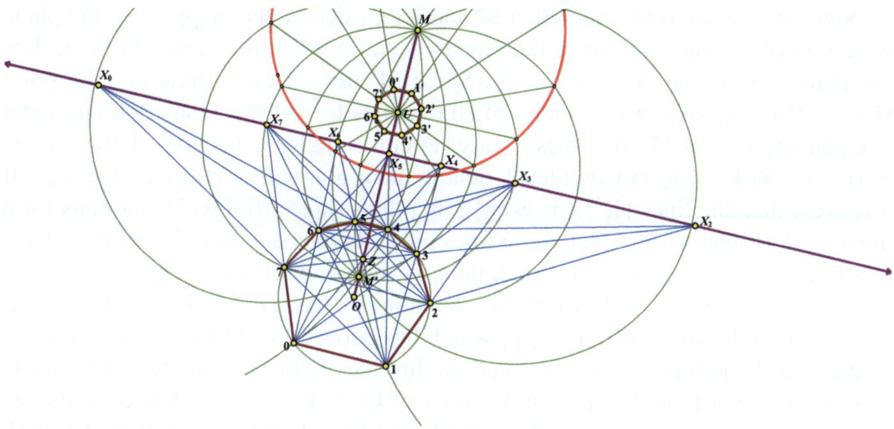

Fig. 11 Perfect octagon is inverse of regular octagon

both j and ω, and so by (E8) cross k in the same two points M, M′. We suppose that M lies in the exterior and M′ in the interior of ω. For each $k \in Z_n$ let ϕ_k denote the mapping which inverts the plane through ω_k.

Claim 1 For each $k \in Z_n$ we have $\phi_k(\omega_{k-1}) = \omega_{k+1}$.

Proof of Claim 1 Suppose k is odd, then $[X_k, (k-1)/2, (k+1)/2]$ where vertex (k-1)/2 lies on ω_{k-1} and vertex (k+1)/2 lies on ω_{k+1}. Since ω_k is orthogonal to ω we must have $\phi_k((k-1)/2) = (k+1)/2$. Also we have $\phi_k(M) = M$ and $\phi_k(M') = M'$ since M, M′ lie on ω_k, thus the three points M, M′, (k-1)/2 of ω_{k-1} are inverted to the three points M, M′, (k+1)/2 of ω_{k+1}. By (E9) we must have $\phi_k(\omega_{k-1}) = \omega_{k+1}$. Now suppose k is even and mark on ω the contact points of the two tangents from each odd perspective point. If such a point lies between vertices k, k+1 mark it as vertex k+(1/2). In accordance with Theorem 11, these points along with the original vertices of P form a perfect 2n-gon on ω, and for each k the lines $\{[x,y] : x+y = k\}$ concur at X_k. We have $[X_k, (k-1)/2, (k+1)/2]$ where (k-1)/2 lies on ω_{k-1} and (k+1)/2 lies on ω_{k+1}, and it follows again that $\phi_k(\omega_{k-1}) = \omega_{k+1}$. This completes the proof of Claim 1.

Claim 2 The n circles ω_k meet at 2n equal angles at M and M′.

Proof of Claim 2 Let θ_k denote the angle at M (and M′) between ω_{k-1} and ω_k. By Claim 1, ω_k is a midcircle for ω_{k-1}, ω_{k+1} and so the angle at M, M′ between ω_{k-1} and ω_{k+1} is bisected by ω_k. Thus the angle at M, M′ between ω_k and ω_{k+1} is also θ_k. Let θ_{k+1} denote the angle at M, M′ between ω_{k+1} and ω_{k+2}. By Claim 1, ω_{k+1} is a midcircle for ω_k and ω_{k+2} and it follows similarly that the angle at M, M′ between ω_k and ω_{k+1} equals the angle between ω_{k+1} and ω_{k+2}, i.e., $\theta_k = \theta_{k+1}$. Since $k \in Z_n$ is arbitrary, Claim 2 follows. As a consequence of Claim 2, we call M, M′ the *equiangular points* of P. If θ is the common angle at M, M′ between consecutive circles ω_k, then we observe also that $\theta = \angle X_k M X_{k+1}$ for each $k \in Z_n$.

Now let α be a circle centered at M and let ϕ_α denote the mapping of the plane which inverts through α. Since the circles ω_k all contain the center M of α, they invert to lines J_k, i.e., $\phi_\alpha(\omega_k) = J_k$. Since M′ lies on each circle ω_k, the point $U = \phi_\alpha(M')$ is a concurrence point of all the lines J_k, and since the circles ω_k meet at equal angles θ at M′, the lines l_k meet at equal angles θ at U (by (E10). Let $\omega' = \phi_\alpha(\omega)$, let $k' = \phi_\alpha(k)$, and let P′ denote the n-gon with vertices $\{k' : k \in Z_n\}$. If n is even, then the lines $\{J_k : k \text{ is even}\}$ join points (k/2)′, ((k+n)/2)′ and thus form the long diagonals of P′. Since these lines meet at equal angles at U, we must have that P′ is a regular n-gon. If n is odd, then we construct the double Q = {0, 0.5, 1, 1.5, 2, 2.5, ..., n-1, n-0.5} of P where the point k+0.5 is defined to be the contact point on ω of the tangent from X_{2k+1} which lies between k and k+1. Each circle ω_k meets ω at the points k/2, (k+n)/2 and the inverses J_k meet ω' at the points (k/2)′, ((k+n)/2)′. Each J_k is thus a long diagonal of $Q' = \{t' : t \in Q\}$. Since the circles ω_k meet at equal angles $\theta = 360°/2n$ at M′, the lines J_k meet at equal angles at U. Thus Q′ is a regular 2n-gon and $P' = \{k' : k \in Z_n\}$ is a regular n-gon which inverts onto P. □

Corollary 13 *Let γ be any conic; let m be a line which does not meet γ, and let Q be a point on γ. Then there exists a unique perfect n-gon on γ with perspective axis m and vertex Q.*

Proof There exists an invertible transformation T which takes γ to a circle ω; m to another line j, and point Q to a point 0. Construct a circle β with center on j and which is orthogonal to ω. Let k denote the line through the center of ω which is perpendicular to j. The circle β meets k at the equiangular points M, M′. Let $0' = [M,0] \cap \omega$ and construct the unique regular n-gon inscribed in ω which contains the vertex 0′. Let P′ denote this regular n-gon and label its vertices 0′, 1′, ..., (n-1)′ consecutively around ω. Let α denote the circle with center M which is orthogonal to ω, and let ϕ_α denote the inversion through α. For each $k \in Z_n$ let vertex k be given by $k = \phi_\alpha(k')$. Since α is orthogonal to ω, the points k form an n-gon P on ω. By Theorem 12, P is a perfect n-gon with perspective axis j. Also P contains the vertex 0 since $0 = \phi_\alpha(0')$. Now apply the inverse transformation T^{-1} to obtain the desired perfect n-gon on the conic γ. □

6 Perfect Polygons with Irreducible Cubic Envelope

In this section we include some basic properties of perfect polygons with an irreducible cubic envelope.

Theorem 14 *Let P be a perfect n-gon with vertex set Z_n, perspective set $\{X_k : k \in Z_n\}$ and nonsingular irreducible cubic envelope γ. Let $(\gamma,*)$ denote the corresponding BHA. If $s,t,p,q \in Z_n$ and $s+t = p+q$, then $X_s * X_t = X_p * X_q$. If no three perspective points of P are collinear, then the perspective points form a perfect n-gon on γ.*

Proof Consider: $X_s*X_t = (p*(s-p))*(0*t) = (p*0)*((s-p)*t) = X_p*X_q$. Thus all lines of the form $\{[X_s,X_t] : s+t = k\}$ meet at the same point U_k on γ. We then obtain, with the further provision that no three perspective points of P are collinear, that the perspective points of P form a perfect n-gon on γ. □

In case P has three collinear perspective points we obtain the following result.

Theorem 15 *Let P be a perfect n-gon with vertex set Z_n, perspective set $X = \{X_k : k \in Z_n$,and nonsingular irreducible cubic envelope γ. Let $(\gamma,*)$ denote the corresponding BHA. If P contains three collinear perspective points, then $(P,*)$ is a subalgebra of $(\gamma,*)$ and $(X,*)$ is a subalgebra of $(P,*)$. Moreover, if $[X_p, X_q, X_r]$ and $p+q+r = m$, then $[X_s, X_t, X_u]$ iff $s+t+u = m$.*

Proof The product of two vertices of P lies in X and the product of a vertex and a perspective point is a vertex, so to show (P,*) is a subalgebra of (γ,*) it suffices to show that X is closed under multiplication. Suppose $[X_p, X_q, X_r]$ and thus $X_p* X_q = X_r$. By Theorem 14, $X_a* X_b = X_r$ for any $a,b \in Z_n$ such that $a+b = p+q$. Then for any $k \in Z_n$ we have $X_k = X_{p+q-k}*(X_{p+q-k}*X_k) = X_{p+q-k}*X_r$. This implies, by Theorem 14 again, that $X_i* X_j = X_k$ whenever $i+j = p+q+r-k$. Consequently both X and P are closed under *. Moreover, $X_i* X_j = X_k$ whenever $i+j+k = p+q+r$, proving the last statement of the theorem. □

If, as in Theorem 15, the perspective points of P form a subalgebra of (γ,*), we call P a *closed perfect polygon.* If P has n points and the subscripts of three collinear perspective points sum to m, then we say that P has *perspective sum m.* The set of perspective points of a closed perfect polygon with perspective sum m provide an example of a *(Z_n,m) geometric triple system.* Let G be an abelian group and $g \in G$, and let $\phi: G \rightarrow \prod$ be an injective mapping from G into the real projective plane $\prod$ such that no four points in $\phi(G)$ are collinear. If for each three element subset {a,b,c} of G, the corresponding points {ϕ(a), ϕ(b), ϕ(c)} are collinear, then ϕ(G) is a *(G,g) triple system.* The group G is the *base* of the triple system, and the element g is the *sum.* Properties of geometric triple systems with base Z or with base Z_n are given in [5]. If S is any subset of (γ,*) and q is any point on γ, then $q*S = \{q*s : s \in S\}$ is called a *translate* of S. The simple relationship between a perfect n-gon and a certain geometric triple system is given in our next theorem.

Theorem 16 *Every perfect n-gon with nonsingular irreducible cubic envelope γ is a translate of a $(Z_n,0)$ geometric triple system.*

Proof Let P be a perfect n-gon on γ and let f be a flex point of γ. Note that f cannot be a vertex of γ, since we would have $f = f*f = X_{2f}$, i.e., a vertex of P would coincide with a perspective point of P, which is not allowed. Let $q = 0*f$ so that $q*0 = f$. We will show that the translate q*P is a $(Z_n , 0)$ triple system. If $a \in Z_n$, (i.e., a is a vertex of P), let a′ denote the point *q*a. Now* suppose $a + b + c = 0$, and consider: $a' * b' = (q*a)*(q*b) = (q*q)*(a*b) = ((0*f)*(0*f))*X_{a+b} = ((0*0)*(f*f))*X_{a+b} = (X_0*f)*(0*(a+b)) = (X_0*f)*(0* -c) = (0*f)*(X_0 * -c) = q*c = c'$. Thus a′, b′, c′ are collinear whenever $a+b+c = 0$, showing that q*P is a $(Z_n, 0)$ triple system. □

In Fig. 12 we illustrate a perfect hexagon along with a translate which is a $(Z_{6,}0)$ triple system. Now we would like to ask the following question. Suppose T is a geometric triple system with base Z_n and irreducible cubic envelope γ. For which points q is q*T another triple system, and for which points q is q*T a perfect n-gon? To answer this question we will use the following results proved in [5]. We will classify irreducible cubic curves according to the classification given in [3]: Type I cubics are those which can be transformed into a cubic with equation $y^2 = x^3$; Type II cubics are those which can be transformed into a cubic with equation $y^2 = x^2(x+1)$; Type III cubics are those which can be transformed into a cubic with equation $y^2 = x^2(x-1)$; Type IV cubics are those which can be transformed into a cubic of the form $y^2 = x(x-1)(x-w)$ where $w>1$; Type V cubics are those which can be transformed into a cubic of the form $x(x^2 + kx + 1)$ where $-2<k<2$. Types I,II,III are singular, each curve with exactly one singular point. In [5] it is shown that no triple systems with base Z_n exist on a Type I or Type II cubic, but we must consider whether it is possible for a perfect polygon to exist on such cubics. Types IV and V are nonsingular, Type V cubics consist of a single connected component

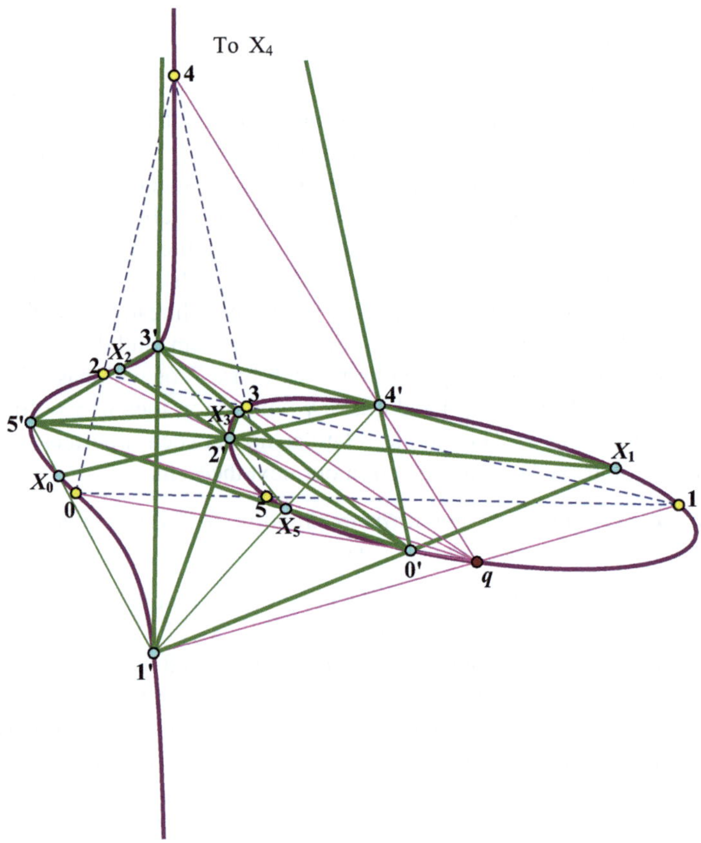

Fig. 12 Perfect hexagon obtained as translate of $(Z_6,0)$ triple system

when regarded in the projective plane, and Type IV cubics consist of two connected components which we call the *bell* and the *oval*. If γ is a Type IV cubic, then the bell of γ is a subalgebra of the BHA (γ,*), and the oval is *anticlosed*, i.e., if a,b $\in$ oval, then a*b lies in the bell. The unique singularity of a Type III cubic is an isolated point; otherwise Type III cubics are like Type V. Type III, IV, V cubics each contain three collinear flex points; Type I, II each have only one flex point. We shall say that two triple systems on a cubic γ are *equivalent* in case they consist of precisely the same set of points on γ, but may have different labels. For example, adding 1 to each vertex of a (Z_n,0) triple system creates an equivalent (Z_n,3) triple system. Here are the useful results from [5]:

(R1) If gcd(n,3) = 1, then every triple system with base Z_n is equivalent to a (Z_n,0) geometric triple system. If 3|n, then every triple system with base Z_n is equivalent to either a (Z_n,0) or a (Z_n,1) triple system. Moreover, if 3|n the (Z_n,0) triple system is not equivalent, or even isomorphic, to the (Z_n,1) triple system.

(R2) Let T be a triple system with base Z_n. If n$\geq$ 10, then the vertices of T lie on an irreducible cubic curve of Type III,IV, or V.

(R3) For each positive integer n, a triple system with base Z_n and any sum can be constructed on any Type III, IV, or V cubic curve.

(R4) Let γ be a Type III, Type V, or bell of a Type IV cubic curve and let f be a flex point of γ. Then for each positive integer n there exists a unique (Z_n,0) triple system on γ with f as vertex 0.

(R5) Let γ be a Type III, Type V, or bell of a Type IV cubic curve. If 3|n then, up to equivalence, there exists a unique (Z_n,0) triple system on γ which contains all three flex points of γ, and there exist exactly two nonequivalent (Z_n,1) triple systems on γ which are translates of each other. If gcd(n,3) = 1, then there exist exactly three mutually nonequivalent (Z_n,0) triple systems on γ, each containing a different flex point of γ.

(R6) If gcd(m,3) = 1, then every triple system with base $Z_2 \times Z_{2m}$ is equivalent to a ($Z_2 \times Z_{2m}$, (0,0)) triple system. If 3|m, then every triple system with base $Z_2 \times Z_{2m}$ is equivalent to either a ($Z_2 \times Z_{2m}$, (0,0)) or a ($Z_2 \times Z_{2m}$, (0,1)) triple system.

In the presence of an irreducible cubic curve γ, we can define perfect polygons with fewer than five points. If we choose any nonflex point of γ for vertex 0, and let X_0 = 0*0, we have a perfect 1-gon. If γ is Type III or V, then for each nonsingular point y on γ there exist exactly two points on γ which satisfy the equation x*x = y. We denote the solutions to this equation by $\sqrt{y}$. So if 0 is any nonflex point of γ, then there exist two points which we label 0,1 such that 0*0 = 1*1 = X_0. We thus obtain a perfect polygon with base Z_2. A (Z_3,0) triple system T consists of the three flex points of γ. It is unique by (R5). If q is any nonflex point of γ then, in accordance with Theorem 16, q*T is a perfect polygon, (i.e., a *perfect triangle*), with base Z_3. First we consider triple systems with a Type III or Type V cubic envelope.

Theorem 17 *Let T be a (Z_n,0) triple system whose vertices lie on a Type V, Type III, or bell of a Type IV cubic curve γ. Let S be the unique (Z_{3n},0) triple system on γ, and let q $\in \gamma$. Then q*T is a geometric triple system with base Z_n iff q $\in$ S. If q is not in S, then q*T is a perfect n-gon.*

Proof By (R5) the vertices of S are uniquely determined. Vertices 0,n,2n correspond to the three collinear flex points of γ. The vertices of S can be partitioned into three subsets: U = {0,3,6, ... , 3n-3}, W = {1,4,7, ... , 3n-2}, and V = {2,5,8, ... ,3n-1} each of which form a subalgebra of (S,*). Also V*W = {v*w : v ∈ V, w ∈ W} = U; V*U = W, and V*W = U. The relabeling of U given by 3k → k turns U into a (Z_n,0) triple system; the relabeling of W given by 3k-2 → k turns V into a (Z_n,2) triple system, and the relabeling of V given by 3k-1 → k turns V into a (Z_n,1) triple system. By (R5), U,V,W constitute the full set of distinct triple systems on γ with base Z_n. In case gcd(3,n) = 1, each of U,V,W is equivalent to a (Z_n,0) triple system, and if 3|n, then U is a (Z_n,0) triple system and V,W are (Z_n, 1) triple systems. In Figs. 13 and 14 we illustrate these cases with n = 6 and n = 5. In either case we can identify T with U, and so we have V*T = W and W*T = V. If q ∈ γ, and q*T is a geometric triple system with base Z_n, then we must have either q*T = T; q*T = V or q*T = W. If q ∈ T, then q*T ⊆ T since T is a subalgebra of (γ,*). Since the mapping ϕ: T → q*T defined by ϕ(t) = q*t is a bijection, we must have |q*T| = |T| and thus q*T = T. We have shown that q*T = T iff q ∈ T. If q ∈ W, then q*T ⊆ V, since W*T = V, and since q*T and V each has n elements, we can conclude that q*T = V. On the other hand, if q*T = V, then q*t = v for some t ∈ T and some v ∈ V. But then q = t*v and so belongs to W. We have shown that q*T = V iff q ∈ W.

Fig. 13 (Z_{18},0) triple system

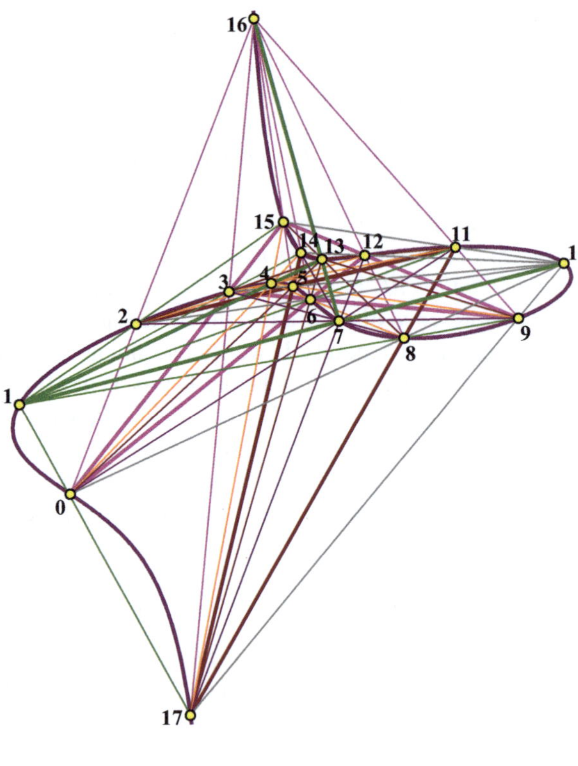

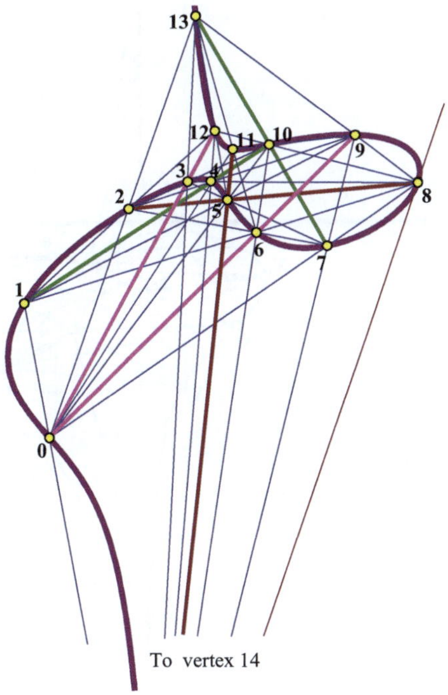

Fig. 14 $(Z_{15},0)$ triple system

Similarly $q^*T = W$ iff $q \in V$. Consequently q^*T is a triple system on γ with base Z_n iff $q \in U \cup V \cup W$, i.e., iff $q \in S$. Then by Theorem 16, if $q \notin S$, then q^*T is a perfect n-gon. □

In Fig. 13 we illustrate a $(Z_{18},0)$ triple system with vertices on a Type V cubic, and in Fig. 14 we illustrate $(Z_{15},0)$ triple system on the same curve. In both figures the key subalgebras U,V,W are emphasized.

Next we will consider *split perfect polygons,* i.e., perfect polygons whose vertices and perspective points lie on both components of a Type IV cubic curve γ. By Theorem 16 every such perfect polygon P on γ is a translate of a geometric triple system T on γ, i.e., $P = q^*T$ for some $q \in \gamma$. Since the oval is anticlosed, the even perspective points of P must lie on the bell of γ. There are several possibilities for the disposition of the vertices and the remaining perspective points of P: (i) all the vertices of P may lie on the oval when all perspective points of P must lie on the bell; (ii) all even vertices of P lie on the oval; all odd vertices of P lie on the bell and all odd perspective points lie on the oval; (iii) all the odd vertices of P lie on the oval; all the even vertices of P lie on the bell, and all odd perspective points lie on the oval. Let us first consider perfect polygons of Type(i). In Fig. 15 such a perfect n-gon P is illustrated on a typical Type IV cubic. By Theorem 16, $P = q^*T$ where T is a $(Z_n,0)$ triple system. Since all the vertices of P lie on the oval, it must be that q lies on the oval and T lies entirely on the bell of the Type IV cubic envelope γ. For every point q on the oval, q^*T must be a perfect n-gon on the oval, and cannot be a geometric triple system. A Type (ii) perfect polygon is illustrated in Fig. 16.

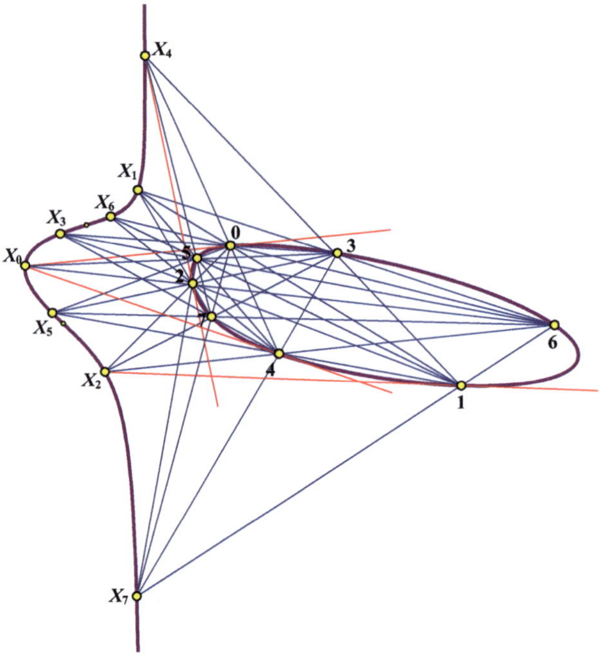

Fig. 15 Perfect octagon on Type IV cubic

We observe that a Type (iii) perfect polygon can be obtained from a Type (ii) perfect polygon by adding 1 to each vertex and 2 to the subscript of each perspective point, so it is unnecessary to consider both Types. Also, we observe that a perfect polygon of either of these types must be realized, in accordance with Theorem 16, as the translate of a split geometric triple system.

Theorem 18 *Let T be a split (Z_n,0) triple system with Type IV cubic envelope γ. and suppose that n = 2m with m even. Let q $\in \gamma$. If gcd(n,3) = 1, let S be the unique split (Z_{3n},0) triple system on γ, then q*T is a (split) triple system on γ iff q $\in$ S. If q $\notin$ S, then q*T is a Type (iii) split perfect polygon. If 3|n, then there is only one split (Z_n,0) triple system T on γ, and q*T is a triple system iff q $\in$ T.*

Proof Suppose first that gcd(*n,3*) = *1*. The vertices of S can be partitioned into three subalgebras U,V,W of (γ,*) given by: U = {0,3,6, ... , 3n-3}; V = {2,5,8, ... , 3n-1}, and W = {1,4,7, ... , 3n-2}. By (R5) each of these is equivalent to a (Z_n,0) triple system, and U,V,W constitute the full set of split (Z_n,0) triple systems on γ. As in the proof of Theorem 17, we may identify T with U, and we then have T*V = W; T*W = V, and V*W = T. It follows similarly that q*T = T iff q $\in$ T; q*T = V iff q$\in$ W, and q*T = W iff q $\in$ V. In short, q*T is a triple system on γ iff q$\in$ S, and if q$\notin$S then, in accordance with Theorem 16, q*T is a Type (iii) split perfect polygon. If 3|n, then by (R5) T is unique and q*T must equal T in order to be a triple system. But q*T = T iff q $\in$ T. □

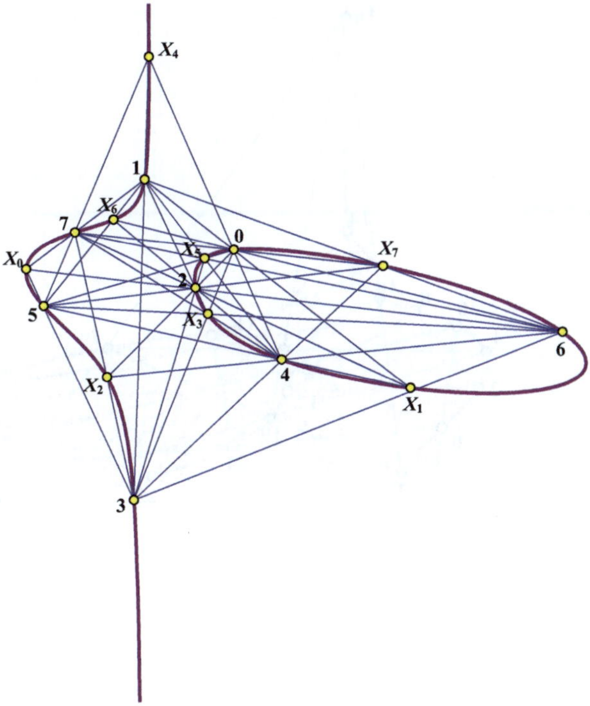

Fig. 16 Perfect octagon: variant

In Fig. 17 we illustrate Theorem 18 using the unique split $(Z_{24},0)$ triple system, (for S), on a typical Type IV cubic curve. In the next section we consider the case in which n is twice an odd number. In case T is a $(Z_n,1)$ triple system, then by (R1) T is equivalent to a $(Z_n,0)$ triple system if gcd(n,3) = 1. But also by (R1) if $3|n$, then T is not equivalent to a $(Z_n,0)$ triple system. Suppose $3|n$ and the cubic envelope γ of T is Type III or Type V. Then by (R5) there are exactly two distinct $(Z_n,1)$ triple systems V, W on the cubic envelope γ. These occur as subalgebras of the unique $(Z_{3n},0)$ triple system S on γ. In fact we can define U,V,W as in Theorem 16, where U is a $(Z_n,0)$ triple system, and we can conclude as in Theorem 16 that q*T is a triple system iff $q \in S$. In case T is split a similar result holds. Similar to (R5) there are exactly two split $(Z_n,1)$ triple systems V,W on a Type IV cubic γ and these occur as subalgebras of the unique split $(Z_{3n},0)$ triple system S on γ along with the unique split $(Z_n,0)$ triple system U. Again we find that q*T is a triple system iff $q \in S$.

7 Twins

When the number n of vertices of a geometric triple system on a Type IV cubic is twice an odd number we will find the following Lemmas useful. If p is a point on the bell of a Type IV cubic γ, then $\sqrt{p}$ contains exactly four points, two on the bell

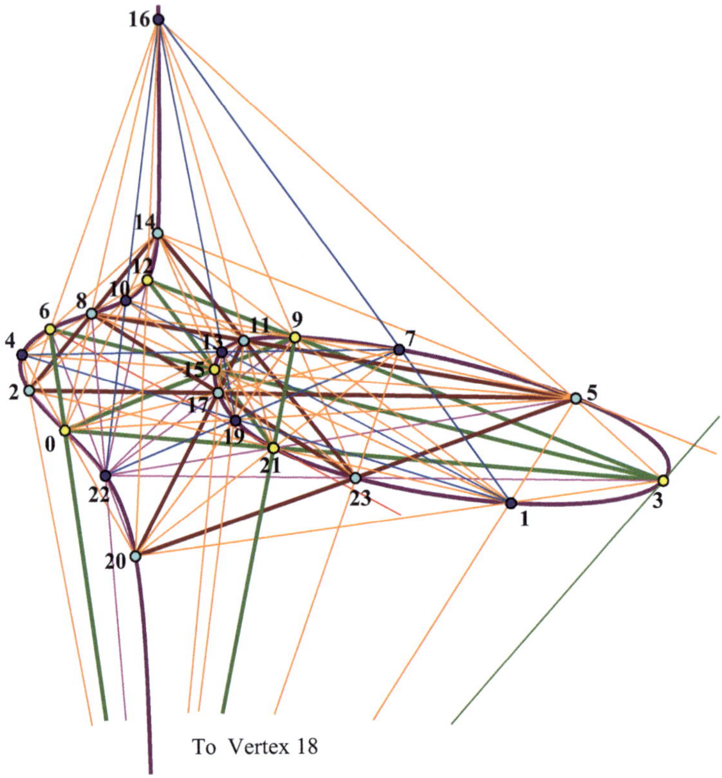

Fig. 17 Three split (Z_8,0) triple systems occur as subalgebras of the split (Z_{24},0) triple system

and two on the oval. If a$\in \sqrt{p}$ and a lies on the oval, let a′ denote the remaining element of $\sqrt{p}$ on the oval, and if b $\in \sqrt{p}$ and b lies on the bell, let b′ denote the remaining element of $\sqrt{p}$ on the bell. In either case we refer to the points x, x′ as *twin points.*

Lemma 19 *Let a,b be points on any irreducible nonsingular cubic curve* γ. *Then* $a{*}b = a'{*}b'$ *and* $a{*}b' = a'{*}b = (a{*}b)'$

Proof Let p = a*a = a′*a′, and let q = b*b = b′*b′. Then the points a*b′, a′*b, a*b, a′*b′ all lie in $\sqrt{(p{*}q)}$. If γ is a Type V cubic, then $\sqrt{(p{*}q)}$ contains only two points so two pairs among these four points must be equal. The point a*b′ cannot be equal to a*b or a′*b′ since cancellation would then require b = b′ or a = a′ which is impossible since no point on γ is equal to its twin. We must then have a*b′ = a′*b and a*b = a′*b′. The same argument can be used in case a,b lie on the bell of a Type IV cubic. Now suppose γ is Type IV and a,b lie on the oval. Then the four points a*b′, a′*b, a*b, a′*b′ all lie on the bell and we can employ the same argument. Finally, if a lies on the bell and b lies on the oval of γ, then the four points all lie on the oval. But $\sqrt{(p{*}q)}$ also contains only two points on the oval with the same result.

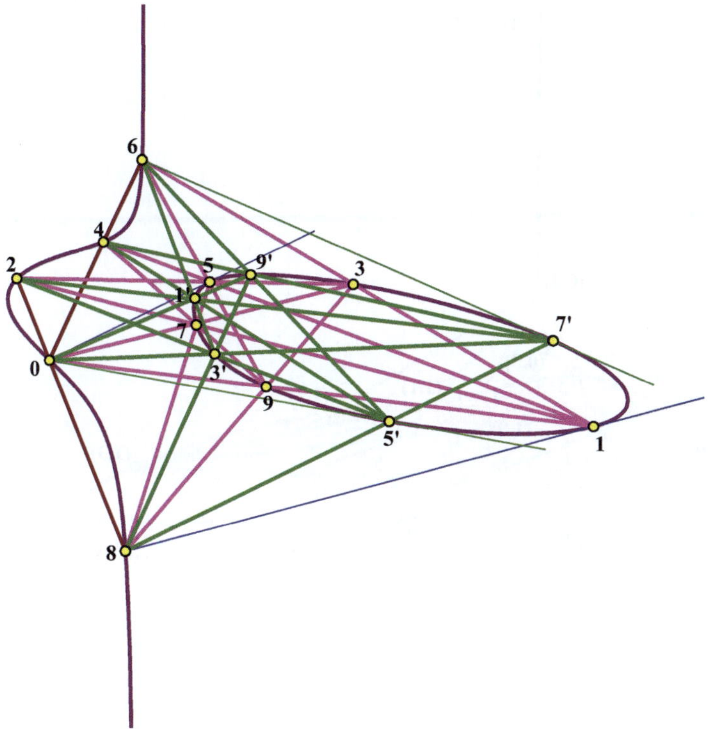

Fig. 18 Twin split $(Z_{10},0)$ triple systems

Since $((a*b)')*((a*b)') = (a*b)*(a*b) = p*q$, the point $(a*b)'$ also lies in $\sqrt{(p*q)}$. No point of γ is equal to its own twin, so we cannot have $(a*b)' = a*b$; so we must have $(a*b)' = a*b' = a'*b$. □

Now let T be a split $(Z_n,0)$ triple system on a Type IV cubic curve γ with $n = 2m$. If S is any subset of γ, we let $S' = \{x' : x \in S\}$. Let Q denote the set of even vertices of T which must lie on the bell, and let W denote the set of odd vertices of T which must all lie on the oval. Let $W' = \{x' : x \in W\}$ and let $\hat{T} = Q \cup W'$. It is shown in [6] that $\hat{T}$ is a split $(Z_n,0)$ triple system. If m is even, then x, x′ both lie in T for each point x in the oval and thus $T = \hat{T}$. But if m is odd, then T, $\hat{T}$ constitute two distinct extensions of Q, (which we call *twin extensions)*, and the extending sets W, W′ are disjoint. In Fig. 18 we illustrate a split $(Z_{10},0)$ triple system T along with its twin $\hat{T}$. Also in [6] the following theorem is proved.

Theorem 20 *Let T be a split $(Z_n,0)$ triple system with $n = 2m$, m odd. Then the points in $T*\hat{T}$ form a $(Z_2 \times Z_n, (0,0))$ triple system.* □

In Fig. 19 we illustrate a $(Z_2 \times Z_{10}, (0,0))$ triple system on a typical Type IV cubic. The twin subalgebras T, $\hat{T}$ of Fig. 18 are emphasized. Now suppose T is a split $(Z_n,0)$ triple system with $n = 2m$, m odd as in Theorem 20, and suppose also

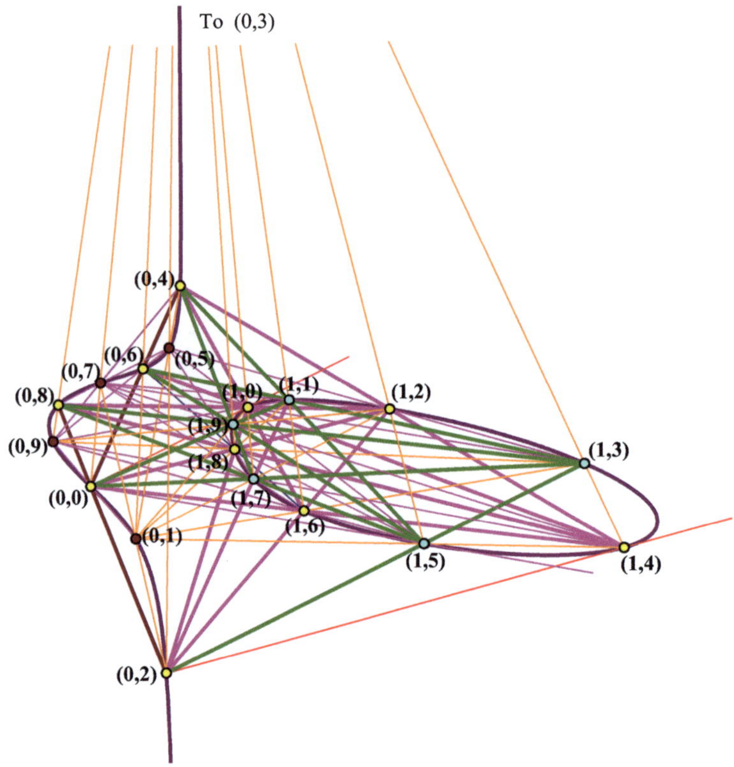

Fig. 19 $(Z_2 \times Z_{10}, (0,0))$ triple system

that gcd(n,3) = 1. Then T is an extension of a $(Z_m,0)$ triple system on the bell of a Type IV cubic γ and, in accordance (R5), there are exactly three distinct $(Z_m,0)$ triple systems on γ. Each of these, in turn, has two twin extensions yielding a total of six distinct split $(Z_n,0)$ triple systems on γ. If $q \in \gamma$ and q*T is a triple system, then q*T must be one of these six split $(Z_n,0)$ triple systems. In our next theorem we determine for which point q this is the case.

Theorem 21 *Let T be a split $(Z_n,0)$ triple system on a Type IV cubic γ with $n = 2m$, m odd and gcd(n,3) = 1. Let e,f,g denote the three flex points of γ and suppose vertex 0 of T coincides with e. If $q \in \gamma$, then q*T is a triple system iff $q \in T \cup (f{*}T) \cup (g{*}T)$. Moreover, $f * \hat{T} = (f *T)\hat{}$ and $g * \hat{T} = (g * T)\hat{}$.*

Proof Let Q denote the set of even vertices of T which all lie on the bell of γ, and let W denote the set of odd points of T which all lie on the oval. Let $b \in f * \hat{T}$, then $b \in f * t'$ for some $t \in T$. By Lemma 19, $b = (f *t)'$ and thus $b \in (f *T)\hat{}$. It follows that $f *\hat{T} = (f *T)\hat{}$ and similarly $g * \hat{T} = (g * T)\hat{}$. Now let S denote the unique $(Z_2 \times Z_{3n}, (0,0))$ triple system on γ. We shall see that S contains all six of the split $(Z_n,0)$ triple systems as subalgebras. It will then follow that if q*T is a triple system, then q must lie in S. Without loss of generality we let e = (0,0); f = (0,n), and g = (0,2n). These are the three flex points of γ which all lie in S. Then T = Q∪W where Q = {(0,6t) : t = 1,2, ... ,m} and W = {(1,6t) : t = 1,2, ... , m}, and T$\hat{}$= Q∪W′ where W′= {(1,6t+3m) : t = 1,2, ... ,m}. Consider: f * T = (f * Q)∪(f * W) where f * Q = {(0, 2n – 6t) : t = 1,2, ... ,m} and f * W = {(1, 2n – 6t) : t = 1,2, ... ,m}; f * T′= (f * Q)∪(f * W′) where f * W′= {(1, m – 6t) : t = 1,2, ... ,m} ; g * T = (g * Q)∪(g * W) where g * Q = {(0,n – 6t) : t = 1,2, ... ,m} and g * W = {(1, n -6t) : t = 1,2, ... ,m}; g * T′= (g * Q)∪(g * W′) where g * W′= {(1, 5m – 6t) : t = 1,2, ... ,m}. The mutually disjoint sets Q, W, W′, f * Q, f * W, f * W′, g * Q, g * W, g * W′ account for all the elements of S except for those in the set M = {(0,t) : t ∈ Z_{3n} is odd}. The sets T, T′, f * T, f * T′, g * T, g * T′ are subalgebras of (S,*) equivalent to each of the six split (Z_n,0) triple systems on γ. We have already remarked that q*T can be a triple system iff q ∈ S, i.e., q lies in one of the sets Q, W, W′, f * Q, f * W, f * W′, g * Q, g * W, g * W′ , M. Note that Q *W = W and Q*W′= W′. If q ∈ T = Q∪W, then q * T = T is of course a triple system. Suppose q ∈ W′, then q * T = q *(Q∪W) contains points in q * W. These points have the form (1, 6t+3m)*(1, 6s) = (0, -6s – 6t -3m) which lies in M since m is odd and not a multiple of 3. None of the six split (Z_n,0) triple systems contain any of the points of M, so q*T cannot be a triple system. If q∈ M, then q * T contains points in q * Q which have the form (0,k)*(0,6t) = (0,-k – 6t) = (0,3n – k – 6t). But this element lies in M since k is odd and 3n is even. Again we can deduce that q * T is not a triple system. Now suppose q ∈ f * Q, then q = f *q_0 for some q_0∈ Q and q * T = (f * q_0)* (Q∪W) = {(f * q_0)* Q}∪{(f * q_0)*W} where {(f * q_0)*(e*Q)} = (e*f)*(q_0*Q) = g * Q, and (f * q_0)*W = (f * q_0)*(e*W) = (f*e)*(q_0*W) = g*W. Thus q*T = g*T is a triple system. Similarly, if q ∈ g*Q, then q*T = f*T is a triple system. Now suppose q ∈f*W, then q = f *w_0 and q*T = (f *w_0)*(Q∪W) = {(f *w_0)*Q}∪{(f *w_0)*W} = {(f *w_0)*(e*Q)}∪{(f *w_0)*(e*W)} = {g * W}∪{g*Q} = g*T. So q*T is a triple system. Similarly, if q∈ g*W, then q*T = f*T is a triple system. We have shown that if q ∈ T∪(f*T)∪(g*T), then q*T is a triple system. If q ∈ f * W′, then q = (1, m – 6t) for some t ∈ {1,2, ... ,m}, and points in q*W have the form (1, m – 6t)*(1,6s) = (0, 5m + 6t – 6s) and so lie in M. We can conclude that q * T is not a triple system. Finally, if q ∈ g*W′, then q = (1, 5m-6t) for some t ∈ {1,2, ... ,m}, and points in q*W have the form (1,5m-6t)*(1,6s) = (0, m + 6t – 6s) which also lie in M, so again, q*T is not a triple system. This completes the proof. □

Theorem 22 *Let T be a split (Z_n, 0) triple system on a Type IV cubic curve γ with $n = 2m$, and m an odd multiple of 3. If $q \in \gamma$, then q*T is a triple system iff $q \in T$.*

Proof If m = 3p, then the points 0, 2p, 4p are idempotent and all occur as vertices of T. The triple systems T, T$\hat{}$ are the only split (Z_n, 0) triple systems on γ. If q*T is a triple system, then it must be that either q*T = T or q*T = T$\hat{}$. So we must have either q∈ T or q∈ T*T$\hat{}$. Since T is contained in T*T$\hat{}$, we can say just q∈ T*T$\hat{}$. By Theorem 20, T*T$\hat{}$ is equivalent to the unique ($Z_2 \times Z_n$, (0,0)) triple system S on γ. Let Q, M, W, W′ denote the partition of S given by Q = {(0,t): t is even}; M = {(0,t): t is odd}; W = {(1,s): s is even} and W′= {(1,s): s is odd}. We can take T equivalent to Q∪W and T$\hat{}$ equivalent to Q∪W′. If q*T is to be a triple system, then it must be that q lies in one of the four sets Q, M, W, W′. Of course, if q ∈T = Q∪W, then q*T = T is a triple system. But if q lies in M or W′, then we obtain q *T = M∪W′ which is not a triple system. □

8 Closed Perfect Polygons

We have previously defined a closed perfect polygon P as a perfect polygon whose perspective points form a subalgebra of (γ,*) where γ is the cubic envelope of P. From Theorem 15 we see that this subalgebra is a geometric triple system T and P is thus a subalgebra of (γ,*) which extends T. In this section we will argue that P itself is a geometric triple system, but first we provide a couple of *perfect* constructions of closed perfect polygons. By *perfect* we mean constructions which are possible with only a straight-edge and compass. In Section 5 we can deduce from Theorem 12 that circular perfect polygons have perfect constructions precisely when regular polygons have such a construction. The constructions given for the Special perfect hexagon (see Theorem 1) and the general perfect hexagon (see Fig. 4 of Section 3) are also perfect. Perfect constructions of perfect polygons with irreducible cubic envelope are relatively rare. We can provide only a few examples. A perfect construction of a closed perfect hexagon can be obtained by following the prescription given in the first paragraph of Section 3, except that we select X_2 to lie at the intersection of [0,2] and [X_1, X_3]. In Fig. 20 we illustrate the two nonisomorphic closed perfect hexagons, one has a perspective sum of 0 and the other has a perspective sum of 1. Only the one, call it P, with perspective sum 0 has a perfect construction. Interestingly the *diagonals* [0, 3], [1, 4] and [2, 5] of P concur at a point Z. This can be proved as follows. The triangles ((0,1),(1,0),(0,5)), ((1,4),(0,3),(1,2)) are in perspective from the line [(1,5),(1,1),(0,0)], and so by

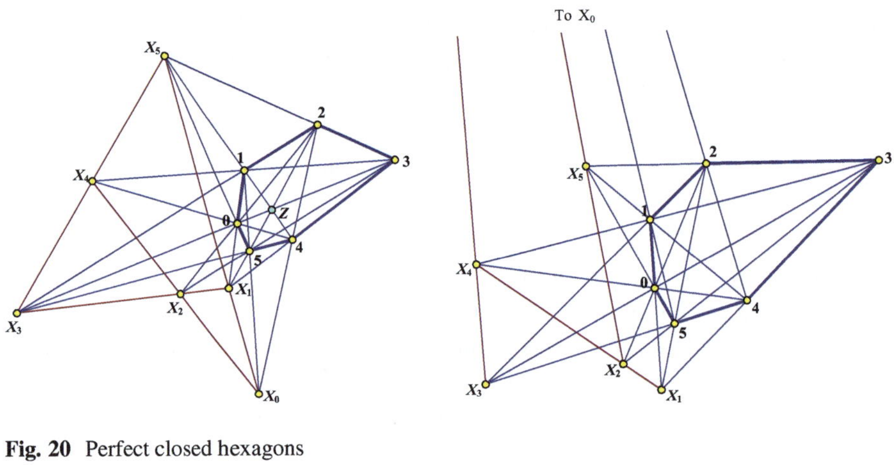

Fig. 20 Perfect closed hexagons

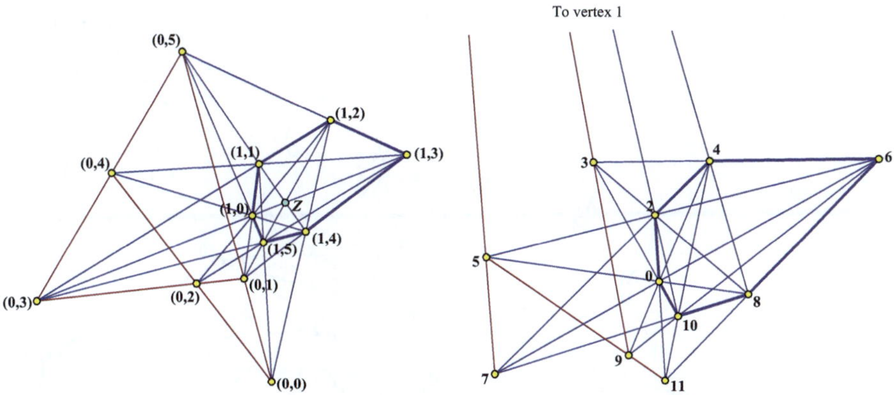

Fig. 21 Perfect closed hexagons relabeled

Desargue's Theorem (E3) the lines [(0,1),(1,4)], [(1,0),(0,3)], and [(0,5),(1,2)] are concurrent (compare with Theorem 6). This little result we might say belongs to the *external geometry* of P along with any concurrence of lines joining vertices of P or any collinearity of points formed by such lines. In Fig. 21 the vertices and perspective points of the two closed perfect hexagons of Fig. 20 have been relabeled to illustrate the equivalence of these perfect polygons with $(Z_2 \times Z_6, (0,0))$ and $(Z_{12}, 1)$ triple systems, respectively. In the following perfect construction the external geometry comes into play in a much more complicated way.

Theorem 23 *There exists a perfect construction of a closed perfect octagon which is equivalent to a* $(Z_2 \times Z_8, (0,0))$ *triple system.*

Proof The construction of a closed perfect octagon P is achieved as follows:

(i) Let vertices 0,1,2 be any three collinear points in the real projective plane and put perspective point X_1 on [0,1] and X_2 on [0,2].
(ii) Let $X_3 = [X_1, X_2] \cap [1, 2]$ and put vertex 3 on $[X_3, 0]$.
(iii) Let vertex 7 = $[X_1, 2] \cap [X_2, 3]$.
(iv) Let $A = [2, 7] \cap [0,3]$; $B = [X_2, 3] \cap [1, A]$; $D = [1, 7] \cap [0,2]$.
(v) Let $X_5 = [B, D] \cap [2, 3]$ and let $X_7 = [X_2, X_5] \cap [0,7]$.
(vi) Let vertex 5 = $[X_7, 2] \cap [X_5, 0]$ and let $X_6 = [X_3, X_5] \cap [1, 5]$.
(vii) Let vertex 6 = $[X_1, 3] \cap [X_7, 1]$ and let vertex 4 = $[X_6, 2] \cap [X_1, 5]$ shows that P.
(viii) Let $X_4 = [0,4] \cap [1, 3]$ and let $X_0 = [1, 7] \cap [2, 6]$.

This construction is illustrated in Fig. 22 where the dashed lines must all be proved. We found the proof of these collinearities too difficult to accomplish by repeated use of the theorems of Desargues or Pappus, and even too difficult to accomplish using cubic curve theory. We could find a proof by analytic means; first transforming vertices 0,1,2,3 to points with coordinates (0,0), (0,1), (1,1), and (1,0) respectively, and then determining coordinates for the remaining vertices, the perspective points, and the points A, B, D given in the construction. The resulting

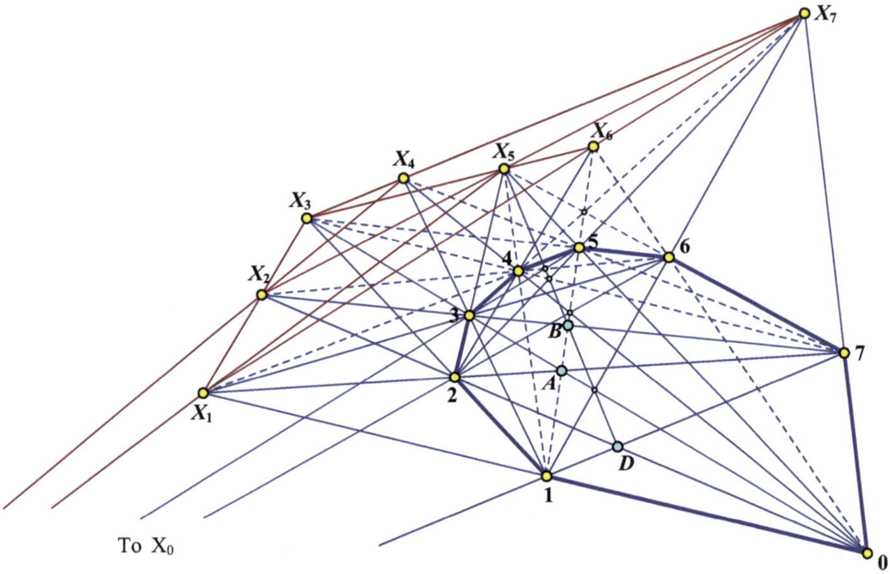

Fig. 22 Perfect construction of perfect closed octagon

demonstration is not particularly informative so we shall omit the details. The fact that points B,D, and X_5 are collinear is a problem involving the *external geometry* of P. Finally, the relabeling of P given by k ←(1, k+3); X_k ← (0, -k+2) for k∈ Z_8 shows that P is equivalent to a (Z_2× Z_8, (0,0)) triple system. □

In the last result of this section we show that every closed perfect polygon P is equivalent to a geometric triple system.

Theorem 24 *Let P be a closed perfect n-gon and let T denote the set of perspective points of P which form a (Z_n, m) triple system. If m is odd, then the relabeling of P given by X_k← 2k (mod 2n) and k ← m – 2k (mod 2n) turns P into a (Z_{2n}, 2m) triple system. If m = 2g is even, then the relabeling of P given by X_k ← (0,k) and k ← (1, g-k) turns P into a (Z_2×Z_n, (0,m)) triple system.*

Proof Let P′ denote the figure resulting from the relabelings. Note that the stipulation that m is odd means that the perspective points of P are relabeled by even vertices of P′ and the vertices of P become the odd vertices of P′. The line [s,t, X_{s+t}] in P corresponds to the line [m-2s, m-2t, 2(s+t)] in P′ whose components sum to 2m (mod 2n), and the line [X_s, X_t, X_q] in P corresponds to the line [2s,2t,2q] in P′ whose components also sum to 2m. These are the only two types of lines in P, so we can conclude that P′ is a (Z_{2n}, 2m) triple system. Now suppose m is even and make the stipulated relabelings. Then the line [s,t, X_{s+t}] in P corresponds to the triple [(1,g-s), (1, g-t), (0, s+t)] in P′, whose components sum to (0,m), and the line [X_s, X_t, X_q] in P corresponds to the triple [(0,s), (0,t), (0,q)] in P′, whose components also sum to (0,m). □

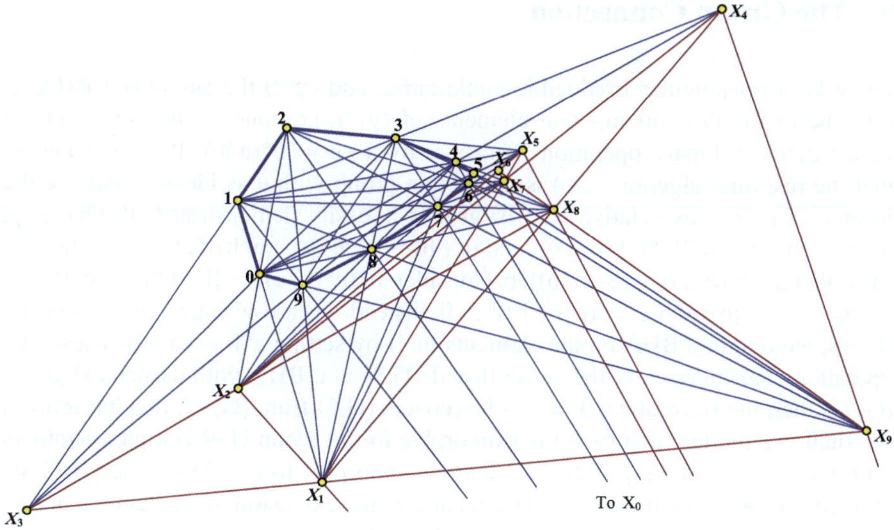

Fig. 23 Closed perfect decagon with perspective sum 3

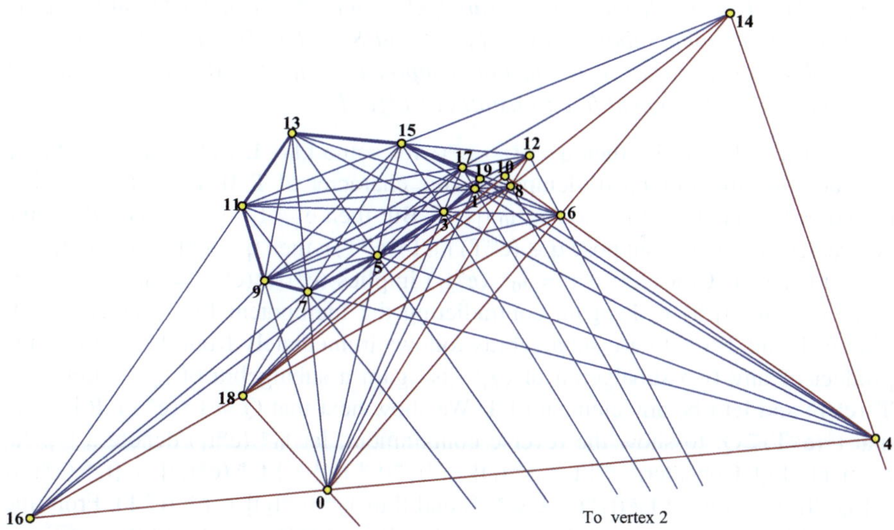

Fig. 24 Closed perfect decagon with perspective sum 3 relabeled to create a $(Z_{20},0)$ triple system

In Figs. 23 and 24 Theorem 24 is illustrated for the case n = 10 and m = 3. In light of (R1) and (R6) Theorem 24 implies that every perfect closed n-gon is equivalent to a $(Z_n,0)$, $(Z_n,1)$, $(Z_2\times Z_n,(0,0))$ or $(Z_2\times Z_n,(0,1))$ triple system.

9 The Group Connection

Let γ be a nonsingular irreducible cubic curve and $(\gamma,*)$ the associated BHA. If e is one of the three idempotent elements of $(\gamma,*)$, i.e., one of the flex points of γ, we define a binary operation $(+)$ on γ by $a+b = e*(a*b)$. It is well known that the resulting algebra $(\gamma,+)$ is an abelian group with e as identity and a*e the inverse of a. The associative law for groups is easily demonstrated: if $a,b,c \in \gamma$, then $a+(b+c) = e*\{a*(e*(b*c))\}$, and $(a+b)+c = e*\{(e*(a*b))*c\}$, so it suffices to show that $a*(e*(b*c)) = (e*(a*b))*c$. Consider: $a*(e*(b*c)) = \{b*(b*a)\}*\{e*(b*c)\} = \{b*(b*c)\}*\{e*(a*b)\} = (e*(a*b))*c$. If instead we are given an abelian group $(G,+)$, we obtain a BHA on the elements of G by setting $a*b = -a - b$. These two operations are inverses in the sense that if $(G,*)$ is a BHA with associated group $(G,+)$, then the operation $a*b = -a -b$ recovers $(G,*)$ from $(G,+)$. In what follows we shall use the fact that $(\gamma,+)$ is isomorphic to the group U of complex numbers with modulus 1 in case γ is Type V, and is isomorphic to $Z_2\times U$ in case γ is Type IV (see [7]). Also the following result regarding the extension of an idempotent free subalgebra of $(\gamma,*)$ to one containing an idempotent will prove useful.

Theorem 28 *Let γ be an irreducible nonsingular cubic curve, let e be a flex point of γ, and let T be an idempotent free subalgebra of $(\gamma,*)$. Then T, e*T, and T*(e*T) are mutually disjoint subalgebras of $(\gamma,*)$, and $S = T \cup (e*T) \cup (T*(e*T))$ is a subalgebra of $(\gamma,*)$ which contains the idempotent element e. Moreover, T and e*T are isomorphic idempotent free translates of T*(e*T).*

Proof If $q \in T \cap (e*T)$, then $q = t = e*t_0$ for some $t,t_0 \in T$. But then $e = t*t_0 \in T$ contradicting the stipulated idempotent free character of T. If $q \in T \cap (T*(e*T))$, then $q = t = t_0* (e*t_1)$ for some $t,t_0,t_1 \in T$. But then $e = t_1* (t_0* t)$ with the same contradiction. Finally suppose $q \in (e*T) \cap (T*(e*T))$, then $q = e*t = t_0* (e*t_1)$ for some $t,t_0,t_1 \in T$. Consider: $e*t = t_0* (e*t_1)$ implies $(e*t_1)*(e*t) = t_0$, $(e*e)*(t_1*t) = t_0$, $e = t_0*(t_1*t) \in T$, again contradicting the idempotent free character of T. That e*T and T*(e*T) are subalgebras follows immediately from the fact that the product of any two subalgebra of $(\gamma,*)$ is again a subalgebra of $(\gamma,*)$. Let $Q = T*(e*T)$ and let t be any element of T. We show next that $Q = t*(e*T)$. It is clear that $t*(e*T) \subseteq Q$; to show the reverse containment, let $t_0* (e*t_1)$ denote a typical element of Q. Consider: $t * \{ t_0* (e*t_1)\} = \{t_1*(t * t_1)\} * \{ t_0* (e*t_1)\} = \{t_1* (e*t_1)\} * \{ t_0*(t_1*t)\} = e * \{ t_0*(t_1*t)\} \in e * T$, and thus $t_0* (e*t_1)\} \in t * (e*T)$. From the just established equation it follows immediately that $t*Q = e*T$ and thus e*T is a translate of Q. To show that T is a translate of Q, again let t be any element of T. We will show that $(e*t)*Q = T$. If $t_0 \in T$, then $t_0 = (e*t)*\{t_0* (e* t)\}$ lies in $(e*t)*Q$. For the reverse containment, suppose $(e*t)* \{t_0* (e* t_1)\}$ is a typical element of $(e*t)*Q$. Then we see immediately that $(e*t)* \{t_0* (e* t_1)\} = (t_0* t) * \{e * (e* t_1)\} = (t_0* t) * t_1$ is an element of T. The mapping ϕ: $T \to e*T$ given by $\phi(t) = e*t$ is easily seen to be an isomorphism. Suppose e*T contains an idempotent element e*t, then $(e*t)*(e*t) = e*t$, $(e*e)*(t*t) = e*t$, and $t*t = t$ by cancellation. But then t is an idempotent element of T which we have not allowed. Note that any translate a*W of a subalgebra W of (γ,*) certainly has the same cardinality as W, but is usually not isomorphic to W, for example, T cannot be isomorphic to Q since Q contains the idempotent element e, and T is idempotent free.

It remains only to show that S = T ∪ (e*T) ∪ (T*(e*T)) is a subalgebra of (γ,*). Since each of the three mutually disjoint components of S are themselves subalgebras, it suffices to verify the equations T*{T*(e*T)} = e*T and (e*T)* {T*(e*T)} = T. We will leave this straightforward verification to the reader. □

A more general result than Theorem 28 is given in [4] where it is shown that an arbitrary idempotent free algebra in the variety of binary hypercommutative algebras can be embedded in a member of the same variety which does contain an idempotent.

Let T be a subalgebra of (γ,*) with finite order n, which contains an idempotent element e of (γ,*). We will argue that T is a triple system. The corresponding abelian group (T,+) is a finite subgroup of (γ,+). If γ is Type V, then it is known that (γ,+) is isomorphic to the group U of complex numbers with modulus 1, and since every finite subgroup of U is cyclic, we can conclude that (T,+) is a cyclic group hence isomorphic to Z_n. Now using the above given formula for converting (Z_n,+) back into a BHA, we obtain a*b = -a – b which clearly realizes T as a (Z_n,0) triple system. In case γ is Type IV we have that (γ,+) is isomorphic to $Z_2\times$ U and thus (T,+) is isomorphic to either Z_n or $Z_2\times Z_{n/2}$ with the consequence that T is either a (Z_n,0) triple system or a ($Z_2\times Z_n$,(0,0)) triple system. Next suppose T is a finite idempotent free subalgebra of (γ,*) with order n. By Theorem 28, T occurs as a subalgebra of S = T ∪ (e*T) ∪ (T*(e*T)) which contains the idempotent element e of (γ,*). If Q = T*(e*T) , then Q is a subalgebra of S which contains e, and consequently both (S,+) and (Q,+) are abelian groups. If t is any element of T, then we have seen that T = (e*t)*Q. The corresponding group expression is T = (-e – t) * Q = e+t – Q = e+t +Q, so we see that T is a coset of Q. If γ is Type V, then (S,+) is cyclic with order 3n, and thus isomorphic to Z_{3n}. We can then represent Q by Q = {0,3,6, ... , 3n-3} and the cosets of Q by {1,4,7, ... , 3n-2} and {2,5,8, ... , 3n-1}. T must be one of these cosets each of which correspond to a (Z_n,1) triple system. If γ is Type IV, then (S,+) is isomorphic to either Z_{3n} or $Z_2\times Z_{3m}$ where n must be even and m = n/2. In the first cast T is again a (Z_n,1) triple system, and in the second case (Q,+) is a subgroup of $Z_2\times Z_{3m}$ with order 2m = n. We must have Q = $Z_2\times$ {0,3,6, ... , 3m-3} and (T,+) must be one of the cosets of Q, namely (0,1)+Q or (0,2)+Q. If T = (0,1)+Q, then the mapping ϕ:T $\rightarrow Z_2\times Z_m$ given by ϕ(0,3k-2) = k and ϕ(1, 3k-2) = (1,k) turns (T,*) into a ($Z_2\times Z_m$, (0,2)) triple system, and if T = (0,2)+Q, the mapping ψ: T $\rightarrow Z_2\times Z_m$ defined by ψ(0,3k-1) = (0,k) and ψ(1,3k-1) = (1,k) turns (T,*) into a ($Z_2\times Z_m$, (0,1)) triple system. If T is ($Z_2\times Z_m$, (0,2)), then by relabeling each vertex of T with its inverse we obtain a ($Z_2\times Z_m$, (0,-2)) triple system. If to this triple system we add (0,1) to each vertex, we obtain a ($Z_2\times Z_m$, (0,1)) triple system equivalent to T. We have proved:

Theorem 29 *Let γ be an irreducible nonsingular cubic curve and $(\gamma,*)$ the corresponding binary hypercommutative algebra. If γ is Type V, then every finite subalgebra of $(\gamma,*)$ of order n is either a $(Z_n,0)$ or a $(Z_n,1)$ geometric triple system. If γ is Type IV every finite subalgebra of $(\gamma,*)$ of order n is either a $(Z_n,0)$, $(Z_n,1)$, $(Z_2{\times}Z_m, (0,0))$ or $(Z_2{\times}Z_m, (0,1))$ geometric triple system where $m = n/2$.* □

10 Conclusion

The notion of perfect polygon can be generalized by allowing the base to be any abelian group G. Let $\phi : G \rightarrow \prod$ be an injective mapping from G into the projective plane such that no three points in $\phi(G)$ are collinear. For each $g \in G$ let W_g denote the set of lines $\{[a,b] : a+b = g\}$. If for each $g \in G$ the lines in W_g are concurrent, then we call the set of points $\phi(G)$ a *perfect G-gon.* Let X_g denote the point of concurrence of the lines in W_g. The set of points $\{X_g : g \in G\}$ is the *perspective set* of the perfect G-gon. We have seen that $(Z_2{\times}Z_n, (0,0))$ triple systems exist with cubic envelope γ. A translate of such a system can produce a perfect $Z_2{\times}Z_n$- gon on γ. If Z denotes the group of integers, then there exists a perfect construction of a perfect Z-gon which we illustrate in Fig. 25. The relabeling of this diagram by $X_k \leftarrow (0,k)$ and $k \leftarrow (1,-k)$ yields a $(Z_2{\times}Z, (0,0))$ triple system.

The vertices and perspective points of a perfect Z-gon must lie on a cubic curve, and it seems likely that every perfect G-gon lies on a cubic curve, thus yielding a generalization of Theorem 4.

It is an interesting open question as to which perfect polygons and geometric triple systems have perfect constructions. Note that our definition of *perfect construction* must be slightly extended to claim that our perfect Z-gon has a perfect construction: the construction is perfect if we can perfectly construct, in principle, any finite number of the vertices and perspective points.

Another further direction for research involves the above-mentioned external geometry associated with a perfect polygon. It sometimes happens that lines joining the vertices of a perfect polygon P (or tangents to the cubic envelope at a vertex) concur at points different from perspective points. We call these *diagonal points* and classify theorems involving the existence of these points, or collinearities among them, as belonging to the *external geometry* of P. A basic result in this direction is the following.

Theorem 30 *Let P be a perfect closed n-gon with Type IV cubic envelope γ and perspective sum m. Suppose n is even and there exists $a \in Z_n$ such that $3a = m \pmod{n}$. If a is even,* i.e., *$a = 2b$, then all points of the form $[X_s, X_t] \cap [X_{m-a-s}, X_{m-a-t}]$ or $[s,t] \cap [a-s, a-t]$ lie on the line $[b, b+(n/2)]$.*

Proof If γ is Type IV, then there exists a transformation which takes γ to a curve of the form $y^2 = x(x-1)(x-w)$ with $w>1$. The envelope γ is transformed to a curve γ' which is symmetric with respect to the x-axis, and P is transformed to a perfect polygon P′ with γ' as its cubic envelope. Since $3a = m$, the perspective point X_a

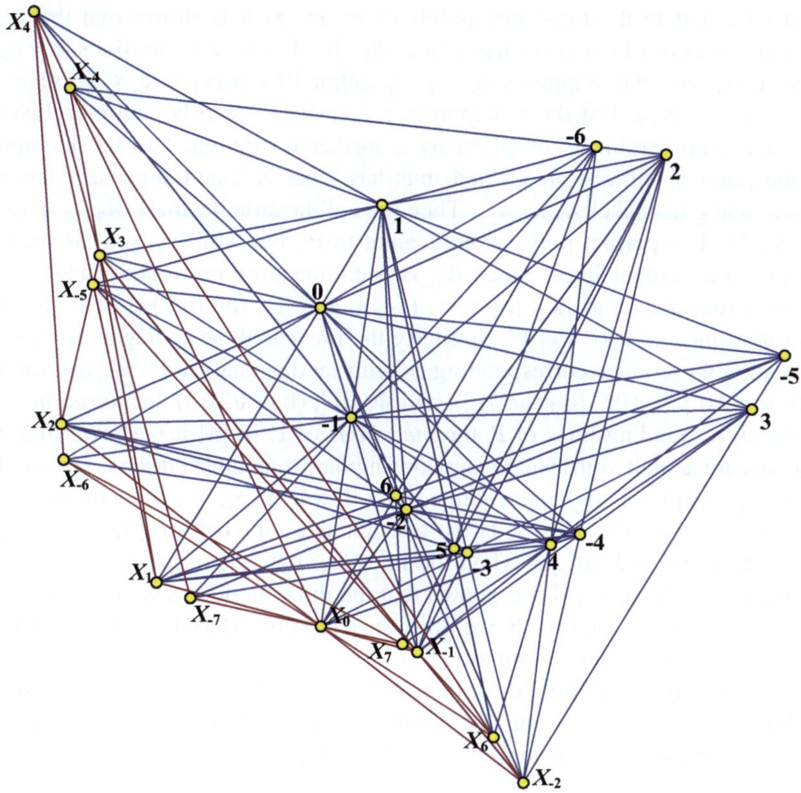

Fig. 25 Perfect Z-gon

of P is an idempotent element of the BHA (γ,*) , and thus its image X_a' under the transformation is an idempotent element of (γ',*). If 3|n, then P contains all three flex points of γ and hence P′ contains all three flex points of γ'. These will have labels X_a', $X_{a+(n/3)}'$, $X_{a+(2n/3)}'$ one of which is the point at infinity on vertical lines, and the other two are reflections of each other across the x-axis. Without loss of generality, we can suppose that X_a' is the point at infinity on vertical lines. Besides X_a' itself, the only three points of γ' which square to X_a' are b′, (b+(n/2))′ and $X_{(a+(n/2))'}$ all of which lie on the x-axis, so the line [b′, (b+(n/2))′] is the x-axis. If s′, t′ are any two vertices of P′, then vertices (a-s)′, (a-t)′ are their reflections across the x-axis, and it follows immediately that the point [s′,t′]∩[(a-s)′, (a-t)′] must lie on [b′, (b+(n/2))′]. Since our original transformation is a collineation, it follows that the corresponding diagonals of P, namely [s,t]. [a-s, a-t] and [b, b+(n/2)] are concurrent. Similarly the points X_a', X_s' and X_{m-a-s}' are collinear since their subscripts sum to m, and they lie on a vertical line since X_a' is the point on γ' at infinity on vertical lines. Since γ' is symmetric with respect to the x-axis, the pairs of points X_s', X_{m-a-s}', and X_t', X_{m-a-t}' are reflections of each other across the x-axis. As a consequence we also have that the lines [X_s, X_t],[X_{m-a-s}, X_{m-a-t}] and [b, b+(n/2)] are concurrent as stipulated. □

Let e,f,g denote the three flex points of γ'. In [8] it is shown that there exist three collineations of the projective plane, Φ_e, Φ_f, Φ_f whose restrictions to γ agree, respectively, with the mappings ϕ_e, ϕ_f, ϕ_g defined by $\phi_e(x) = e^*x$, $\phi_f(x) = f^*x$, $\phi_g(x) = g^*x$. These last three mappings are easily seen to be automorphisms of $(\gamma',^*)$ and automorphisms of $(P',^*)$ by a further restriction. Say the idempotent e is the point at infinity on γ', and thus has label $X_a{}'$, and suppose f has label $X_{a+(n/3)}{}'$ and g has label $X_{a+(2n/3)}{}'$. Then the collineation Φ_g maps $X_{a+(n/3)}{}'$ to $X_a{}'$. Now $\Phi_g(P')$ is a perfect polygon equivalent to P′, but with $X_{a+(n/3)}{}'$ as the point at infinity on vertical lines. Since Φ_g is a collineation, the images under Φ_g of the concurrent diagonals [s′,t′], [(a-s)′, (a-t)′], and [b′, (b+(n/2))′] of P′ are again concurrent lines in $\Phi_g(P') = P'$. Similarly the images of these diagonals under the collineation Φ_f produce corresponding concurrent diagonals in P′. We call the lines [b′, (b+(n/2))′], Φ_g([b′, (b+(n/2))′]), and Φ_f ([b′, (b+(n/2))′]) the *meridians* of P′, and the corresponding lines of P, the *meridians* of P. In [8] it is shown that these meridians are actually intrinsic to any nonsingular irreducible curve γ. For our Type IV cubic γ, the meridians can be defined as follows. Let e_1, e_2 denote the two points in $\sqrt{e}$ which lie on the bell; f_1, f_2 the two points on the bell in $\sqrt{f}$, and g_1, g_2 the two points on the bell in $\sqrt{g}$. Then the lines $[e_1, e_2]$, $[f_1, f_2]$, and $[g_1, g_2]$ are the meridians of γ. Also, in [8] it is shown that the three meridians concur at a point Z called the *center* of γ. In Fig. 26 we illustrate a closed perfect 12-gon with diagonal points on three concurrent meridians.

In case n is not a multiple of 3, then $3a = m \pmod{n}$ has at most one solution and thus P contains at most one flex point of γ. In Fig. 27 we have a perfect closed octagon P with perspective sum m = 6. The perspective point X_2 is the only flex

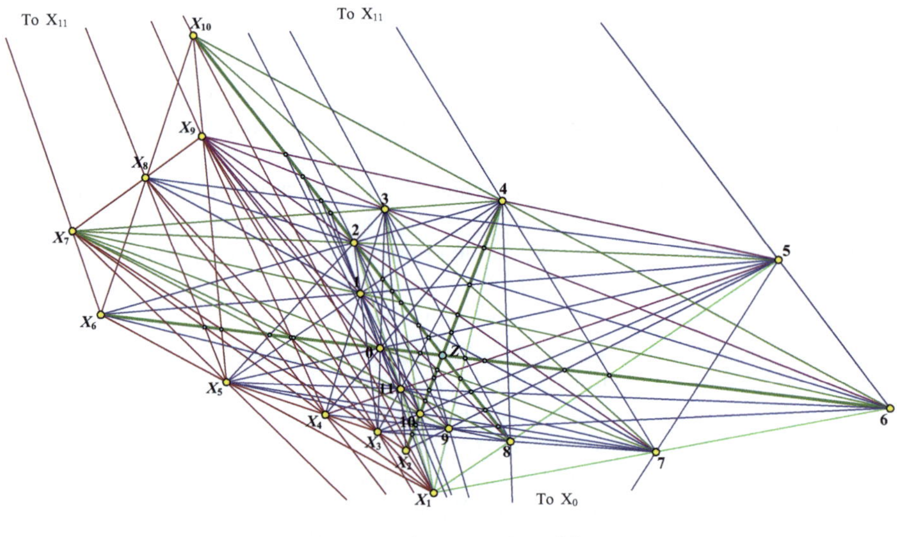

Fig. 26 Perfect closed 12-gon with diagonal points and meridians

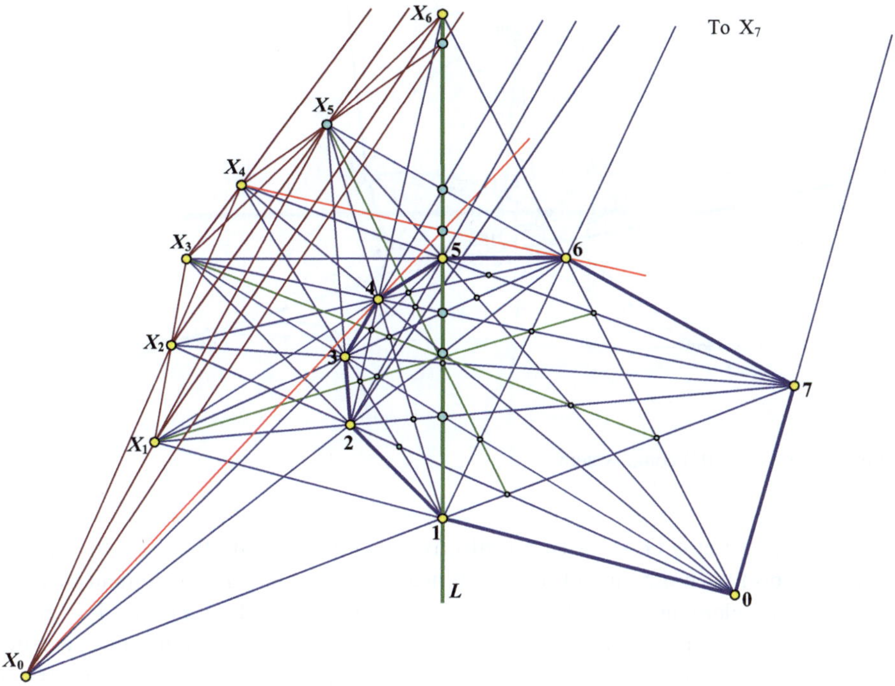

Fig. 27 Perfect closed octagon with meridian and diagonal points

point on P and the line [1, 5] is the only meridian of γ which contains the full complement of diagonal points stipulated by the theorem.

There are several interesting analogs of geometric triple systems which involve circles, conics, and planes. If G is an abelian group, then a *G-circle system* can be defined as follows. Let ϕ:G → ∏ be an injective map from G into the projective plane such that no 5 points in ϕ(G) are collinear, and let g be a fixed element of G. If for every four element subset {a,b,c,d} of G such that a+b+c+d = g, the corresponding points {ϕ(a), Φ(b), ϕ(c), Φ(d)} are cocyclic or possibly collinear, we call the points in ϕ(G) and the associated circles or four point lines a *(G,g) circle system*. These circle systems have already been studied in [9, 10]. In [10] the (Z,0) circle system is used to gain general structural information for an irreducible self-inversive cubic curve.

Suppose ψ: G→Λ is an injective map from the abelian group G into 3-dimensional projective space such that no four points in ψ(G) are coplanar, and let g be a fixed element of G. If for every four element subset {a,b,c,d} of G such that a+b+c+d = g, the corresponding points {ψ(a), ψ(b), ψ(c), ψ(d)} are coplanar or possibly collinear, we call the points in ψ(G) and the associated planes a *(G,g) plane system*. Does a (Z,0) plane system exist? If so does there exist a space curve which contains its points?

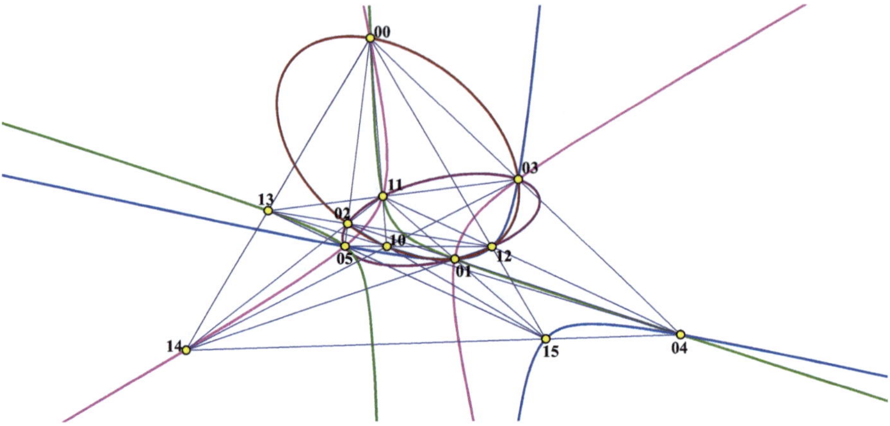

Fig. 28 ($Z_2 \times Z_6$, 02) conic system

Finally, let Let ϕ:G → $\prod$ be an injective map from G into the projective plane such that no four points in ϕ(G) are collinear, and let g be a fixed element of G. If for every six element subset {a,b,c,d,e,f} of G such that a+b+c+d+e+f = g, the corresponding points {ϕ(a), Φ(b), ϕ(c), Φ(d), ϕ(e), Φ(f)} lie together on the same conic, or form a pair of collinear triples, then we call the points in ϕ(G) a *(G,g) conic system.* One interesting result which relates to geometric triple systems is the following.

Theorem 31 *Every geometric triple system is a conic system.*

In Fig. 28 we depict a ($Z_2 \times Z_6$, (0,1)) triple system which is equivalent to a ($Z_2 \times Z_6$, (0,2)) conic system. All 6-tuples which sum to (0,2) lie either on a conic or form two sets of collinear triples. We reserve the proof of Theorem 31 for a future work.

References

1. Coxeter, H.S.M., Greitzer, S.L.: Geometry Revisited. MAA, Washington, D.C. (1967)
2. Eves, H.: Fundamentals of Modern Geometry. Jones and Bartlett, Boston (1992)
3. Bix, R.: Conics and Cubics, a Concrete Introduction to Algebraic Curves, 2nd edn. Springer Sci. and Business Media, LLC, New York (2006)
4. Harris, M.: Thirdpoint Groupoids. Master's Thesis, Virginia State University (2008)
5. Fletcher, R.: Geometric Triple Systems with Base Z and Z_n, submitted for publication in Springer Proceedings in Mathematics and Statistics (2015)
6. Fletcher, R.: Geometric Triple Systems with Base $Z_2 \times Z_n$, unpublished
7. Walker, W.: Algebraic Curves. Springer Verlag, New York (1949)

8. Fletcher, R.: Perfect Hexagons, Elementary Triangles and the Center of a Cubic Curve. In: Springer Proceeding in Mathematics and Statistics, vol. 24. Bridging Mathematics, Statistics, Engineering and Technology, Springer Science + Business Media, New York (2012)
9. Robinson, D.: Group Circle Systems. Master's Thesis, Virginia State University (2015)
10. Fletcher, R.: Self-inversive Cubic Curves, Accepted for publication in Springer Proceedings in Mathematics and Statistics (2016)

Geometric Triple Systems with Base Z and Z_n

Raymond R. Fletcher III

Abstract In this paper we discuss triple systems whose elements come from an abelian group G. Every triple of elements with the same sum in G will correspond to three collinear points in the projective plane. We will show that the set of all points in such a triple system must lie on a cubic curve γ which we call the envelope of the system. We will determine the number and location of equivalent triple systems which have the same cubic envelope. These results will be used to determine exactly which finite abelian groups can be used to construct a geometric triple system.

Keywords Triple system • Cubic curve • Abelian group • Hypercommutative groupoid • Equilateral triangle

1 Introduction

Let G be an abelian group and $g \in G$. Let $\phi : G \to \Pi$ be an injective map from G into the projective plane Π such that no four points of $\phi(G)$ are collinear. If for each three element subset $\{a, b, c\}$ of G with $a + b + c = g$, the corresponding points $\{\phi(a), \phi(b), \phi(c)\}$ are collinear, then we call the set of points $\phi(G)$ and the associated collinear triples $\{\phi(a), \phi(b), \phi(c)\}$, a *geometric triple system with base G and sum g,* or more shortly, a *(G,g) triple system.* In Figs. 1 and 2 we illustrate a $(Z_9, 0)$ and a $(Z_9,1)$ triple system. The *size* of a triple system is the number of triples and the *order* of a triple system is the number of vertices (or points). Thus the triple systems in Figs. 1 and 2 both have order 9 but the $(Z_9, 0)$ system has size 10 and the $(Z_9,1)$ system has size 9. A (G,g) triple system is *equivalent* to a (H,h) triple system if there exists a bijection $\psi : G \to H$ such that whenever $a, b, c \in G$ and $a + b + c = g$, then $\psi(a) + \psi(b) + \psi(c) = h$. In case $H = G$ we refer to ψ as

R.R. Fletcher III (✉)
Department of Mathematics and Economics, Virginia State University, Petersburg, VA 23806, USA
e-mail: rfletcher210@gmail.com

B. Toni (ed.), *New Trends and Advanced Methods in Interdisciplinary Mathematical Sciences*, STEAM-H: Science, Technology, Engineering, Agriculture, Mathematics & Health, DOI 10.1007/978-3-319-55612-3_2

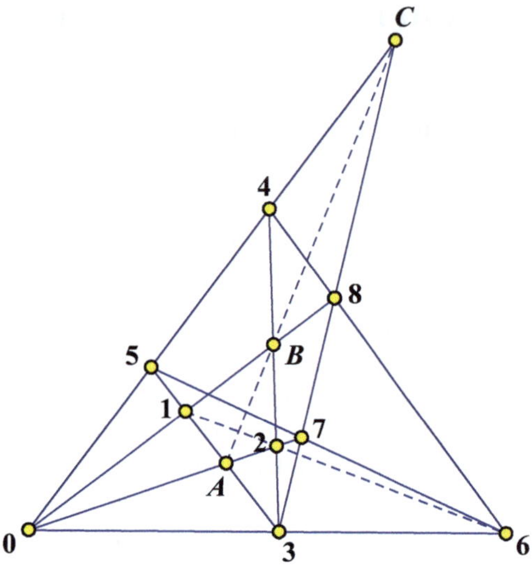

Fig. 1 (Z_9,0) triple system

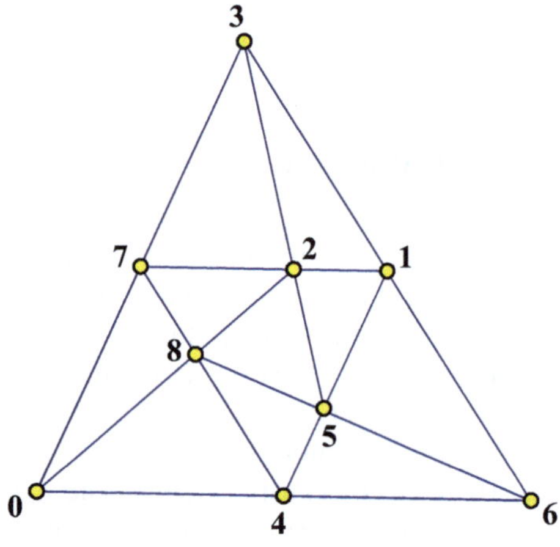

Fig. 2 (Z_9,1) triple system

a *relabeling* of the points of G. Two equivalent triple systems must have the same order and the same size, so we see that the triple systems in Figs. 1 and 2 are not equivalent. We will show in the next section, that up to equivalence, the illustrated triple systems are the only ones with base Z_9. It is convenient to represent the points of a triple system by the labels assigned to them from the base group. If a, b are two elements from the base group then [a,b] denotes either the line through a, b or the segment with endpoints a,b. We use the notation [a,b,c] for a triple. The (Z_9, 0) triple system was constructed starting with the triangle (0,4,6), then putting 5, 8, 3 on the appropriate sides. Then vertex 1 is put at the intersection of segments [3,5] and [0,8] and vertex 7 is put at the intersection of [5,6] and [3,8]. Finally vertex 2 is put at the intersection of [3,4] and [0,7]. When this is done the remaining triple [1,2,6], indicated by the dashed line, is automatically obtained. The dashed line is proved as follows. Let A $= [1,5] \cap [2,7]$; B $= [1,8] \cap [2,4]$ and C $= [0,5] \cap [3,7]$. The triangles (0,A,3), (8,C,4) are in perspective from line [5,7,6] and so, by Desargues Theorem, are in perspective from a point. Since $[0,8] \cap [3,4] =$ B, it must be that B is the point, and it then follows that points A,B,C are collinear. Then triangles (3,2,7), (0,1,5) are in perspective from line [A,B,C] and so, by Desargues Theorem again, these triangles are in perspective from a point. Since $[5,7] \cap [0,3] = 6$, we can conclude that points 1, 2, 6 are collinear. We thus have what we call a *perfect construction* of a (Z_9,0) triple system, i.e., a construction that can be accomplished, in principle, by straightedge and compass alone. The *independent points* 0,4,6 can be put anywhere in the plane, so long as they are not collinear, then vertices 3,5,8 can be put anywhere on the sides of triangle (0,4,6), and the positions of the remaining vertices are forced. Interestingly no such perfect construction is available for the (Z_9,1) system in Fig. 2. The triangle (0,3,6) is chosen freely and the points 7,1,4 can be placed anywhere on the sides, but vertex 2 must be *slid* along segment [1,7] until the line [6,5,8] is achieved. It can easily be proved that this is possible, but it is not a perfect construction.

Perfect constructions also exist for (Z_{10},0) and (Z_{12},0) triple systems which we illustrate in Figs. 3 and 4. In each case only the dashed lines need to be proved. The reader may try to prove these using the Theorem of Desargues repeatedly. The remaining single line in the (Z_{10},0) construction is not difficult to prove, but the four dashed lines in the (Z_{12},0) triple system are quite difficult to prove in this manner, and the difficulty increases in unmanageable proportions as the order of the triple system increases. It is clear then, that the general existence of geometric triple systems cannot be proved in this ad hoc manner, even if we insist that the base group be Z_n. A clue to a general procedure is afforded by the (not perfect) construction of a (Z_{18},0) triple system illustrated in Fig. 5. (Vertex 14 is constructed as the intersection of the red segment with line [5,17]. The red segment is a *lever* which allows us to move vertex 14 by very small increments until it obtains the position required.) It begins to be clear that the vertices all lie on a curve which has

Fig. 3 (Z_{10},0) triple system

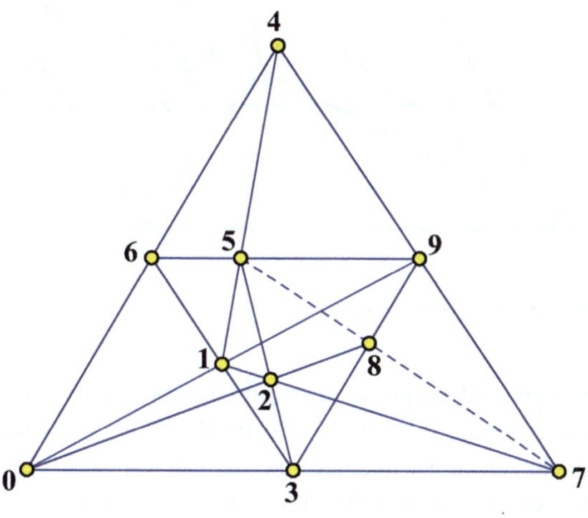

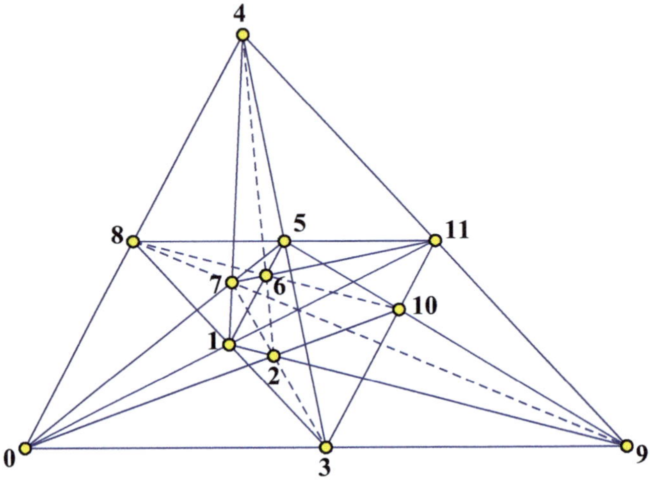

Fig. 4 $(Z_{12},0)$ triple system

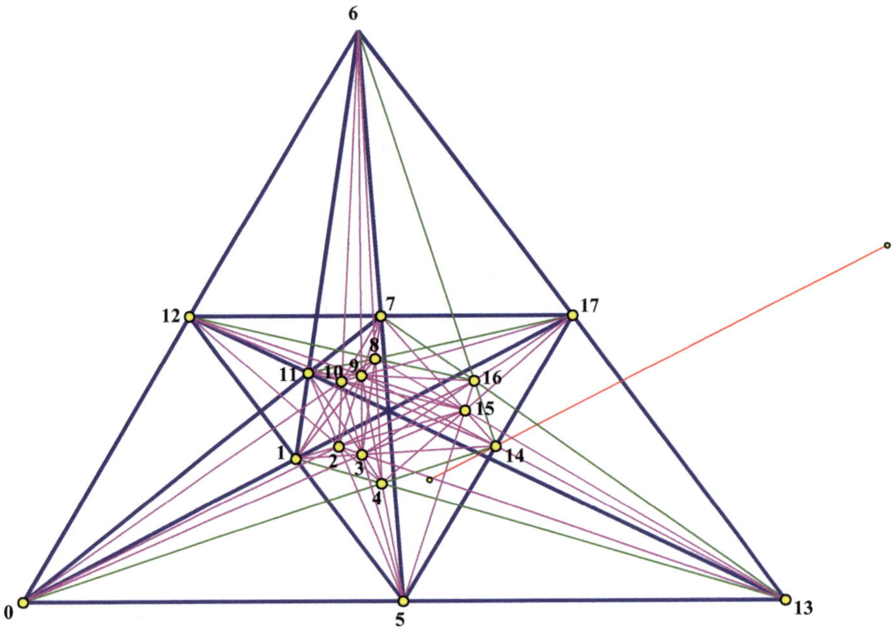

Fig. 5 $(Z_{18},0)$ triple system

the general shape of a certain species of cubic. In our construction in Fig. 5 we even observe that the vertices occur along the supposed cubic curve in consecutive order as the curve is traversed in one direction: starting at vertex 0 we travel consecutively to vertex 5 then follow the cubic to infinity and then back from infinity to vertex 6, then we follow to vertex 12 and go off to infinity again and coming back to vertex 13. Then we continue along the conjectured cubic to vertex 17 and off to infinity once again and back from infinity to the starting point at vertex 0. We will show that all geometric triple systems with base Z or Z_n lie on irreducible cubic curves.

2 Equivalence

The simplest of all groups is the cyclic group and every cyclic group of order n is isomorphic to Z_n, so it is natural to begin our investigation by considering geometric triple systems with base Z_n. Our first Theorem allows us to focus our research on Z_n triple systems with sum 0 or 1.

Theorem 1 *If gcd(n,3) = 1, then every Z_n triple system is equivalent to a (Z_n,0) geometric triple system. If 3|n then a Z_n triple system is equivalent to a (Z_n,0) or (Z_n,1) triple system. Moreover, if 3|n then the (Z_n,0) system is not equivalent to the (Z_n,1) triple system.*

Proof The bijection $\phi : Z_n \rightarrow Z_n$ defined by $\phi(x) = x + k$ is an equivalence between a (Z_n,m) and a (Z_n, m + 3k) triple system, and the bijection $\psi : Z_n \rightarrow Z_n$ defined by $\psi(x) = -x$ is an equivalence between a (Z_n,m) and a (Z_n, −m) triple system. Thus a (Z_n,3k) triple system is equivalent to a (Z_n,0) system, and a (Z_n, 3k + 1) triple system is equivalent to a (Z_n,1) system. Finally, a (Z_n, 3k − 1) is equivalent to a (Z_n, −1) system, which is equivalent, in turn, to a (Z_n,1) system by means of the equivalence ψ. Suppose 3|n and $\xi : Z_n \rightarrow Z_n$ is an equivalence between a (Z_n,0) triple system T and a (Z_n,1) triple system S, then $0+0+0 = 0$ in T would imply $\xi(0) + \xi(0) + \xi(0) = 1$ in S which is not possible. □

If 3|n we can also prove nonequivalence of a (Z_n,0) triple system T, and a (Z_n,1) triple system S by determining the sizes of these triple systems. Let p denote the size of T and let q denote the size of S. There are n/3 multiples of 3 in Z_n and so pn/3 is the count of three element subsets of Z_n which sum to a multiple of 3. There are 2n/3 nonmultiples of 3 in Z_n, so 2qn/3 counts the number of three element subsets of Z_n which sum to a nonmultiple of 3. The total number of 3 element subsets of Z_n is given by $(n(n-1)(n-2))/6$, so we have $(pn/3) + (2qn/3) = (n(n-1)(n-2))/6$ which simplifies to $p+2q = (n-1)(n-2)/2$. Now if $p = q$ this would imply that $3 \mid (n-1)(n-2)$ which is impossible since n is a multiple of 3. We conclude that $p \neq q$, and thus that S is not equivalent to T.

In the next section we prove the existence of geometric triple systems with base the full set of integers Z. These are equivalent to a (Z,0) or a (Z,1) triple system. We omit the proof which is similar to the proof used in Theorem 1.

Theorem 2 *Let Z denote the group of integers under addition. Then every triple system on Z is equivalent to a (Z,0) or a (Z,1) triple system and these two triple systems are nonequivalent.* □

If G is a finite abelian group, x ∈ G, and T is a (G,g) triple system, let δ(x) denote the number of triples of T which contain x. Then:

$$\text{size (T)} = (1/3)\sum_{x\in G}\delta(x) \tag{1}$$

Now for x ∈ G, let ρ (x) denote the number of two element subsets of G which sum to x, and let ξ (x) denote the cardinality of the set {g ∈ G : g + g = x}. Then:

$$\rho(\mathrm{x}) = (1/2)\left(|\mathrm{G}| - \xi(\mathrm{x})\right). \tag{2}$$

The explanation of (2) is as follows. In the Cayley table for G, the element x occurs exactly once in each row and column. The expression ξ (x) counts the number of times that x occurs on the main diagonal of the table. So x occupies | G | −ξ (x) nondiagonal locations. Each pair of unequal elements a,b which sum to x yield two x entries, one in row a and column b, and one in row b and column a. Thus there are (1/2) (|G| − ξ (x)) two element subsets of G which sum to x. Let ε (x) = 0 if 3x = g and ε (x) = 1 otherwise, then:

$$\delta(\mathrm{x}) = \rho(\mathrm{g}-\mathrm{x}) - \varepsilon(\mathrm{x}). \tag{3}$$

The explanation of (3) is as follows. A triple containing x consists of x and a pair of points which sum to g − x. The expression ρ (g − x) counts these pairs. But from this count we must remove the unique pair {x, g − 2x} since the element x would be repeated. If it happens that g = 3x, the {x, g − 2x} is not a pair and no removal is required. In the first case ε (x) = 1 and in the second case ε (x) = 0. Now combining (1), (2) and (3) we obtain:

$$\begin{aligned}\text{size (T)} &= (1/3)\sum_{x\in G}[\rho(g-x)-\varepsilon(x)] = (1/3)\sum_{x\in G}\rho(x) - (1/3)\sum_{x\in G}\varepsilon(x)\\ &= (1/3)\sum_{x\in G}\left(\tfrac{1}{2}\right)(|G|-\xi(x)) - \left(\tfrac{1}{3}\right)\sum_{x\in G}\varepsilon(x)\\ &= (1/6)\sum_{x\in G}|\,G\,| - (1/6)\sum_{x\in G}\xi(x) - \left(\tfrac{1}{3}\right)\sum_{x\in G}\varepsilon(x),\end{aligned}$$

and finally:

$$\text{Size (T)} = (1/6)\,|\mathrm{G}|^2 - (1/6)\sum_{x\in G}\xi(x) - \left(\tfrac{1}{3}\right)\sum_{x\in G}\varepsilon(x). \tag{4}$$

We saw in Figs. 1 and 2 that a $(Z_9,0)$ triple system and a $(Z_9,1)$ triple system have different size. In our next result we determine precisely the size of triple systems with base Z_n.

Theorem 3 *The size of a* $(Z_n, 0)$ *triple system is* $(n-1)(n-2)/6$ *if* $\gcd(n,3)=1$. *If* $3|n$ *then the size of a* $(Z_n, 0)$ *triple system is* $\left(n^2-3n+6\right)/6$ *and the size of a* $(Z_n, 1)$ *triple system is* $\left(n^2-3n\right)/6$.

Proof Suppose gcd (n, 3) = 1 and let p denote the size of a $(Z_n, 0)$ triple system. By Theorem 1, every triple system in $S = \{(Z_n, k) : k \in Z_n\}$ has size p, and so there are np distinct three element subsets which constitute the triples belonging to the systems in S. Since every three element subset of Z_n is represented among the triples in the systems in S, it must be that $np = \binom{n}{3}$. This yields $p = (n-1)(n-2)/6$. Now suppose $3|n$ and let T be a $(Z_n, 0)$ triple system. The equation $3x = 0$ has exactly three solutions: $\{0, n/3, 2n/3\}$, so $\sum_{x \in G} \varepsilon(x) = n - 3$. If n is odd, then $\xi(x) = 1$ for all $x \in G$, so $\sum_{x \in G} \xi(x) = |G| = n$. Then (4) yields: size (T) $= (1/6)n^2 - (1/6)n - (1/3)(n-3) = \left(n^2 - 3n + 6\right)/6$. If n is even then $\xi(x) = 0$ if x is odd, and $\xi(x) = 2$ if x is even, so $\sum_{x \in G} \xi(x) = 2(n/2) = n$, so we obtain the same value for size(T). If q denotes the size of a $(Z_n, 1)$ triple system, then we have:

$$p + 2q = (n-1)(n-2)/2 \qquad (5)$$

from the discussion following Theorem 1. Substituting $p = \left(n^2 - 3n + 6\right)/6$ into (5) and solving for q yields $q = \left(n^2 - 3n\right)/6$ as stipulated. □

3 Triple Systems with Base Z

In this section we provide perfect constructions of triple systems with base Z and show that the vertices of such systems must all lie on a cubic curve. We find that certain properties of irreducible cubic curves greatly facilitate the proof of existence for these triple systems. If γ is an irreducible cubic curve, then a binary operation * can be defined on the nonsingular points of γ by setting a*b equal to the third point on the line [a,b] and on γ. Here we use the fact that every line which meets γ in two points, also meets γ in a third point, counting multiplicities. The product a*a refers to the point besides a on γ and on the tangent to γ at a. If a*a = a, then a is called a *flex point* of γ. The tangent to γ at a flex point a, intersects γ with multiplicity 3 at a. If γ is nonsingular then it is known that γ has exactly three collinear flex points in the projective plane [1]. The following identities are satisfied by *:

(i) a*b = b*a
(ii) a*(a*b) = b
(iii) (a*b)*(c*d) = (a*c)*(b*d)

Identities (i), (ii) are clear from our definition of *. Identity (iii), which we call the *hypercommutative identity* is a remarkable property of irreducible cubic curves which we illustrate in Fig. 6. An algebra which satisfies identities (i), (ii), (iii) we call a *hypercommutative groupoid;* thus the nonsingular points on any irreducible cubic curve γ form a hypercommutative groupoid which we denote by (γ,*). If a*b = a*c then b = a*(a*b) = a*(a*c) = c, so *cancellation* holds in any hypercommutative groupoid. If T is a (G,t) triple system then a *t-square for T* is 3 × 3 matrix of elements from G arranged so that each row and each column represents a triple belonging to T.

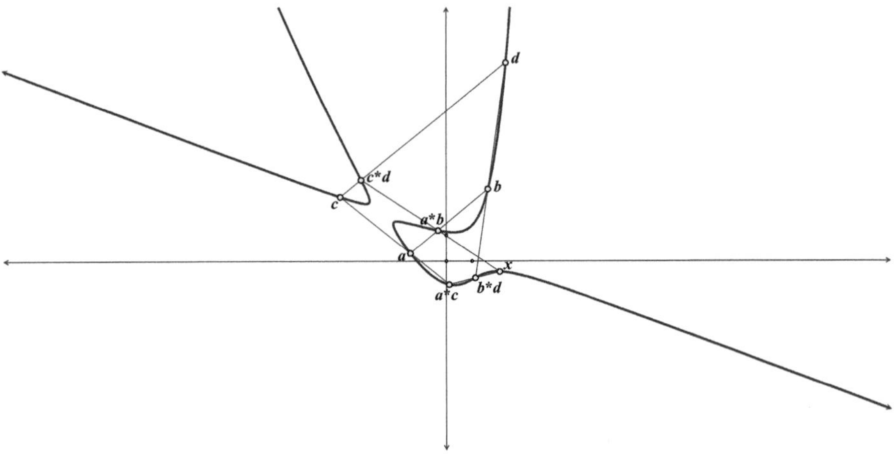

Fig. 6 (a*b)*(c*d) = (a*c)*(b*d) = x

Lemma 4 *Let T be a (G,t) triple system. If eight of the entries of a t-square for T lie on an irreducible cubic curve γ, then so does the ninth.*

Proof Let $\begin{bmatrix} a & b & c \\ d & e & f \\ g & h & k \end{bmatrix}$ be a t-square for T. Since the rows and columns of a t-square can be permuted independently without affecting the row or column sums, we can assume that vertex a is the only element of G not known to lie on γ. Then the second two rows and columns represent collinear triples each point of which lies on γ. Thus the product of any two members of these triples is the third member of the triple. Then we have:

$$b^*c = \left(e^*h\right)^* \left(f^*k\right) = \left(e^*f\right)^* \left(h^*k\right) = d^*g$$

Consequently b*c = d*g is a point q on γ which must lie on both lines [b,c], [d,g]. But then $q = [b, c] \cap [d, g] = a$ lies on γ. □

Theorem 5 *There exists a perfect construction of a (Z,0) triple system. The vertices of such a system must lie on an irreducible cubic curve.*

Proof The starting configuration S for our construction involves the nine points $\{0, \pm 1, \pm 2, \pm 3, 4, -5\}$ and is illustrated by the bold lines in Fig. 7. Any nine points in the projective plane lie on some cubic curve, so let γ denote a cubic which contains these nine points. No three lines cover all the points in S so γ cannot consist of three lines. The removal of any line from S, except for $[3, 0, -3]$ leaves a configuration which still contains three collinear points. This configuration cannot be covered by any conic since no line can meet a conic in more than two points. There is a unique conic which contains the five points $\{\pm 1, \pm 2, -5\}$. If we select

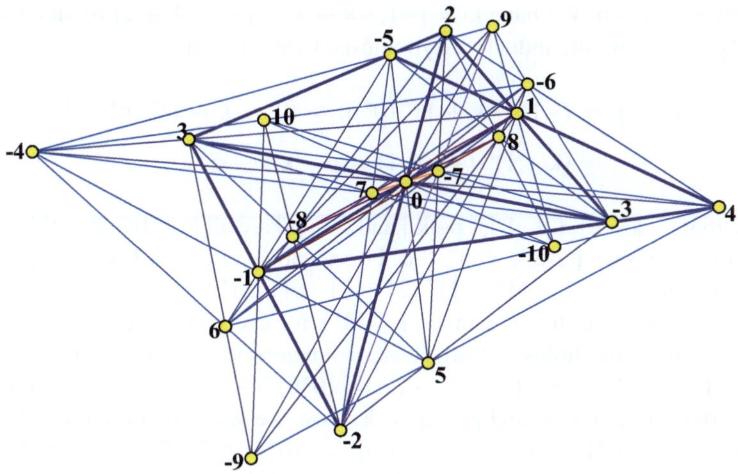

Fig. 7 (Z,0) triple system

vertex 4 on line $[-1, -3]$ so that it does not lie on this conic, then also in case we cover $\{3, 0, -3\}$ with a line, the remaining six points of S cannot be covered by a conic. So we can suppose that γ is irreducible, and since every point in S lies on a line with two other points of S, we can also suppose that each of the nine points in S is a nonsingular point of γ. (A line through a singular point of a cubic cannot meet the cubic in two additional points.) Define $-4 = [1, 3] \cup 0, 4]$ and $5 = [0, -5] \cup [-2, -3]$. The 0-squares:

$$\begin{bmatrix} -4 & 1 & 3 \\ 4 & -3 & -1 \\ 0 & 2 & -2 \end{bmatrix}, \begin{bmatrix} 5 & -2 & -3 \\ -5 & 3 & 2 \\ 0 & -1 & 1 \end{bmatrix}$$

imply by Lemma 4 that -4 and 5 lie on γ. Now suppose $k \geq 5$ and assume inductively that points $\{0, \pm 1, \pm 2, \pm 3, \ldots, \pm k\}$ have been defined and shown to lie on γ. Define vertices $\pm (k + 1)$ by: $-k - 1 = [1, k] \cup [2, k - 1]$ and $k + 1 = [-1, -k] \cup [-2, -k + 1]$. Then the 0-squares:

$$\begin{bmatrix} -k-1 & 1 & k \\ 2 & 0 & -2 \\ k-1 & -1 & -k+2 \end{bmatrix}, \begin{bmatrix} k+1 & -1 & -k \\ -2 & 0 & 2 \\ -k+1 & 1 & k-2 \end{bmatrix}$$

imply, by Lemma 4, that the points $\pm (k + 1)$ lie on γ. It now follows by induction that every integer has been assigned to a point in the plane which lies on γ. It remains to show that the resulting configuration is a (Z,0) triple system. Suppose $k \geq 5$, $0 < p < q < k$ and $p + q = k$. Assume inductively that if $|a|$, $|b|$, $|a + b| < k$, then we have a*b = −a − b. We have k = [−1, −k + 1] ∪ [−2, −k + 2] by definition and also (0*p) = −p by the inductive hypothesis. Consequently:

$$k^* - p = (-1^* - k + 1)^* (0^*p) = (0^* - k + 1)^* (p^* - 1)$$
$$= (k-1)^* (-p+1) = p - k = -q.$$

Note that | −k+1 |, | p−1 |, | p−k | and | k−p | are all less than k so this equation is hypothesis. So we have k = p* − q. For negative values of k we can proceed as follows. Consider: 0*0 = (1* − 1)* (2* − 2) = (1*2)* (−1* − 2) = (−3* 3) = 0. We will use this result to show that −k = 0*k for every positive integer k. Assume inductively that this holds for all positive integers less than k. Consider: −k = (1* (k − 1)) = (0* − 1)* (0* − k + 1) = (0*0)*(−1* (−k + 1) = 0*k. Again suppose 0 < p < q < k and p + q = k. Now we want to show that −k = p*q. Consider: −k = 0*k = (0*0)* (−p* − q) = (0* − p)* (0* − q) = p*q. We can now conclude that a*b = −a − b for all a, b ∈ Z with a, b, −a − b distinct, and thus our construction results in a (Z,0) triple system. □

Theorem 6 *There exists a perfect construction of a (Z,1) triple system. The vertices of such a system must lie on an irreducible cubic curve.*

Proof The starting configuration S for our construction involves the nine points {0, ±1, ±2, ±3, ±4} and is illustrated by the bold lines in Fig. 8. Let γ denote a

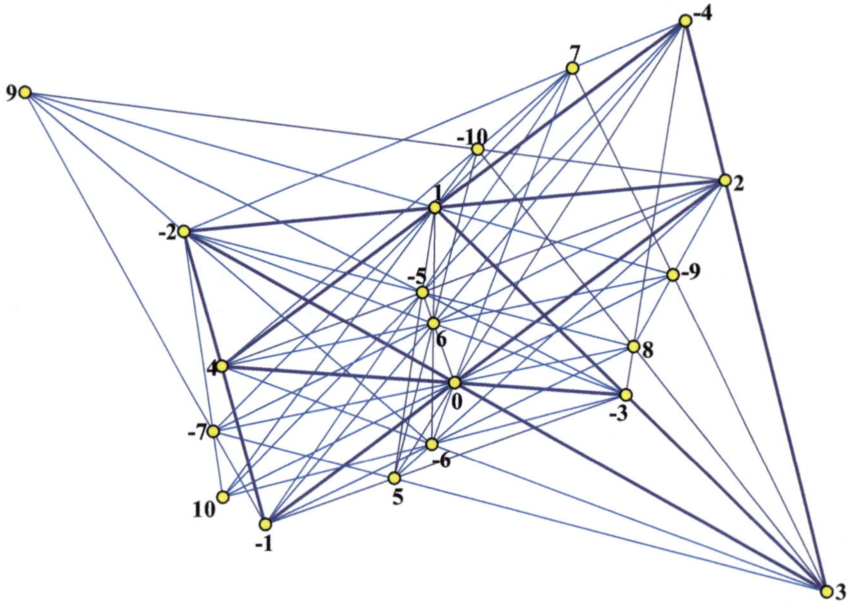

Fig. 8 (Z,1) triple system

cubic curve containing the points in S. As in Theorem 5, γ must be irreducible and each point in S must be nonsingular. We can then define the binary operation * on the points of γ which satisfies identities (1), (2), (3). Define vertex 5 by $5 = [-1,-3] \cup [0,-4]$ and vertex -5 by $-5 = [1,5] \cup [2,4]$. The 1-squares:

$$\begin{bmatrix} 5 & -1 & -3 \\ 0 & -2 & 3 \\ -4 & 4 & 1 \end{bmatrix} \begin{bmatrix} -5 & 5 & 1 \\ 2 & -1 & 0 \\ 4 & -3 & 0 \end{bmatrix}$$

Along with: $0^*0 = (2^* - 1)^* (3^* - 2) = (2^*3)^* (-1^* - 2) = -4^* 4 = 1$ imply that the points 5 and -5 lie on γ. For $k \geq 6$ we define the points recursively by: $k = [-1,-k+2] \cup [-2,-k+3]$ and $-k = [1,k] \cup [2,k-1]$. Assume inductively that the points $\{0, \pm 1, \pm 2, \ldots, \pm (k-1)\}$ as defined have been shown to lie on γ. Suppose $k \geq 6$, then the 1-squares:

$$\begin{bmatrix} k & -1 & -k+2 \\ -2 & 2 & 1 \\ -k+3 & 0 & k-2 \end{bmatrix} \begin{bmatrix} -k & 1 & k \\ 2 & 0 & -1 \\ k-1 & 0 & -k+2 \end{bmatrix}$$

imply by Lemma 4 that k and $-k$ lie on γ. We have now shown by induction that the entire set of integers correspond to points which lie on the cubic curve γ. Let T denote this collection of points. It remains to show that T is a (Z,1) triple system. Suppose $k \geq 6$, $-k < p < q \leq 0$ and $k + p + q = 1$. We must show that k,p,q are collinear. Consider: $k^* p = (-1^* (-k+2))^* (1^* - p) = ((-k+2)^* 1)^* (-1^* - p) = (k-2)^* (p+2) = -k - p + 1 = q$. Here we can assume $p \neq -1$ since we know from the definition of k that $k^* - 1 = -k+2$. Since $k^*p = q$ it must be that k.p.q are collinear as desired. Now suppose $k \leq -6, 0 \leq p < q < -k$ and $k + p + q = 1$. Consider: $k^*p = (1^* - k)^* (0^* (-p+1)) = (0^*1)^* (-k^* (-p+1)) = 0^* (p+k) = -p - k + 1 = q$ and so we have again that k,p,q are collinear. □

In Figs 7 and 8, (Z,0) and (Z,1) triple systems are (partially) illustrated. In the (Z,0) triple system constructed in Theorem 5 we showed that $0^*0 = 0$. Since the construction is perfect, this property holds in every (Z,0) triple system. If $a \neq 0$, then $a^*a = (0^* - a)^* (1^* (-a-1)) = (0^*1)^* (-a^* (-a-1)) = -1^* (2a+1) = -2a$, so we can say that in a (Z,0) triple system $a^*b = -a - b$ for all $a, b \in Z$. In the (Z,1) triple system we showed that $0^*0 = 1$. If $a \neq 0$ then $a^*a = (0^* (-a+1))^* (1^* - a) = (0^*1)^* ((-a+1)^* - a) = 0^*2a = -2a + 1$, so we can say that in a (Z,1) triple system $a^*b = 1 - a - b$ for all $a, b \in Z$. Recall that a^*a denotes the point besides a which lies on the cubic γ and on the tangent to γ at a. The cubic curve γ containing the points of a geometric triple system T is called the *cubic envelope* of T. A nonempty subset of γ which is closed under * is called a *subalgebra* or a *subgroupoid* of the hypercommutative groupoid (γ,*). It is clear from the above that the vertices of a (Z,0) or (Z,1) triple system form a subalgebra of (γ,*). Let T be any (G,0) triple system and H a subgroup of G. If $a, b \in H$, then $a^*b = -a-b$ also lies in H and so H is a subgroupoid of T and (H,0) is a triple system in its own right. We call H a *subsystem* of T. It is not true that every subalgebra of T is a subsystem as we see in our next result.

Theorem 7 *Let T be a (Z,0) triple system with irreducible cubic envelope γ and associated hypercommutative groupoid (γ,*). Then (T,*) contains a subalgebra equivalent to a (Z,1) triple system.*

Proof Let $S = \{3k - 1 : k \in Z\}$. If $3s - 1, 3t - 1 \in S$ then $(3s - 1)^* (3t - 1) = -3s - 3t + 2 = 3(-s - t + 1) - 1 \in S$, and so S is a subalgebra of T. We cannot say that S and the associated triples form a subsystem of T since S is not an abelian group. It is rather a coset of the subsystem of T formed by the multiples of 3. Now apply the relabeling $\phi(3k - 1) = k$ to the points in S. If $3s - 1, 3t - 1, 3q - 1$ are points in T which sum to 0 then it must be that the relabeled points sum to 1. Thus the *coset subsystem* S of T is equivalent to a (Z,1) triple system. □

We now know that the points of a triple system on Z must lie on an irreducible cubic curve. It is possible to construct such a triple system on any given irreducible cubic, as we show next. We use the fact [1] that every irreducible cubic curve contains at least one flex point.

Theorem 8 *Let γ be an irreducible cubic and let e be a flex point of γ. The following construction yields a (Z,0) triple system on γ. Assign vertex 0 to e, and let vertex 1 be any nonsingular point on γ which is not a flex point and such that $1^*1 \neq 0$. Let $-1 = 0^*1$; $2 = -1^* - 1$; $-2 = 0^*2$; $3 = -1^* - 2$ and $-3 = 0^*3$. If points $\{0, \pm 1, \pm 2, \ldots, \pm(k-1)\}, k \geq 3$, have been defined, let (i) $k = -1^*(-k+1)$, and (ii) $-k = 0^*k$. We assume that vertex 1, (the generating vertex), is chosen on γ so that no point on γ is assigned more than one integer label.*

Proof Note that vertex 0 must be a flex point since we must have $0^*0 = 0$ in a (Z,0) triple system. Observe that (ii) holds for every integer k. We will show first that (i) also holds for every integer k. It holds for $k = -2, -1, 0$ and all positive integers. Now suppose $k < 0$. Then $-k > 0$ and so by (i) $-k = -1^* \ (k+1)$ which yields $k + 1 = -1^* - k$. Replacing k by $k - 1$ in this last equation yields (i), and so (i) holds for every integer k. Let T denote the set of points defined in the Theorem and let a, b be two such points. We must show that a^*b is the point $-(a + b)$ of T. Now using (i), (ii) we obtain: $a^*b = (-1^* \ (-a+1))^* \ (0^* - b) = (0^* \ (-a+1))^* \ (-1^* - b) = (a-1)^* \ (b+1)$. This yields $a^*b = (a-k)^* \ (b+k)$ for any integer k. If we take $k = a$ we obtain $a^*b = 0^* \ (a+b) = -(a+b)$ as desired. □

The construction of a (Z,1) triple system on a given irreducible cubic curve is accomplished similarly:

Theorem 9 *Let γ be an irreducible cubic curve. The following construction yields a (Z,1) triple system with cubic envelope γ. Let vertex 0 be a nonsingular nonflex point of γ and let vertex 1 be the point 0*0. Let* $-1 = 1^*1$*; let* $2 = 0^* - 1$ *and* $-2 = 1^*2$*. If* $k \geq 3$ *and vertices* $\{0, \pm 1, \pm 2, \ldots, \pm (k-1)\}$ *have been assigned to points on γ, let (i)* $k = 0^* (-k+1)$*, and (ii)* $-k = 1^*k$*. We assume that vertex 0, (the generating vertex), is chosen so that no point on γ is assigned more than one integer label.*

Proof First we show that (i) holds for every integer. It clearly holds for k = 0,1,2 and thus for every nonnegative integer. Now suppose $k < 0$. Then $-k > 0$ and so by (i) we have: $-k = 0^* \ (k+1)$ and thus $k + 1 = 0^* - k$. Replacing k by $k - 1$ in this last equation yields (i). Let T denote the set of integer labeled points as defined in the Theorem statement and let a,b be two points of T. We must show that $a^*b = 1 - a - b$. Consider: $a^*b = (1^* - a)^* \ (0^* - b + 1) = (0^* - a)^* \ (1^* \ (-b+1)) = (a+1)^* \ (b-1)$. From this we again obtain that $a^*b = (a+k)^* \ (b-k)$ for every integer k. Letting $k = b$ we obtain $a^*b = (a+b)^*0 = -a - b + 1$ as desired. □

In Fig. 9 we illustrate the construction of a (Z,0) triple system given in Theorem 8. For the cubic envelope γ we have selected the curve: $xy^2 + x^3 + x^2y + 2y = 0$ which is symmetric with respect to the origin and has an easily located flex point at the origin.

In Section 5 we show that triple systems with base Z_n can and must be constructed on an irreducible cubic curve. The perfect constructions given in Theorems 5, 6 provide the basis for *nicer* constructions of such triple systems, which we will outline without proof. Let Q denote the points $\{0, \pm 1, \pm 2, \pm 3, 4\}$ used in the starting configuration for our perfect construction of a (Z,0) triple system given

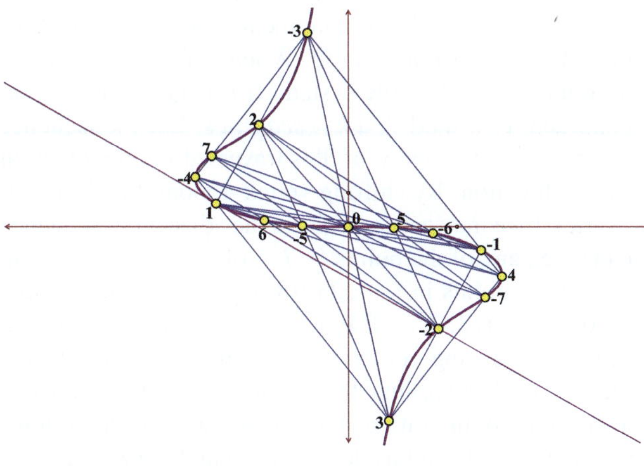

Fig. 9 (Z,0) triple system on Type V cubic

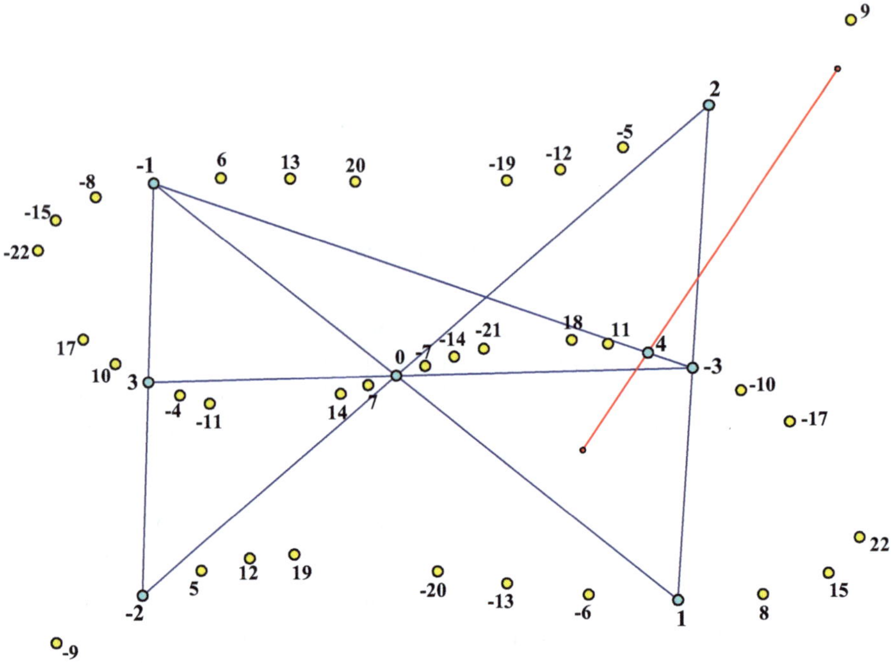

Fig. 10 $(Z_{4+7k}, 0)$ triple system generator

in Theorem 5. Of these we will hold the points $\{0, \pm 1, \pm 2, \pm 3\}$ as fixed and allow vertex 4 to move along line $[-1, -3]$. In accordance with Theorem 5, any position of vertex 4 on this line which does not result in the same point being assigned more than one integer label, will yield a (Z,0) triple system T. An example of a possible starting configuration is given in Fig. 10. In the construction of T we note that as vertex 4 is shifted towards vertex -3, all points which are congruent to k mod 7 converge towards vertex k, i.e., to one of the eight points in Q. When vertex 4 is made to coincide with vertex -3, we obtain a $(Z_7,0)$ triple system. If $k \in Q$ we call the set of points of T which are congruent to k mod 7, a *k – sequence*. Each k-sequence has an *increment* $\eta = 7$ as the cubic envelope γ of T is traversed in one direction which we shall call the *positive* direction. By shifting vertex 4 along line $[-1, -3]$ away from -3 and towards vertex 1 we may cause points in the adjacent 0- sequence and -3 – sequence to coincide, i.e., any of the points $\{-7, -14, -21, -28, \ldots\}$ can be made to coincide with any of the points in $\{\ldots, 25, 18, 11, 4\}$ resulting in triple systems of the form $\{(Z_{4+7k}, 0) : k = 1, 2, 3, \ldots\}$. In Fig. 11 a $(Z_{39},0)$ triple system is illustrated which was created by shifting vertex 4 along line $[-1, -3]$ until the points 18 and -21 coincide. By further varying the position of vertex 4 on line $[-1, -3]$ or by varying the starting configuration many more such triple systems with base Z_n can be obtained. It may be that all triple systems with base Z_n, $(n \geq 7)$, can be constructed in this fashion, but this we leave as an open question.

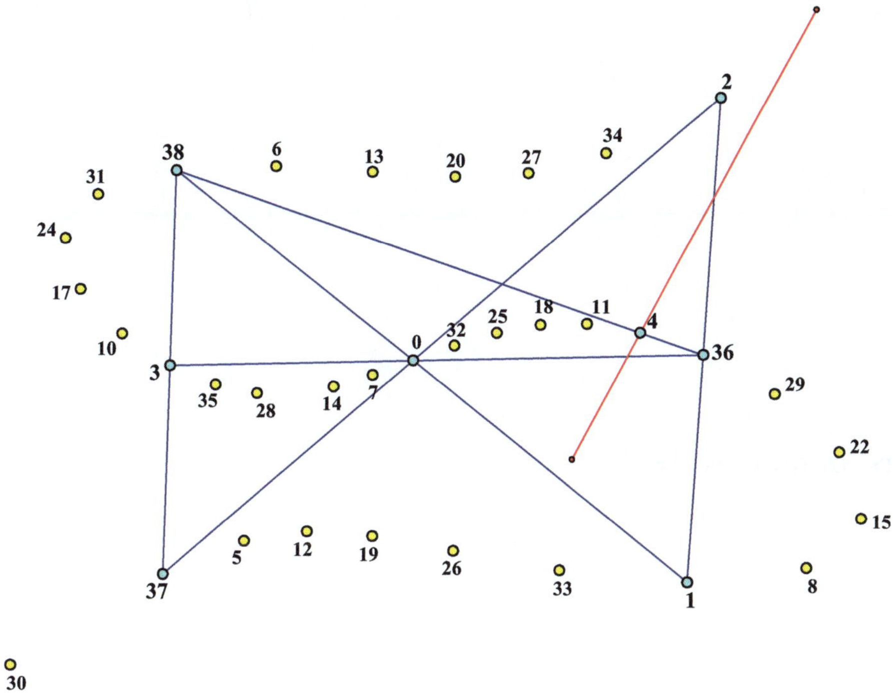

Fig. 11 $(Z_{39}, 0)$ triple system

4 Types of Cubic Curves

According to Bix [1] there are three types of singular irreducible cubic curves: those which can be transformed into:

(I) $y^2 = x^3$,
(II) $y^2 = x^2\,(x + 1)$, or
(III) $y^2 = x^2\ (x - 1)$.

These are illustrated in Figs. 12, 13 and 14 Each has one singular point at the origin and one flex point at infinity on vertical lines. Again, in accordance with Bix, there are two types of nonsingular irreducible cubic curves; those which can be transformed into:

(IV) $y^2 = x\,(x - 1)\,(x - w);\ w > 1$, or
(V) $y^2 = x\left(x^2 + kx + 1\right);\ -2 < k < 2$.

These are illustrated in Figs. 15 and 16. Each of these has three collinear flex points, one of which is the point at infinity on vertical lines. Type V cubics consist of one connected component when regarded in the projective plane, and Type IV cubics have two connected components which we call the *bell* and the *oval*. If γ is a

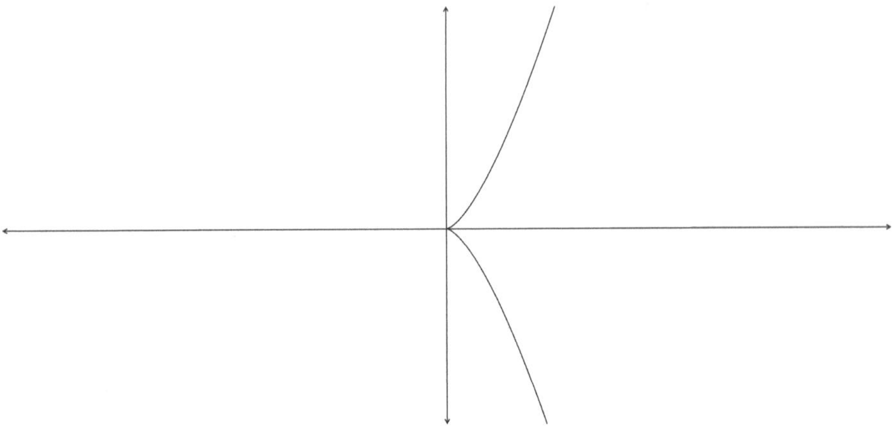

Fig. 12 Type I cubic $y^2 = x^3$

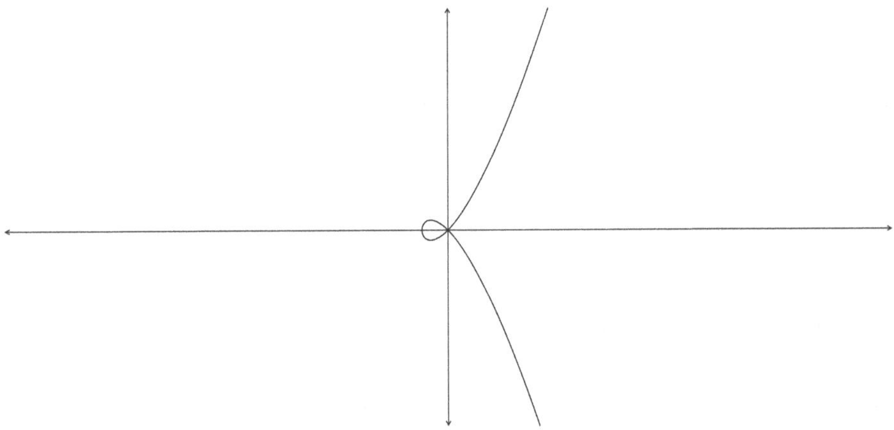

Fig. 13 Type II cubic $y^2 = x^3 (x + 1)$

Type IV cubic then the bell of γ is a subalgebra of (γ,*). A subset S of an irreducible cubic curve with the property that if a, b $\in$ S then a*b lies in the complement of S is called *anticlosed*. A *maximal anticlosed set* is an anticlosed set which is not a proper subset of any other anticlosed subset of γ. In a Type I cubic the unique singularity separates two maximal anticlosed sets. In a Type II cubic the unique singularity at the self-crossing point separates the curve into three maximal anticlosed sets one of which is the loop section of the curve. In Type III and Type V cubics the three flex points separate the curve into three maximal anticlosed sets and in Type IV cubics the three flex points separate the bell into three maximal anticlosed sets, and additionally, the oval is a maximal anticlosed subset of the cubic. Any of the five Types of irreducible cubic curves can serve as the envelope for a (Z,0) or (Z,1) triple system. In Fig. 17 a (Z,0) triple system T is (partially) represented on a Type II cubic

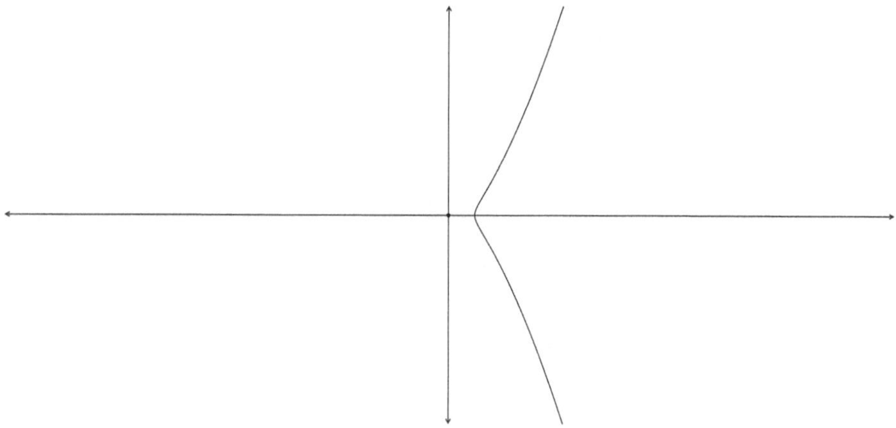

Fig. 14 Type III cubic $y^2 = x^2\ (x - 1)$

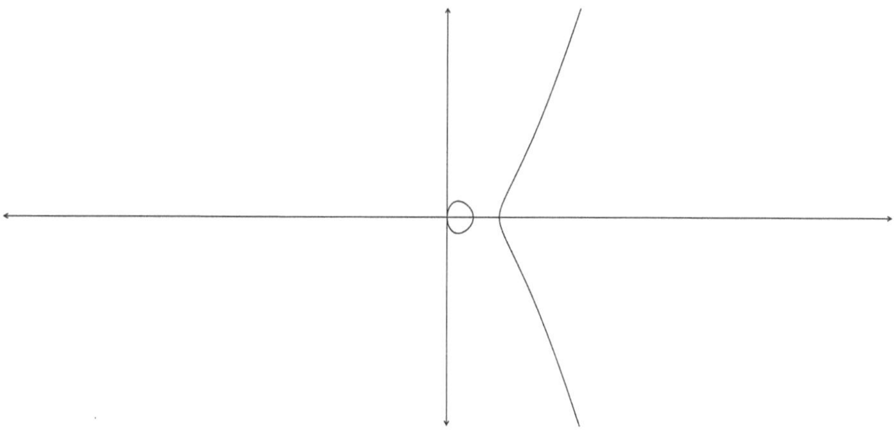

Fig. 15 Type IV cubic $y^2 = x\ (x - 1)\ (x - 2)$

γ; note that the singularity at the self-crossing point is an accumulation point for the vertex set of the triple system. The loop portion of the curve is anticlosed and the remainder of the curve is a subalgebra. For this reason, if x is any nonsingular point of γ, then x*x must lie in the subalgebra and thus all even vertices $2m = -m^* - m$ must lie in the subalgebra and only odd vertices can lie on the loop. Similarly, if T is a (Z,0) triple system with a Type IV cubic envelope γ, then it may be that all the vertices of T lie on the bell of γ, or it may be that all the odd vertices lie on the oval and all the even vertices lie on the bell of γ as in Fig. 18. In either case, since $0^*0 = 0$ vertex 0 must be a flex point of γ. If S is a (Z,1) triple system with Type IV cubic envelope μ then all the vertices of S may lie on the bell of μ or it may be that all the even vertices of S lie on the oval and all the odd vertices on the bell of μ as in Fig. 19. No flex point of μ is a vertex of S.

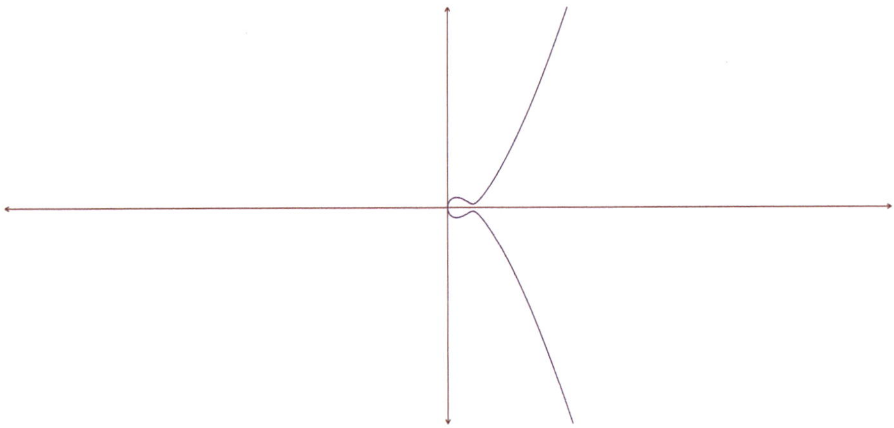

Fig. 16 Type V cubic $y^2 = x\left(x^2 - 1.98x + 1\right)$

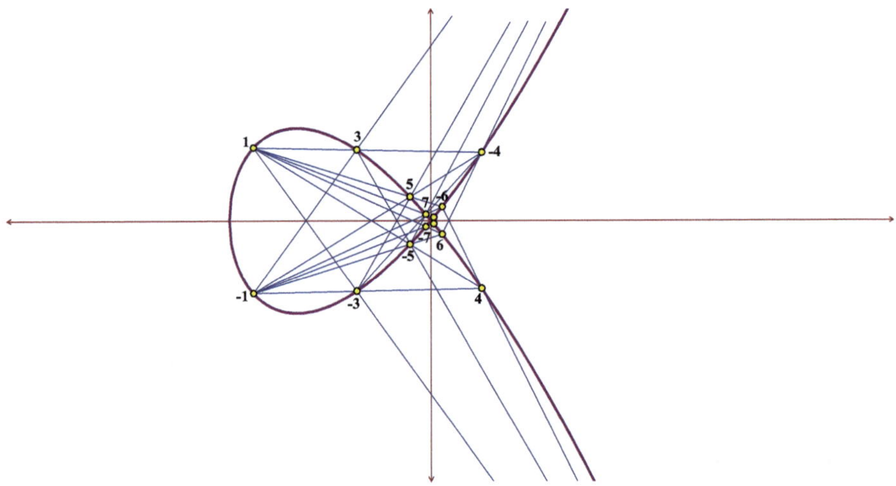

Fig. 17 (Z,0) triple system on Type 2 cubic

5 Triple Systems with Base Z_n

The constructions given in Theorems 8, 9 provide the basis for our general construction of triple systems on Z_n. First we assume the existence of such triple systems and show that they must lie on an irreducible cubic curve of Type III, IV or V.

Theorem 10 *Let T be a triple system with base Z_n. If $n \geq 10$ then the vertices of T lie on an irreducible cubic curve of Type III, IV or V.*

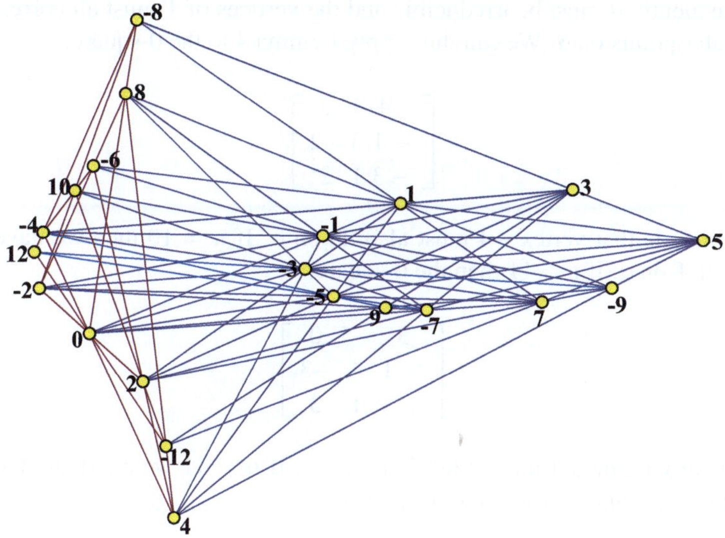

Fig. 18 (Z,0) triple system on Type IV cubic

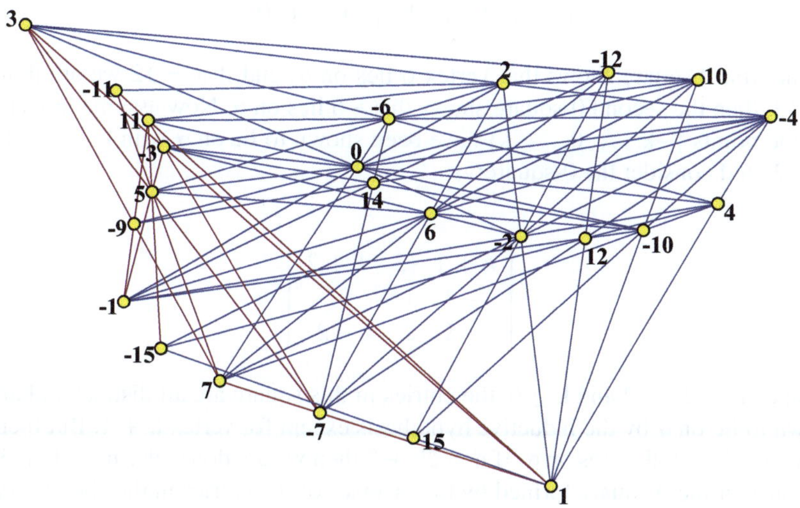

Fig. 19 (Z,1) triple system with odd vertices on bell and even vertices on oval

Proof First suppose T is a $(Z_n,0)$ triple system and let $S = \{0, 1, -1, 2, -2, 3, -3, 4, -5\}$. Any set of nine points lies on some cubic so let α denote a cubic curve which contains these nine vertices of T. No three triples of T contain all nine vertices of S so α cannot consist of three lines. If any triple of T in S is removed there remains at least one triple of T among the remaining vertices of S. No conic can contain three collinear points so α cannot consist of a line and a conic.

Consequently α must be irreducible and the vertices of T must all correspond to nonsingular points on α. We can thus apply Lemma 4 to the 0-square:

$$\begin{bmatrix} -4 & 1 & 3 \\ -1 & 3 & -2 \\ -3 & 1 & 2 \end{bmatrix}$$

To conclude that vertex -4 must also lie on α. If $n = 10$ then we have all ten vertices of T on α. If $n \geq 11$ then the 0-square:

$$\begin{bmatrix} 5 & -5 & 0 \\ -1 & 4 & -3 \\ -4 & 1 & 3 \end{bmatrix}$$

Implies by Lemma 4 that vertex 5 lies on α. If $n = 11$ we are done. Otherwise suppose $n \geq 12$, and consider the 0-squares:

$$\begin{bmatrix} 6 & -4 & -2 \\ -5 & 3 & 2 \\ -1 & 1 & 0 \end{bmatrix} \begin{bmatrix} -6 & 4 & 2 \\ 5 & -3 & -2 \\ 1 & -1 & 0 \end{bmatrix}$$

The first 0-square shows that vertex 6 lies on α, and if $n = 12$ we are done. If $n \geq 13$, then the second 0-square shows that -6 lies on α. Now assume inductively that the points $\{-k, \ldots, 0, \ldots, k\}$ have been shown to lie on α with $k \geq 6$ and $n \geq 2k + 2$, and consider the 0-square:

$$\begin{bmatrix} k+1 & -3 & -k+2 \\ -1 & 1 & 0 \\ -k & 2 & k-2 \end{bmatrix}$$

Since $n \geq 2k + 2$ and $k \geq 6$, the entries in this square are all distinct and all are known to lie on α by the inductive hypothesis except for vertex $k + 1$. But then, by Lemma 4, $k + 1$ also lies on α. If $n = 2k + 2$ then we are done, else $n \geq 2k + 3$ and we consider the 0-square formed by the inverses of the entries in the above square:

$$\begin{bmatrix} -k-1 & 3 & k-2 \\ 1 & -1 & 0 \\ k & -2 & -k+2 \end{bmatrix}$$

The entries in this square are all distinct and all except for $-k - 1$ lie on α by the inductive hypothesis. By Lemma 4 vertex $-k - 1$ must also lie on α and so the induction is complete.

Now suppose T is a $(Z_n,1)$ triple system. We have $n \geq 10$ by hypothesis, and if 3 does not divide n, then T is equivalent to a $(Z_n,0)$ triple system and we are done by the first part of the proof. So we can assume that $n \geq 12$. Let β denote the irreducible cubic curve which contains the nine points $\{0, \pm 1, \pm 2, \pm 3, \pm 4\}$ of T. Consider the 1-squares:

$$\begin{bmatrix} 5 & -4 & 0 \\ -1 & 4 & -2 \\ -3 & 1 & 3 \end{bmatrix} \begin{bmatrix} -5 & 5 & 1 \\ 4 & 0 & -3 \\ 2 & -4 & 3 \end{bmatrix}$$

The first of these shows that vertex 5 lies on β, and then the second shows that vertex -5 lies on β. The 1-square:

$$\begin{bmatrix} 6 & -5 & 0 \\ -1 & 4 & -2 \\ -4 & 2 & 3 \end{bmatrix}$$

Shows that vertex 6 lies on β. If $n = 12$ we are done, otherwise, (since we need only consider multiples of 3), we can assume that $n \geq 15$. Consider the 1-squares:

$$\begin{bmatrix} -6 & 6 & 1 \\ 4 & -1 & -2 \\ 3 & -4 & 2 \end{bmatrix} \begin{bmatrix} 7 & -2 & -4 \\ -1 & 2 & 0 \\ -5 & 1 & 5 \end{bmatrix} \begin{bmatrix} -7 & 7 & 1 \\ 2 & -6 & 5 \\ 6 & 0 & -5 \end{bmatrix}$$

These show successively that vertices $-6, 7, -7$ lie on β. If $n = 15$ we are done. Now assume inductively that the points $\{-k, \ldots, 0, \ldots, k\}$ of T have been shown to lie on β, that $k \geq 7$ and $n \geq 2k + 2$. Then the entries in the 1-square:

$$\begin{bmatrix} k+1 & -k & 0 \\ -1 & k-1 & -k+3 \\ -k+1 & 2 & k-2 \end{bmatrix}$$

are all distinct and all lie on β by the inductive hypothesis, except for vertex k which by Lemma 4 must also lie on β. If $n = 2k + 2$ we are done, otherwise $n \geq 2k + 3$ and the entries in the 1-square:

$$\begin{bmatrix} k+1 & -k-1 & 1 \\ -1 & k-1 & -k+3 \\ -k+1 & 3 & k-3 \end{bmatrix}$$

are all distinct and show that $-k-1$ also lies on β. This completes the induction and so we have shown that a geometric triple system with base Z_n has all its vertices on an irreducible cubic curve.

It remains to show that Type I and Type II cubics cannot contain triple systems with base Z_n. Consider the generic Type II cubic γ illustrated in Figs. 13 and 17 and suppose A is a nonsingular, nonflex point on this curve. The curve has three sections: the loop in the second and third quadrants; the section in the first quadrant and the section in the fourth quadrant. If A lies on the loop then A*A lies either in the first or the fourth quadrant section, since the loop is anticlosed. The combined first and fourth quadrant sections form a subalgebra of $(\gamma,*)$. The first as well as the fourth quadrant sections are anticlosed: if A lies in the first quadrant then A*A lies in the fourth, and if A lies in the fourth then A*A lies in the first. More significantly, if A lies in either the first or fourth section, and A is not the point ∞ at infinity on γ, then A*A lies closer to the singularity at the origin. If we let $\phi(x) = x^*x$, then the sequence $\{A, \phi(A), \phi^2(A), \phi^3(A), \ldots\}$ is infinite. We conclude that $\{\infty\}$ and $\{\infty, (-1,0)\}$ are the only finite subalgebras of $(\gamma,^*)$. If $n > 2$ then a triple system with base Z_n on γ would be a finite nontrivial subalgebra of $(\gamma,^*)$ which as we see is not possible. If γ is a Type I cubic, a similar analysis shows that $\{\infty\}$ is the unique finite subalgebra of $(\gamma,^*)$, and thus no nontrivial finite triple systems can be constructed on γ. □

Having shown that triple systems with base Z_n must lie on irreducible cubic curves, we now start with an irreducible cubic and construct the triple system on it. A triple system as defined does not have much meaning when $n = 1$ or 2 since three points are required to make even a single triple. But in the context of an irreducible cubic curve γ, the three collinear flex points count as, (trivial), Z_1 triple systems, and a two element set {a,e} where e is a flex point of γ and $a^*a = e$, is a Z_2 triple system. The three flex points of γ form a Z_3 triple system. Ultimately it will be important for us to describe exactly how many distinct but equivalent triple systems with base Z_n can be accommodated on a given Type IV or Type V cubic curve, but first we give the constructions:

Theorem 11 *For each positive integer n, a triple system with base Z_n can be constructed on an irreducible cubic curve.*

Proof First suppose n is odd and let γ denote the Type V cubic with equation: $xy^2 + x^3 + x^2y + 2y = 0$ illustrated in Fig. 9. We will construct a $(Z_{2n}, 0)$ triple system on γ by following the procedure given in Theorem 8 and then shifting the generating vertex, (vertex 1), so that the points $\{(n-1)/2, (n+1)/2\}$ lie on a vertical line. Let ∞ denote the point at infinity on vertical lines and let 0 denote the flex point of γ at the origin. Let A, B denote the two points on γ which satisfy $A^*A = B^*B = \infty$. (We call A,B the *square-roots of* ∞.) Choose vertex 1 in section (0,A) of γ and near vertex 0. Following the construction given in Theorem 8, we find that the points $\{1, 2, 3, \ldots\}$ occur sequentially on γ as we traverse γ from 0 in a *positive* direction. The reflections $\{-1, -2, -3, \ldots\}$ occur sequentially on γ as γ is traversed from 0 in a *negative* direction. By choosing vertex 1 sufficiently close to vertex 0 we can arrange for the section (0,A) of γ to contain precisely the points $\{1, 2, 3, \ldots, (n-1)/2\}$. Then the first point of section (A,∞) will be $(n+1)/2$. By further shifting of vertex 1 we can arrange for the points $\{(n-1)/2, (n+1)/2\}$ to lie on a vertical line. In case n is even we follow the same procedure, but shift vertex 1 until point n/2 coincides with A. If n is odd then $\infty = \frac{n+1}{2}^* \frac{n-1}{2} = -n$ and if n is even then $-n = \frac{n}{2}^* \frac{n}{2} = A^*A = \infty$, so in either case $0 = \infty^*\infty = -n^* - n$. But also $n^* - n = 0$, so we have $n = -n = \infty$. The coincidence of the points $-n, n$ forces any two points of the (Z,0) triple system which are congruent mod 2n to coincide: if $a \in Z$ then $-a^*\, n = a - n = a + n$ implies $a = a + 2n$. Thus the (Z,0) triple system constructed in accordance with Theorem 8 collapses to a $(Z_{2n},0)$ triple system T on γ. Now let $S = \{0, 2, 4, \ldots, 2n - 2\}$ denote the set of even vertices of T and relabel them by the mapping $\phi : S \to Z_n$ defined by $\phi\,(2k) = k$. Then the points $\phi\,(S)$ form a $(Z_n, 0)$ triple system on γ.

To construct a $(Z_n, 1)$ triple system on γ we first construct a $(Z_{3n}, 0)$ triple system T on γ using the method in the first paragraph. Let $S = \{3k - 1 : k = 1, 2, \ldots, n\}$, then S is easily seen to be a subalgebra of (T,*) and, in fact, a coset subsystem. The relabeling $\phi : S \to Z_n$ given by: $\phi\,(3k - 1) = k$ (mod n) turns S into a $(Z_n, 1)$ triple system. □

Note that a geometric triple system is a pure incidence structure and so is preserved by transformations of the projective plane. Our construction of Z_n triple systems on a particular Type V cubic therefore means that such triple systems can be constructed on any Type V cubic. Similar procedures can be used to construct Z_n triple systems on a Type IV cubic curve γ. If the generating vertex lies on the bell then the entire triple system will lie on the bell. If the generating vertex (vertex 1) is chosen on the oval then we have the following possibilities. If n is even and T is a (Z,0) triple system on γ then every even vertex 2k of T is equivalent to $-k^* - k$ and must lie on the bell since the oval is anticlosed and so cannot contain the square of any of its vertices, and the bell is a subalgebra which consequently contains the square of any of its vertices. The oval must then be populated with all the odd vertices of T. In Fig. 31, (see Appendix) we illustrate a $(Z_{18},0)$ triple system on the Type IV cubic with equation: $xy^2 = (x - 1)\left(16 - x^2\right)$. This curve is symmetric with respect to the x-axis and has a flex point at the point at infinity on vertical lines. We let vertex 0 be this point at infinity, then as a consequence vertices $k, -k$ are reflections of each other across the x-axis for each integer k. Let A,B denote the two points on the oval and on the x-axis as in Fig. 31, and select the generating vertex, (vertex 1) on the oval near point A. As the construction proceeds, the odd vertices form consecutively on the oval with increment 2 and the even vertices fall consecutively on the bell also with increment 2. By shifting vertex 1 on the oval we can arrange for vertex n/2 to coincide with point B, resulting in a $(Z_n,0)$ triple system on γ. If a is a vertex of T on the oval, (and vertex 1 lies on the oval), then the mapping $\phi\,(a) = 1^*a$ is a bijection between the points of T on the oval and the points of T on the bell. And so T must have an even number of vertices. From this we have:

Theorem 12 *For any positive integer n, a $(Z_n,0)$ triple system T can be constructed on the bell of a Type IV cubic curve. If n is odd then every vertex of T must lie on the bell. If n is even then T can be constructed with its odd vertices on the oval and its even vertices on the bell.* □

The situation is reversed for a $(Z_n,1)$ triple system T with vertices on the oval of a type IV cubic curve γ. T must again have an even number n of vertices but this time every odd vertex is a square and so must lie on the bell. The even vertices must all lie on the oval along with the generating vertex (vertex 0). That such a triple system exists can be argued by first constructing a $(Z_{3n},0)$ triple system T_0 according to the above prescription and then taking points of the form 3k-1 from T_0 and relabeling by the mapping $\phi\,(3k-1) = k$. A $(Z_{18},1)$ triple system of this type with increment 2 is illustrated in Fig. 32, (see Appendix). We thus have:

Theorem 13 *For any positive integer n, a $(Z_n,1)$ triple system T can be constructed on a Type IV cubic curve. If n is odd then every vertex of T must lie on the bell. If n is even then T can be constructed with its even vertices on the oval and its odd vertices on the bell.* □

We shall refer to a triple system which has half of its vertices on the bell and the other half on the oval of a Type IV cubic as a *split triple system.*

6 Construction of Z_n Triple Systems Using Equilateral Triangles

If T is a triple system with base Z_n and $3|n$ then we have found a beautiful construction of T using concentric equilateral triangles. The construction will be accomplished by putting the vertices of T on an irreducible cubic curve which has a rotational symmetry of order 3. We first determine the formula for such a curve.

Theorem 14 *After a suitable rotation an irreducible cubic curve which rotates onto itself clockwise about the origin through an angle of 120° has an equation of the form:*

$$(e-3ax)\,y^2+\left(ax^3+ex^2+j\right)=0. \tag{6}$$

Proof The general equation of a cubic is given by:

$$ax^3+bx^2y+cxy^2+dy^3+ex^2+fxy+gy^2+hx+ky+j=0, \tag{7}$$

and the equations:

$$x'=\left(-\frac{1}{2}\right)x-\left(\frac{\sqrt{3}}{2}\right)y \quad y'=\left(\frac{\sqrt{3}}{2}\right)x-\left(\frac{1}{2}\right)y \tag{8}$$

effect a clockwise rotation of 120° of the point (x,y) to the point (x′, y′). Now substituting x′ for x and y′ for y in (7), we obtain:

$$a\left\{\left(-\frac{1}{2}\right)x-\left(\frac{\sqrt{3}}{2}\right)y\right\}^3+b\left\{\left(-\frac{1}{2}\right)x-\left(\frac{\sqrt{3}}{2}\right)y\right\}^2\left\{\left(\frac{\sqrt{3}}{2}\right)x-\left(\frac{1}{2}\right)y\right\}$$
$$+c\left\{\left(-\frac{1}{2}\right)x-\left(\frac{\sqrt{3}}{2}\right)y\right\}\left\{\left(\frac{\sqrt{3}}{2}\right)x-\left(\frac{1}{2}\right)y\right\}^2$$
$$+d\left\{\left(\frac{\sqrt{3}}{2}\right)x-\left(\frac{1}{2}y\right)\right\}^3+e\left\{\left(-\frac{1}{2}\right)x-\left(\frac{\sqrt{3}}{2}\right)y\right\}^2$$
$$+f\left\{\left(-\frac{1}{2}\right)x-\left(\frac{\sqrt{3}}{2}\right)y\right\}\left\{\left(\frac{\sqrt{3}}{2}\right)x-\left(\frac{1}{2}\right)y\right\}+g\left\{\left(\frac{\sqrt{3}}{2}\right)x-\left(\frac{1}{2}\right)y\right\}^2$$
$$+h\left\{\left(-\frac{1}{2}\right)x-\left(\frac{\sqrt{3}}{2}\right)y\right\}+k\left\{\left(\frac{\sqrt{3}}{2}\right)x-\left(\frac{1}{2}\right)y\right\}+j=0. \tag{9}$$

Since (7), (9) must represent the same cubic curve γ, we can equate corresponding coefficients. By equating the coefficients of x we obtain:

$$3h=k\sqrt{3}, \tag{10}$$

And by equating the coefficients of y we obtain:

$$3k=-h\sqrt{3}. \tag{11}$$

Solving (10), (11) yields $h=k=0$. Equating the coefficients of x^2 yields:

$$3e=3g-f\sqrt{3}, \tag{12}$$

And equating the coefficients of xy we obtain:

$$f\sqrt{3}=e-g. \tag{13}$$

The Eqs. (12) and (13) yield $e=g$, and then from (12) we obtain $f=0$. With these results the Eq. (7) of γ reduces to:

$$ax^3+bx^2y+cxy^2+dy^3+e\left(x^2+y^2\right)+j=0. \tag{14}$$

By equating coefficients of y^3 in (9), (14) we obtain:

$$9d=\left(-3\sqrt{3}\right)a-3b-\left(\sqrt{3}\right)c, \tag{15}$$

and by equating the coefficients of xy^2 in (9), (14) we obtain:

$$3c = -9a + \left(\sqrt{3}\right)b + \left(3\sqrt{3}\right)d. \tag{16}$$

Solving (15), (16) together yields $b = -3d$, and substituting this for b in (15) yields $c = -3a$. Equating the coefficients of x^3 or of x^2y yields only redundant equations. Equation (14) can then be rewritten:

$$ax^3 - 3dx^2y - 3axy^2 + dy^3 + e\left(x^2 + y^2\right) + j = 0. \tag{17}$$

To further simplify the equation of γ, we can rotate γ about the origin until it contains the point (0,1,0) at infinity on vertical lines. The homogenization of (17) is:

$$ax^3 - 3dx^2y - 3axy^2 + dy^3 + ez\left(x^2 + y^2\right) + jz^3 = 0. \tag{18}$$

Putting $x = 0$, $y = 1$ and $z = 0$ in (18) yields $d = 0$, and so (17) reduces to:

$$ax^3 - 3axy^2 + e\left(x^2 + y^2\right) + j = 0. \tag{19}$$

Since y can be replaced by $-y$ without changing Eq. (19), γ is symmetric wrt the x-axis. In (19) clearly a cannot be 0, so we can divide by a and rename coefficients to obtain:

$$x^3 - 3xy^2 + e\left(x^2 + y^2\right) + k = 0. \tag{20}$$

□

In the next result we develop important algebraic properties pertaining to a cubic curve with a 120° rotational symmetry.

Lemma 15 *Let γ be an irreducible cubic curve with rotational symmetry ξ of order three about the origin. Then (i) ξ is an automorphism of $(\gamma,^*)$ of order 3, and (ii) $b^*\xi(c) = \xi(b)^*c = \xi^2(b)^*\ \xi^2(c)$* for any points *b*, *c* on γ.

Proof We may assume that the equation of γ is given by (20) and so consists of three *petals* each of which have an axial symmetry as in Fig. 20. Let $\infty_v, \infty_u, \infty_w$ denote the three points at infinity on γ. These also are the three flex points of γ and form a subalgebra I of $(\gamma,^*)$. Let b be a point of the Euclidean plane on γ, and let α denote the circle with center at the origin and which contains the points $\Delta_b = \left\{b, \xi(b), \xi^2(b)\right\}$. (In Fig. 20 we use the notation $b' = \xi(b)$, and $b'' = \xi^2(b)$). This circle meets γ in three more points, counting multiplicities, which are given by $\{b^*\infty_u, b^*\infty_v, b^*\infty_w\}$ due to the aforementioned symmetry of the petals of γ. We also see that the following equations hold:

$$b^*\infty_u = \xi(b)^*\infty_v = \xi^2(b)^*\infty_w, \tag{21}$$

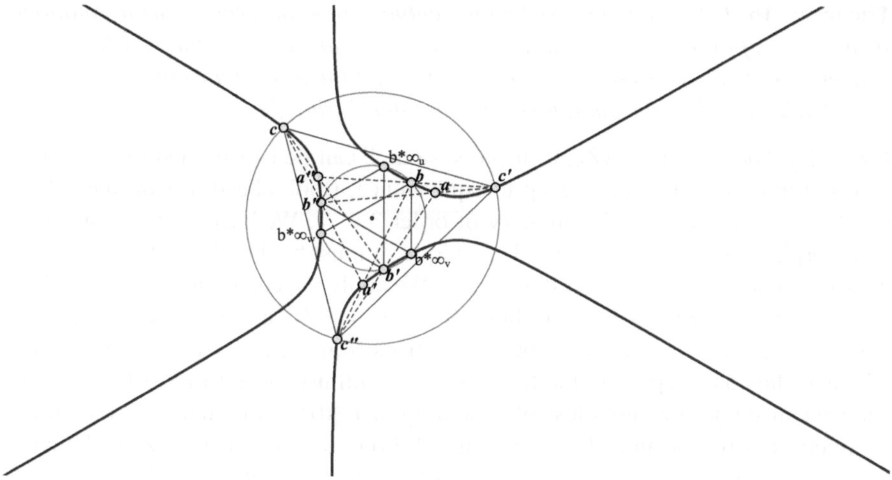

Fig. 20 Cubic curve with 120° rotational symmetry

$$b^*\infty_v = \xi(b)^*\infty_w = \xi^2(b)^*\infty_u, \tag{22}$$

$$b^*\infty_w = \xi(b)^*\infty_u = \xi^2(b)^*\infty_v. \tag{23}$$

If we let $q = b^*\infty_v$ then we see that $q^*\ I = \Delta_b$, i.e., every equilateral triangle with its vertices on γ is a translate of I. If c is a second point on γ, then $\Delta_c = p^*I$ for some $p \in \gamma$, and $\Delta_b{}^*\Delta_c = (q^*I)^*\ (p^*I) = (q^*p)^*\ (I^*I) = (q^*p)^*I = \Delta_{b^*c}$. We may call Δ_b a *coset* of the subalgebra I, then, as in the theory of groups, the product of two cosets of I is a coset of I. General properties of cosets of a subalgebra of a hypercommutative algebra are given in [2]. If s,t are any two points on γ, then of course the points $\{s,t,s^*t\}$ are collinear. Since ξ preserves lines it must be that the points $\{\xi\,(s)\,,\xi\,(t)\,,\xi\,(s^*t)\}$ are collinear, when it follows immediately that $\xi(s)^*\xi\,(t) = \xi\,(s^*t)\}$, and thus ξ is an automorphism of $(\gamma,^*)$. The points of Δ_{b^*c} must be given by $\left\{b^*c, b^*\xi\,(c)\,, b^*\xi^2\,(c)\right\}$. If we set $a = b^*c$, then we have $\xi\,(a) = \xi(b)^*\xi\,(c)$, and $\xi^2\,(a) = \xi^2(b)^*\xi^2\,(c)$. Then we must have $\xi\,(a) \in \left\{b^*\xi\,(c)\,, b^*\xi^2\,(c)\right\}$. We cannot have $\xi\,(a) = b^*\xi\,(c)$ since this would imply $\xi(b)^*\xi\,(c) = b^*\xi\,(c)$ which yields $b = \xi\,(b)$ by cancellation. So we must have $\xi(b)^*\xi\,(c) = \xi\,(a) = b^*\xi^2\,(c)$, and $\xi^2(b)^*\xi^2\,(c) = \xi^2\,(a) = b^*\xi\,(c)$. In the equation $\xi(b)^*\xi\,(c) = b^*\xi^2\,(c)$, if we replace c by $\xi^2\,(c)$, we obtain $\xi(b)^*c = b^*\xi\,(c)$ as desired. □

Now we are prepared to give our intended construction of a Z_n triple system using equilateral triangles.

Theorem 16 *Let γ be an irreducible cubic curve of Type V with rotational symmetry ξ of order three about the origin. If $3|n$, say $n = 3m$, then a $(Z_n,0)$ triple system with its vertices on γ has the property that the three points of every coset of $I = \{0,m,2m\}$, different from I, form an equilateral triangle.*

Proof By Theorem 11, a $(Z_n, 0)$ triple system T can be constructed on γ. Since n is a multiple of 3, T is unique up to equivalence, (look ahead to Corollary 24 !). Let ξ denote the rotational symmetry of order 3 of γ. We have shown that ξ is an automorphism of $(\gamma,*)$, and since T is unique, it must be that the restriction of ξ to the subalgebra T, is an automorphism of T. We apply the equation $b^*\xi(c) = \xi(b)^*c$ from Lemma 15 with $b = 0$ to obtain $0^*\xi(c) = \xi(0)^*c$ which yields $-\xi(c) = -m - c$ or $\xi(c) = c + m$. Note that the vertices $\{0,m,2m\}$ of T must be flex points of γ and thus they represent the three points at infinity. So $\xi(0)$ must be m or 2m, and we can suppose without loss of generality that $\xi(0) = m$. Since ξ is a rotational symmetry of order 3, and $\xi(c) = c + m$, it follows that for each nonzero element c of Z_n, the points $\{c, c^*m, c + 2m\} = c + I$ form an equilateral triangle. □

In Fig. 21 we illustrate a $(Z_{18}, 0)$ triple system constructed in accordance with Theorem 16. The result of Theorem 16 also applies to the case in which γ is a Type IV cubic and T is a split $(Z_n,0)$ triple system. In Fig. 22 we illustrate a split $(Z_{18},0)$ triple system with its vertices on a cubic curve with 120° rotational symmetry. If $3|n$ then, in accordance with Theorem 1, a $(Z_n,1)$ triple system is not equivalent to a $(Z_n,0)$ triple system, but such a triple system also has a construction involving equilateral triangles. A $(Z_{18},1)$ triple system is illustrated in Fig. 23 in which each three element subset of Z_{18} of the form $\{q, q + m, q + 2m\}$, $(m = n/3)$ corresponds to an equilateral triangle. With the help of Lemma 15 we can prove the following interesting result concerning nested equilateral triangles.

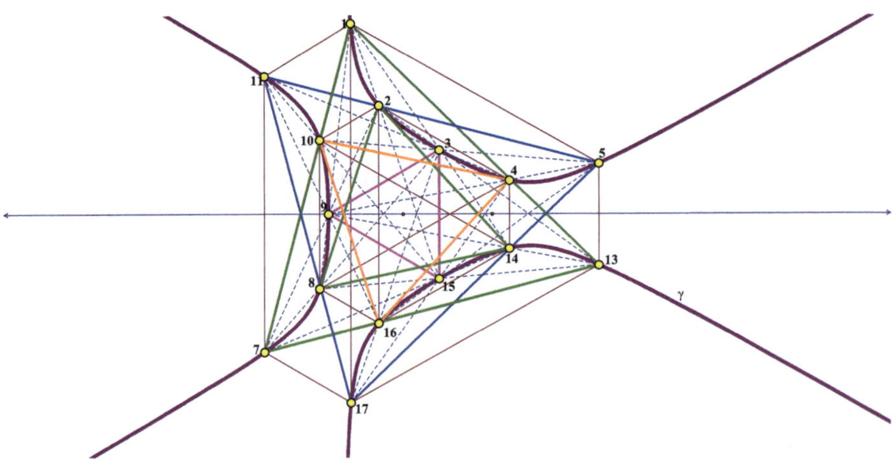

Fig. 21 Equilateral triangle construction of $(Z_{18}, 0)$ triple system

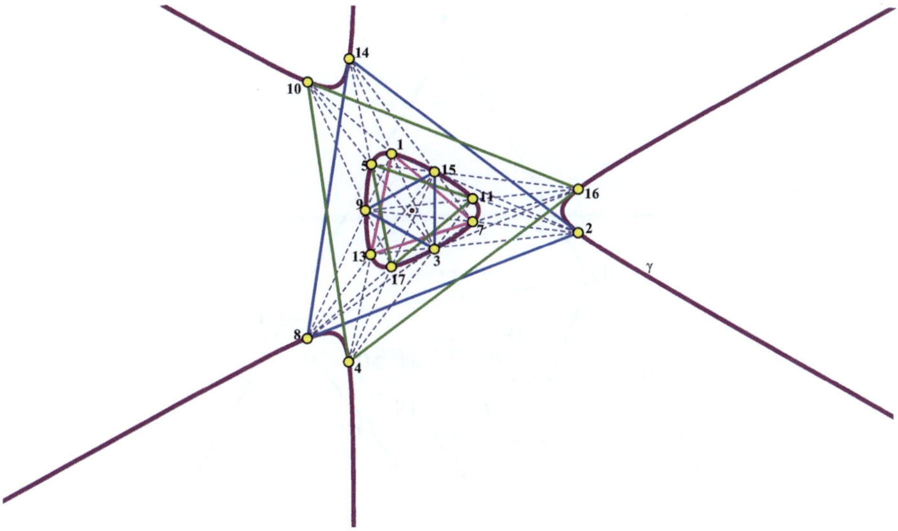

Fig. 22 (Z_{18}, 0) triple system on Type 2 cubic: equilateral construction with vertices 0,6,12 at infinity

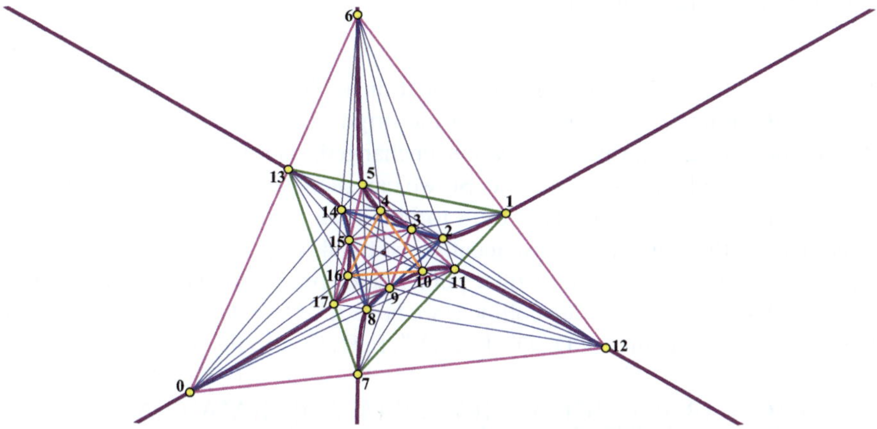

Fig. 23 Equilateral construction of (Z_{18},1) triple system

Theorem 17 *Let $\Delta_B = (B, B', B'')$ be an equilateral triangle with one vertex on each side of an equilateral triangle $\Delta_A = (A, A', A'')$, and let $\Delta_C = (C, C', C'')$ be an equilateral triangle with one vertex on each side of Δ_B as in Fig. 24. Let $D = [A, C'] \cap [A', C]$; $D' = [A', C''] \cap [A'', C']$; $D'' = [A'', C] \cap [A, C'']$; $E = [B, D''] \cap [B'', D]$; $E' = [B, D'] \cap [B', D]$; $E'' = [B', D''] \cap [B'', D']$; $F = [E, A'] \cap [A, E']$; $F' = [E', A''] \cap [E'', A']$; $F'' = [E'', A] \cap [E, A'']$. Then F lies on $[C, C']$; F' lies on $[C', C'']$, and F'' lies on $[C'', C]$.*

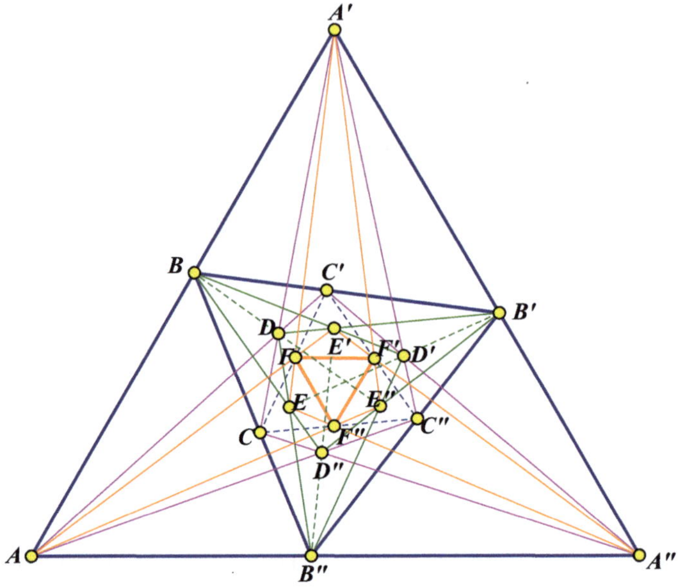

Fig. 24 Nested equilateral triangle Theorem

Proof Note that (17) is the general equation of an irreducible cubic curve γ with a 120° rotational symmetry about the origin. In this equation interchanging the coefficients a,d leaves the equation unchanged, and a,d cannot both be zero otherwise we no longer have a cubic. So we can suppose without loss of generality that $a \neq 0$, and then divide (17) by a to obtain an equation for γ with only the three free coefficients d,e,j. Consequently just the three points A,B,C determine γ, and thus γ contains the nine distinct points belonging to the equilateral triangles $\Delta_A, \Delta_B, \Delta_C$. Since A,B, A′ are collinear we have $A^* A' = B$; $A'^* A'' = B'$, etc. By Lemma 15, all the points $D, D', D'', E, E', E'', F, F', F''$ also lie on γ. Then:

$$\begin{aligned} C^*C' &= (B^*B'')^* (B^*B'') = \{(A^*A')^* (A^*A'')\}^* \{(A^*A')^* (A'^*A'')\} \\ &= \{(A^*A')^* (A'^*A'')\}^* \{(A^*A)^* (A'^*A'')\} \\ &= [A^* (A^*A)]^* [\{ A^* ((A^*A')^* (A'^*A'')\} ^* (A'^*A'') \}] \\ &= A^* [\{A^* (B^*B')\}^* \{A'^*A''\}]= A^*[(A^*C')^* (A'^*A'')] \\ &= A^* (D^*B') = A^*E' = F. \end{aligned}$$

Thus F lies on [C, C′], and by 120° rotations we obtain F′ on [C′, C″], and F″ on [C″,C]. □

7 Increments

We have observed in all our constructions of triple systems on Z_n that there is a constant increment between consecutive vertices as the cubic envelope is traversed in one direction. This property is proved in our next result.

Theorem 18 *Let T be a triple system with base Z_n and irreducible cubic envelope γ. If s is any point of T, let $\sigma(s)$ denote its successor on γ as γ is traversed in one direction. Then there exists an integer k such that $\sigma(s) = s + k$ for all s in T.*

Proof Let $\rho(s)$ denote the predecessor of vertex s on γ. Since $\sigma(s)$ is the successor of s, it must be that $s * s = \rho(s) * \sigma(s)$. Suppose $\sigma(s) = s+k$ for some integer k. If T is a (Z_n,0) triple system then $-\rho(s) - \sigma(s) = \rho(s)*\sigma(s) = -2s$ implies $\rho(s) = s - k$. If T is a (Z_n,1) triple system then $1 - \rho(s) - \sigma(s) = \rho(s) * \sigma(s) = s * s = 1 - 2s$ implies similarly that $\rho(s) = s - k$. thus for any triple system with base Z_n on γ and for any three consecutive vertices of such a triple system, the increment (k) between the first and second vertex is the same as the increment between the second and third. This implies $\sigma(s) = s + k$ for every vertex s of T. □

We call the value k stipulated by Theorem 18 the *increment* of T. The construction given in Theorem 11 shows that triple systems on Z_n can have an increment of 1. Suppose T is a (Z_n,0) triple system on a Type V cubic curve with increment 1, and suppose t is an integer relatively prime to n. Then the relabeling of T given by $\phi(s) = ts$ produces a triple system on Z_n with sum 0 which is equivalent to T and which has increment t. On the other hand, If t is not relatively prime to n, then the multiples of t form a proper subgroup of Z_n and so the full triple system T cannot be formed. If we start with a (Z_n,0) triple system with increment t and relabel according to the inverse of ϕ, we obtain an equivalent triple system with increment 1. In accordance with Theorem 1, a (Z_n,1) triple system is equivalent to a (Z_n,0) triple system in case gcd $(n, 3) = 1$. The relabelings used, namely adding a fixed integer k to each vertex or replacing a vertex by its inverse, preserve the set of possible increments. Finally if T is a (Z_n,1) triple system with Type V cubic envelope γ and with n a multiple of 3, let t be an integer relatively prime to n. The relabeling of T given by $\psi(s) = ts$ transforms T into a (Z_n,t) triple system S equivalent to T but with increment t. In accordance with Theorem 1, S is in turn, equivalent to a (Z_n,1) triple system via relabelings which either preserve the increment or negate it. So we have:

Theorem 19 *Let T be a triple system with base Z_n on a Type V cubic curve. Then T can have increment t for any $t \in \{1, 2, \ldots, n-1\}$ which is relatively prime to n. In any case T is equivalent to a triple system on γ with increment 1.* □

In Theorem 19 we can replace "Type V" with "Type III" or "bell of a Type IV", but for triple systems on a Type IV cubic which have vertices both on the bell and the oval the increment between consecutive vertices must be a multiple of 2 since either the odd (even) vertices lie together on the bell and the even (odd) vertices lie together on the oval. Such triple systems must have an even number of vertices. We have seen, as in the discussion following Theorem 11, that a $(Z_n,0)$ triple system can be constructed with *clockwise* (around the oval) increment 2, (See Appendix Fig. 31).

Theorem 20 *Let T be a $(Z_n,0)$ triple system on a Type IV cubic curve γ with even vertices on the bell and odd vertices on the oval of γ. Then T can have clockwise increment 2t for any $t \in \{1, 2, \ldots, \frac{n}{2} - 1\}$ with $gcd\left(t, \frac{n}{2}\right) = 1$. In any case T is equivalent to a $(Z_n,0)$ triple system on γ with increment 2.*

Proof Let T be a $(Z_n,0)$ triple system as described in the Theorem statement which has increment 2. Such a triple system exists as per our discussion following Theorem 11 where it is also made clear that T must have an even number of vertices which we classify as odd and even. The even vertices must lie on the bell and the odd ones on the oval. This explains why only even increments are possible. It is also not possible for $\gcd\left(t, \frac{n}{2}\right) > 1$, since such increments would yield only the points belonging to a subgroup of Z_n. Now suppose first that n/2 is odd and define the following mapping $\phi : Z_n \to Z_n$

$$\text{If t is even let } \phi(x) = \{tx \text{ if } x \text{ is even}\,;\left(\frac{n}{2} + t\right) x \text{ if x is odd}\}$$

$$\text{If t is odd let} \phi(x) = tx \text{ for each } x \in Z_n$$

Note that whether t is even or odd, ϕ is a bijection. Suppose t is odd and a, b, c $\in$ T with $a + b + c = 0 \pmod{n}$. Then $\phi(a) + \phi(b) + \phi(c) = ta + tb + tc = t(a + b + c) = 0 \pmod{n}$. Thus $\phi(T)$ is a $(Z_n,0)$ triple system equivalent to T which clearly has increment 2t. Now suppose t is even and a, b, c $\in$ T with $a + b + c = 0$ (mod n). There are essentially two cases to consider: (I) a,b,c are all even and $\phi(T)$ is a $(Z_n, 0)$ triple system equivalent to T with increment 2t as in the previous line. (II) a,b are odd and c is even. Then $\phi(a) + \phi(b) + \phi(c) = \left(\frac{n}{2} + t\right) a + \left(\frac{n}{2} + t\right) b + tc = t(a + b + c) + \frac{n}{2}(a + b) = 0$. Note that $\frac{n}{2}(a + b) = 0 \pmod{n}$ since a + b is even and so $\frac{n}{2}(a + b)$ is a multiple of n.

Next suppose n/2 is even. Then $\gcd\left(t, \frac{n}{2}\right) = 1$ implies that t is odd. Let $\psi : Z_n \to Z_n$ be defined by $\phi(x) = tx \pmod{n}$. This relabeling of T yields a $(Z_n,0)$ triple system equivalent to T with increment 2t. In Fig. 25 we illustrate a $(Z_{16},0)$ triple system with increment 2. The construction technique is essentially the same as the one used, for example in Fig. 31, for a sum 0 triple system on Z_n where n/2 is odd, except that we shift the generating vertex, (vertex 1), on the oval until vertices $\frac{n}{2} - 1, \frac{n}{2} + 1$ lie on a vertical line. Now we consider the case for split $(Z_n,1)$ triple systems on a Type IV cubic curve.

Theorem 21 *A split $(Z_n,1)$ triple system can have clockwise increment 2t for any $t \in \{1, 2, \ldots, n - 1\}$ with $gcd(t, n) = 1$. All such triple systems are equivalent to a split $(Z_n,1)$ triple system with clockwise increment 2.*

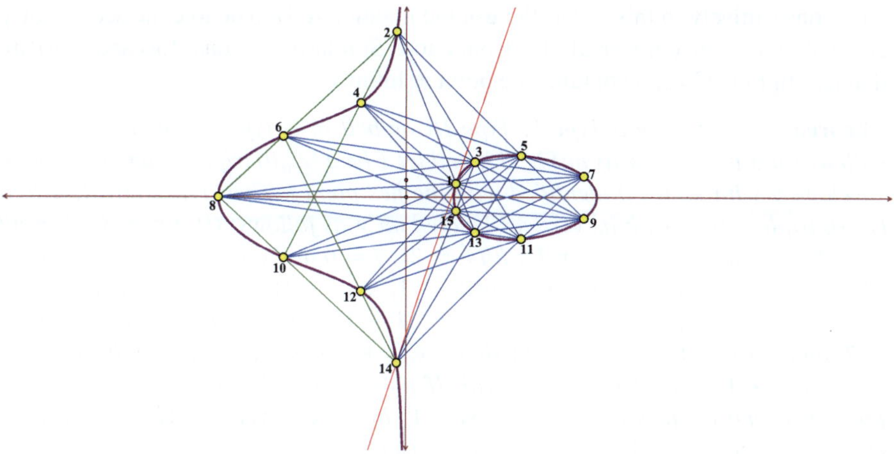

Fig. 25 (Z_{16},0) triple system on Type IV cubic

Proof Let T be a split (Z_{3n},0) triple system, (n even), with clockwise increment 2 and let $S = \{3k - 1 : k \in Z_n\}$ denote the coset subsystem of T consisting of points of the form $3k - 1$. The relabeling of S given by $\phi(3k - 1) = k$ turns $\phi(S)$ into a split (Z_n,1) triple system Q on γ with clockwise increment 2. Note that the relabeling replaces odd vertices with even ones and even vertices with odd ones. Let $t \in \{1, 2, 3, \ldots, n - 1\}$ with $\gcd(t, n) = 1$ and define $\psi : Q \to Z_n$ by $\psi(x) = tx$. This relabeling turns Q into a (Z_n,t) triple system R on γ with clockwise increment 2t (mod n). Suppose n is not a multiple of 3. Then $\gcd(t, 3) = 1$, and R is equivalent by Theorem 1 to a (Z_n,1) triple system by means of relabelings which preserve the increment. So in this case we obtain a (Z_n,1) triple system with clockwise increment 2t (mod n) which is equivalent to Q. If $3|n$ and t has the form $3k + 1$ then, again by Theorem 1, R is equivalent to a (Z_n,1) triple system with clockwise increment 2t (mod n). If $3|n$ and t has the form $3k - 1$ then R is again equivalent to a (Z_n,1) triple system but a relabeling is required which replaces each vertex by its inverse resulting in a *counterclockwise* increment of 2t. However, $\gcd(t, n) = 1$ implies $\gcd(n - t, n) = 1$, and if t has the form $3k - 1$ and $n = 3m$, then $n - t$ has the form $3(m - k) + 1$ and so we obtain anyway a (Z_n,1) triple system equivalent to Q with a clockwise increment of 2t. □

8 Vertex Location and Uniqueness

Our objective in this section is to show that for each integer n, up to equivalence, there is a unique (Z_n,0) triple system on a Type V, Type III or bell of a Type IV cubic, which has a given flex point identified with vertex 0. To facilitate this objective we first determine precise locations for the vertices of such a triple system. In Fig. 33, (see appendix), we illustrate a typical Type V cubic curve γ with collinear flex points e,f,g. We shall consider a *positive* traversal of γ to be the traversal which encounters e,f,g consecutively in this order. We use the notation (e,f) to denote the section of γ covered in a positive traversal of γ from e to f. Similarly sections (f,g) and (g,e) are defined. In Fig. 33 (g,e) contains the point at infinity.

Theorem 22 *Let γ be a Type V, Type III or bell of a Type IV cubic curve with collinear flex points e,f,g as in Fig. 33, and let T be a (Z_n,0) triple system with cubic envelope γ with vertex 0 identified with the flex point f. Then T is equivalent to a (Z_n,0) triple system with its vertices located on γ as follows: (i) if $n = 3k$ then all three flex points are vertices of T with $e = 2k$; $f = 0$ and $g = k$. Moreover vertices $\{2k+1, \ldots, 3k-1\}$ lie in (e,f); vertices $\{1, \ldots, k-1\}$ lie in (f,g) and vertices $\{k+1, \ldots, 2k-1\}$ lie in (g,e). (ii) If $n = 3k+1$ then $f = 0$ is the only flex point of T, and vertices $\{2k+1, \ldots, 3k\}$ lie in (e,f); vertices $\{1, \ldots, k\}$ lie in (f,g) and vertices $\{k+1, \ldots, 2k\}$ lie in (g,e). (iii) If $n = 3k+2$ then $f = 0$ is the only flex point of T, and vertices $\{2k+2, \ldots, 3k+1\}$ lie in (e,f); vertices $\{1, \ldots, k\}$ lie in (f,g) and vertices $\{k+1, \ldots, 2k+1\}$ lie in (g,e).*

Proof Let (γ,*) be the usual thirdpoint groupoid defined on the points of γ. Then (T,*) is a subalgebra of (γ,*) which satisfies $a^*b = -(a+b)$. The flex points e,f,g of γ are precisely the idempotent elements of (γ,*). If n = 3k then points {0,k,2k} satisfy $x = x^*x = -2x$ and so all three flex points of γ must belong to T. If n is not a multiple of 3 then vertex 0 is the unique flex point of T. By Theorem 19 we may regard T as having increment 1 in the positive direction. If n = 3k then we must have e = 2k, f = 0 and g = k. This immediately forces the vertices of T to align themselves as indicated in (i) of the Theorem statement. Now suppose n = 3k + 1 and define $\psi : (f, g) \to (e, f)$ by $\psi(x) = f^* x = 0^*x$. This mapping is a bijection which induces a one to one correspondence between the points of T in (f,g) and the points of T in (e,f). Note that $k^*k = k+1$ and k is the only point of T which squares to its successor in the positive direction along T. As a consequence the vertices k, k + 1 cannot lie together in any of the intervals (e,f), (f,g), (g,e), since these are anticlosed subsets of γ. Also f = 0 is the only flex point belonging to T so neither k nor k + 1 can be a boundary point for these intervals. It must then be that k is the last point of T in one of these intervals, and k + 1 is the first point of T in the succeeding interval. The points 3k and 1 directly precede and follow f and neither can k, k + 1 precede and follow the flex point e: If k + 1 follows e on γ then the 2k points {k + 1, ... ,3k} must all lie in (e,f). But then due to the bijection ψ, the interval (f,g) must contain another 2k points of T, which is impossible since 4k exceeds the number of vertices in T. So we must have that k immediately precedes g and vertex k + 1 immediately follows g. As a consequence the vertices {1, ... , k} lie in (f,g) and the images {3k, ... , 2k + 1} of these under ψ must constitute the points of T which lie in (e,f). The remaining points {k + 1, ... , 2k} must then all lie in (g,e) in accordance with the stipulation of the Theorem. In Fig. 33 we illustrate a triple system of this type with n = 3(4) + 1. Finally suppose n = 3k + 2. The only point of T which satisfies $x^*x = x + 1$ is x = 2k + 1. Thus the points 2k + 1, 2k + 2 cannot lie in the same anticlosed set. Since they are adjacent points of T, 2k + 1 must be the last point of T in (g,e) and 2k + 2 must be the first point of T in (e,f). As a consequence the points {2k + 2, ... , 3k + 1} must lie in (e,f) and due to the bijection ψ, the points $\{1, \ldots, k\}$ must lie in (f,g). The remaining points of T, (different from f = 0), namely $\{k + 1, \ldots, 2k + 1\}$ must then lie in (g,e). □

In Fig. 34, (see appendix), we illustrate a (Z_n, 0) triple system with n = 3(4) + 2. Now, still focusing on (Z_n,0) triple systems we will consider the question of uniqueness. Let γ be a Type V, Type III or bell of a Type IV cubic curve. We may consider any one of the three flex points of γ to be a (Z_1,0) triple system with the unique element 0 identified with the flex point. For any point a on γ the equation x*x = a has exactly two solutions for x, we may call these the *square-roots of a.* Each flex point has itself and one other point as a square root. These two points form a subalgebra of (γ,*) which we may consider to be a (Z_2,0) triple system. Thus there are exactly three (Z_1,0) triple systems and exactly three (Z_2,0) triple systems on γ. The three flex points {e,f,g} constitute the unique (Z_3,0) triple system on γ. A (Z_4,0) triple system T must contain a flex point of γ for vertex 0, the remaining square-root of this flex point for vertex 2, and since 1*1 = 3*3 = 2, the remaining two points of T must be the square-roots of 2. The three flex points of γ yield three distinct (Z_4,0) triple systems. Thus we see that for n = 1,2,3,4 there is a unique (Z_n,0) triple system on γ which contains a given flex point. To consider the general case it is useful to establish some *motion principles* for points on γ:

*M1: Let a be a fixed point of γ and x a moving point of γ. If x moves along γ in one direction then a*x moves in the opposite direction.*
*M2: If a, b move in the same direction along γ, then a*b moves in the opposite direction.*

Let T be a (Z_n,0) triple system with n > 4 and suppose first n = 3k + 1. We can assume that T has increment 1 in the positive direction along its cubic envelope γ and that the vertices of T lie in the intervals specified by Theorem 22. As vertex 1 is shifted along γ from f = 0 in the positive direction towards g the vertices 2, ..., k + 1 move ahead of it. This is a consequence of M1. By Theorem 22 vertex k is confined to the interval (f,g) and by M2, as k moves from f to g in the positive direction, point k*k moves from f to g in the negative direction. Vertex k + 1 is the first point of T in (g,e) and it moves in a positive direction ahead of vertex k in this interval. Since points k + 1, k*k move in opposite directions, there is a unique position for vertex 1 at which these points coincide. They must coincide in a triple system with sum 3k + 1, so we conclude that T is uniquely determined. If n = 3k + 2 we argue similarly: as vertex 1 moves in the positive direction from f to g, vertex k + 1 also moves in a positive direction from f. By M2 point (k + 1)*(k + 1) moves in the negative direction from f to g. Since points k and (k + 1)*(k + 1) move in opposite directions there is a unique position on γ for which these points coincide, and once again T is uniquely determined. The final case is when n = 3k. In this case we must have f = 0; g = k and e = 2k since 0, k, 2k are idempotent in (T,*). As vertex 1 is shifted in the positive direction starting at f, vertex k also moves in the positive direction starting at f. By M2 point k*k moves in the negative direction along γ, starting at f. Again, since the points k and k*k move in opposite directions, there is a unique position for vertex 1 in (f,g) for which points k and k*k coincide and thus T is uniquely determined. In this case we must have k = k*k = g. We have proved:

Theorem 23 *Let γ be Type V, Type III or bell of a Type IV cubic curve and let f be a flex point of γ. Then for each positive integer n there exists, up to equivalence, a unique $(Z_n,0)$ triple system on γ with $f = 0$.* □

This result will ultimately be very useful in the determination of exactly which finite abelian groups can serve as bases for triple systems. If n is a multiple of 3 then a $(Z_n,0)$ triple system must contain all three flex points of its cubic envelope γ. Since there is a unique such triple system containing any one of the flex points, we conclude that, up to equivalence only one such triple system exists on γ. If n is not a multiple of 3 then a $(Z_n,0)$ triple system contains only one of the flex points of γ. Since γ contains three flex points, we conclude that there exist exactly three distinct but equivalent triple systems of this type on γ. Thus we have:

Corollary 24 *Let γ be a Type V, Type III or bell of a Type IV cubic curve. If n is a multiple of 3 then, up to equivalence, there exists a unique $(Z_n,0)$ triple system on γ. If n is not a multiple of 3 then there exist exactly three distinct $(Z_n,0)$ triple systems on γ.* □

Next we would like to obtain a similar determination for $(Z_n,1)$ triple systems. In accordance with Theorem 1, we may assume that n is a multiple of 3. In the proof of Theorem 11 we saw that a $(Z_{3n},0)$ triple system T contains a coset subsystem equivalent to a $(Z_n,1)$ triple system. The equivalence was obtained by first taking vertices $U = \{3k - 1 : k = 1, 2, \ldots, n\}$ and then relabeling by $\phi\,(3k - 1) = k$. The subset $W = \{3k + 1 : k = 0, 1, \ldots, n - 1\}$ of T yields a second coset subsystem of T equivalent to a $(Z_n,1)$ triple system under the relabeling $\psi\,(3k + 1) = -k$ (mod n). In Fig. 26 we depict two distinct $(Z_{12},1)$ triple systems, one with bold and the other with dashed lines, as subalgebras of a $(Z_{36},0)$ triple system. It should be observed that $W = 0^*\ U = \{0^*u : u \in U\}$,i.e., W,U are translates of each other. By Corollary 24, there is a unique $(Z_{3n},0)$ triple system with a given cubic envelope

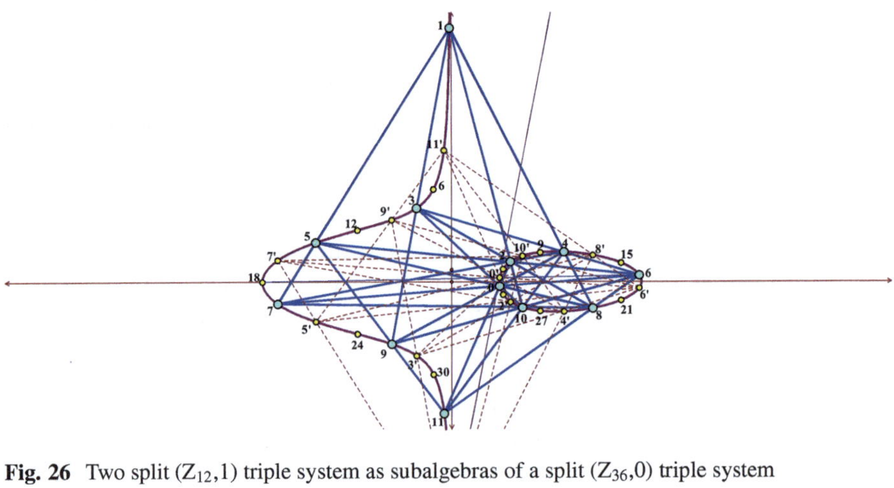

Fig. 26 Two split $(Z_{12},1)$ triple system as subalgebras of a split $(Z_{36},0)$ triple system

γ of Type V, Type III or bell of Type IV. So it would appear that for each n there are exactly two distinct (Z_n,1) triple systems on γ. It must, however, be shown that every (Z_n,1) triple system on γ arises in this way, i.e., as a subsystem of a (Z_{3n},0) triple system. For this purpose the next Theorem will suffice:

Theorem 25 *Let γ be a nonsingular irreducible cubic curve, and let T be a (Z_n,1) triple system on γ. Then T can be extended to a (Z_{3n},0) triple system on γ.*

Proof If $\gcd(n, 3) = 1$ then T is equivalent to a (Z_n,0) triple system by Theorem 1. So we may as well regard T as a (Z_n,0) triple system. The vertex 0 must label one of the flex points, say f, of γ. By Theorem 23, T is uniquely determined. Now let Q be a (Z_{3n},0) triple system on γ with vertex 0 identified with f. Let $R = \{0, 3, 6, \ldots., 3n - 3\}$ denote the subgroup of Z_{3n} consisting of the multiples of 3. The points of Q with labels in R form a subsystem of Q equivalent to a (Z_n,0) triple system. By Theorem 23, the triple system R is uniquely determined and must therefore be equivalent to T. Thus Q is the required extension of T. Now suppose $3|n$. Then T is not equivalent to a (Z_n,0) triple system, and in fact, contains none of the three flex points of γ. We will build a (Z_{3n},0) triple system W which contains the points of T. Relabel each point of T by $\phi : T \to Z_{3n}$ defined by $\phi(k) = 3k - 1 \pmod{3n}$ and select a flex point, say f for vertex 0. Let $W_{-1} = \phi(T)$. Let $W_1 = \{s^*0 : s \in W_{-1}\}$, and for each $s \in W_{-1}$ let s^*0 have label $-s$. Since $-1 \in W_{-1}$, we now have a point with label 1 in W_1. Finally let $W_0 = \{ s^*1 : s \in W_{-1}\}$, and for each $s \in W_{-1}$ label the point s^*1 with $-s - 1$. Let $W = W_{-1} \cup W_1 \cup W_0$. For $i = 0, -1, 1$ the points in W_i are congruent to i mod 3 and the elements of Z_{3n} constitute precisely the set of labels assigned to points in W. We will show that the points of W, as labeled, form a (Z_{3n},0) triple system. Let $a, b \in W$ and consider the following cases. In each case we must show that $a^*b = -a - b$.

Case (i): $a, b \in W_{-1}$. We have $a = 3s - 1$ and $b = 3t - 1$ for some $s, t \in Z_n$. The points $s, t, -s - t + 1$ are collinear in T and so the points $\phi(s) = a, \phi(t) = b$ and $\phi(-s - t + 1) = 3(-s - t + 1) - 1 = -3s - 3t + 2 = -a - b$ are collinear in W and consequently $a^*b = -a - b$. Note that W_{-1} is a relabeling of the points of T and thus is a subalgebra of $(\gamma,^*)$.

Case (ii): $a, b \in W_1$. In this Case we have $a = s^*0 = -s$ and $b = q^*0 = -q$ for some $s, q \in W_{-1}$. So $a^*b = (s^*0)^*(q^*0) = (s^*q)^*(0^*0) = (-s - q)^*0 = s + q = -a - b$. Since $s \equiv -1 \bmod 3$ and $q \equiv -1 \mod 3$, we have $a^*b = s + q \equiv -2 \equiv 1 \mod 3$.So $a^*b \in W_1$ and thus W_1 is also a subalgebra of $(\gamma,^*)$.

Case (iii): $a \in W_1, b \in W_0$. Here we have $a = 0^*s = -s$ for some $s \in W_{-1}$, and $b = 1^*t = -1 - t$ for some $t \in W_{-1}$. Then $a^*b = (0^*s)^*(1^*t) = (0^*1)^*(s^*t)$. Now since $0^*(-1) = 1$, we have $0^*1 = -1$, and by Case (i) we have $s^*t = -s - t$, and thus $a^*b = -1^*(-s - t) = 1 + s + t$. The last equality follows also from Case (i). Since $1 + s + t = -a - b$ we have the desired result.

Case (iv): $a \in W_{-1}, b \in W_0$. In this case $b = 1^*s = -1 - s$ for some $s \in W_{-1}$, so: $a^*b = (0^*(-a))^*(1^*s) = (-a^*1)^*(0^*s)$. Since $-a = 0^*a$ we have $-a \in W_1$. Also $1 = 0^*(-1) \in W_1$, so by Case(ii) $-a^*1 = a - 1$. So $a^*b = (a - 1)^*(-s) = -a + 1 + s$. The last equality holds again by Case(ii) since both $a - 1$ and $-s$ lie in W_1. Then since $-a + 1 + s = -a - b$ we have the desired result.

Case (v): $a \in W_1$, $b \in W_{-1}$. Here we have $a = 0^*s = -s$ and $b = 1^*t = -1 - t$ for some $s, t \in W_{-1}$. Then $a^*b = (0^*s)^* (1^*t) = (0^*1)^* (s^*t) = -1^* (-s - t) = 1 + s + t$. The last equality follows from Case(i) since both -1 and $-s - t$ both lie in W_{-1}. Since $1 + s + t = -a - b$ we have achieved once again the desired result.

Case (vi): $a, b \in W_0$. In this case $a = 1^*s = -1 - s$ and $b = 1^*t = -1 - t$ for some $s, t \in W_{-1}$. So $a^*b = (1^*s)^* (1^*t) = (1^*1)^* (s^*t) = (-2)^* (-s - t)$. This last equality follows from Cases (i), (ii). Now since $-2 \in W_1$ and $(-s - t) \in W_{-1}$, we have by Case(iv) that $(-2)^* (-s - t) = 2 + s + t = -a - b$ as desired. □

Theorem 25 along with the preceeding discussion yields the following result:

Theorem 26 *Let γ be a Type III, Type V or bell of a Type IV cubic curve. If gcd(3,n) = 1 then a (Z_n,1) triple system on γ is equivalent to a (Z_n,0) triple system and contains exactly one of the three flex points of γ. So in this case there are exactly three distinct (Z_n,1) triple systems on γ, each one determined by the particular flex point it contains. If $3|n$ then there are exactly two distinct (Z_n,1) triple systems on γ which are translates of each other and which contain none of the flex points of γ.* □

Next we describe the different possibilities for split triple systems. We have already remarked that a split triple system must have an even number of vertices with half of its vertices on the oval and the other half on the bell of a Type IV cubic curve. Let T be a split (Z_n,0) triple system with $n = 2m$. The odd vertices of T lie on the oval and the even vertices on the bell of a Type IV cubic γ. The even vertices of T form a subsystem equivalent to a (Z_m,0) triple system which lies entirely on the bell and is uniquely determined by Theorem 23 once a flex point of γ is selected for vertex 0. Let Q denote this subsystem and suppose the flex point f has been selected for vertex 0. Now we distinguish two cases depending on whether m is even or odd. Suppose first that m is even. The position of vertex 1 certainly determines T, and we must have $1^*1 = -2$. There are exactly two points a,b on the oval which serve as square-roots of -2 and both of these belong to T. One of these must be labeled 1 and the other $m + 1$. Suppose $a = 1$ and $b = m + 1$ and consider the relabeling of T given by $\phi(x) = x$ if x is even, and $\phi(x) = x + m$ if x is odd. This relabeling does not affect triples whose vertices lie entirely in the bell. The remaining triples of T have the form {x,y,z} where x lies in the bell and y,z lie on the oval. Then $\phi(x) + \phi(y) + \phi(z) = x + (y + m) + (z + m) = x + y + z = 0$. So the relabeling produces a triple system equivalent to T but now the vertices 1, $m + 1$ have exchanged positions, i.e., now $a = m + 1$ and $b = 1$. We conclude, (in the case m even), that Q has a unique, (up to equivalence), extension to a split (Z_n,0) triple system on γ. If gcd(n,3) = 1, then, in accordance with Corollary 24, there are exactly three distinct ways to situate the subsystem Q on the bell and thus exactly three distinct split (Z_n,0) triple systems on γ. If $3|n$, then in accordance with Corollary 24, Q is uniquely determined and so its extension to a split (Z_n,0) triple system is uniquely determined. These results are summarized in:

Theorem 27 *Let $n = 2m$ with m even and let γ be Type IV cubic curve. If gcd(n,3) = 1 then, up to equivalence, there are exactly three distinct split (Z_n,0) triple systems on γ. If $3|n$ then, up to equivalence, there is a unique split (Z_n,0) triple system on γ.* □

Now for the case n = 2m where m is odd we will find the following Lemma useful. If p is a point on a Type IV cubic γ then p has exactly four square-roots on γ, two on the bell and two on the oval. If a is a square- root of p on the oval, let a′ denote the remaining square –root of p on the oval, and if b is a square-root of p on the bell, let b′ denote the remaining square-root of p on the bell. In any case we refer to the points x, x′ as *twin points.*

Lemma 28 *Let a, b be points on the oval of a Type IV cubic curve γ. Then a * b = a′ * b′ and a * b′ = a′ * b.′*

Proof Let p = a * a = a′ * a′, and q = b * b = b′ * b′. Then the points a * b′, a′ * b, a * b, a′ * b′ all lie on the bell and all square to p * q. There are only two square-roots of p * q on the bell, so two pairs from among these four points must be equal. The point a * b′ cannot be equal to a * b or a′ * b′ since cancellation would then require b = b′ or a = a′ which is impossible since no point on γ is equal to its twin. We must then have a * b′ = a′ * b and similarly a * b = a′ * b′. □

The result of Lemma 28 is illustrated in Fig. 27.

Theorem 29 *Let n = 2m with m odd and let γ be Type IV cubic curve. If gcd(n,3) = 1 then there are exactly six distinct split (Z_n,0) triple systems on γ which occur in three twin pairs. If 3|n then there are exactly two distinct split (Z_n,0) triple systems on γ and these are twins.*

Proof Let T be a split (Z_n,0) triple system on a Type IV cubic curve γ with n = 2m. Let Q denote the set of even vertices of T which must lie on the bell, and let W denote the set of odd vertices of T which must lie on the oval. Let W′ = {x′ : x ∈ W} and let T′ = Q ∩ W′. We claim that T′ is a split (Z_n,0) triple system. There are essentially only two kinds of triples to check: those which contain two points in the bell, and those which contain two points in the oval. If a, b ∈ Q, then a* b = −a − b as in T. If a, b ∈ W′, then a′, b′ ∈ W and thus a′, b′ ∈ T and a′* b′ = −a − b as in T.

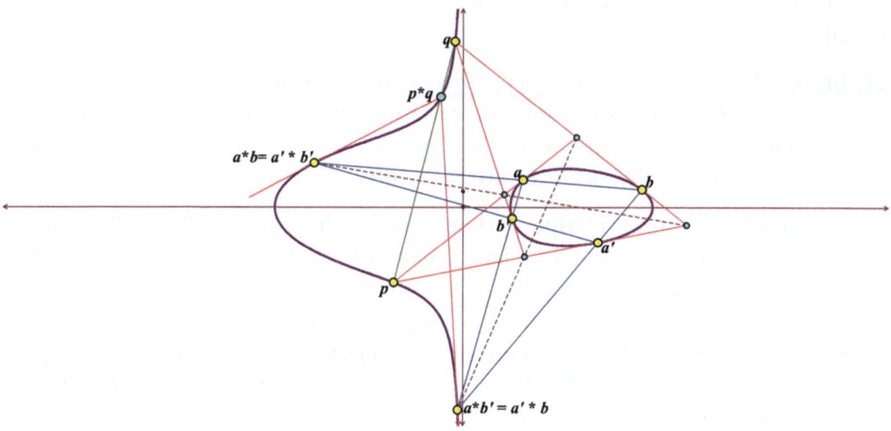

Fig. 27 a * b = a′ * b

Then by Lemma 28, a* b = a′* b′ = −a − b as desired. We call T′ the *twin of T.* If m is even then x, x′ both lie in T for each point x in the oval. In this case we can say that T is *self-twin* and we obtain the result of Theorem 27. But if m is odd as in our current Theorem, then T, T′ constitute two distinct extensions of the set Q of even points. In fact, the extending sets W, W′ are disjoint. (If 2s is any point of Q, then the points, −s, −s + m are the only two square-roots of 2s in T or T′. Since m is odd, exactly one of these square-roots is odd and lies on the oval. If b is the unique odd point of W such that b*b = 2s, then b′ cannot lie in W since b′* b′ is also equal to 2s. It follows that W ∩ W′ is empty.) If gcd (n, 3) = 1 then the subsystem Q can occur in three different locations on the bell, each containing a different one of the three flex points on the bell. Each of these has two distinct extensions and thus there are a total of six distinct split (Z_n,0) triple systems on γ. In case 3|n, then, in accordance with Corollary 24, the set Q of points is uniquely determined. Its two twin extensions then account for all possible split (Z_n,0) triple systems on γ. □

Similar results hold for split (Z_n,1) triple systems:

Theorem 30 *Let n = 2m with m even and let γ be a Type IV cubic curve. If gcd(3,n) = 1 then there are exactly three distinct split (Z_n,1) triple systems on γ. If 3|n then there are exactly two distinct split (Z_n,1) triple systems on γ.*

Proof Let T be a split (Z_n,1) triple system with n = 2m and m even and let γ denote the Type IV cubic envelope of T. Let Q denote the set of odd vertices of T which all lie on the bell, and let W denote the set of even vertices of T which all lie on the oval of γ. If gcd(3,n) = 1 then by Theorem 1, T is equivalent to a (Z_n,0) triple system, and by Theorem 29 we can deduce that there are exactly three distinct split (Z_n,1) triple systems on γ. Now suppose 3|n. The relabeling given by ξ (2k + 1) = −k turns Q into a (Z_m,1) triple system. Since 3|n, we also have 3|m, so by Theorem 26 there are exactly two distinct (Z_m,1) triple systems on the bell. Thus there are exactly two ways to arrange the odd vertices of T on the bell. Let W′ denote the set of twins of points in W. Since m is even we have W = W′ and thus each of the two choices for Q has, up to equivalence, a unique extension to a split (Z_n,1) triple system. □

Theorem 31 *Let n = 2m with m odd and let γ be a Type IV cubic curve. If gcd(3,n) = 1 then there are exactly six distinct split (Z_n,1) triple systems on γ, which occur in three twin pairs. If 3|n then, there are exactly four distinct (Z_n,1) triple systems on γ, which occur in two twin pairs.*

Proof Let T be a split (Z_n,1) triple system with n = 2m and m odd. Let Q denote the set of odd points of T which all lie on the bell of the cubic envelope γ of T, and let W denote the set of even points all of which lie on the oval. If gcd(3,n) = 1, then by Theorem 1, T is equivalent to a split (Z_n,0) triple system, and so by Theorem 29 there are exactly six mutually nonequivalent split (Z_n,1) triple systems on γ. Now suppose 3|n and apply the relabeling ξ (2k + 1) = −k to the odd points of T to obtain a (Z_m,1) triple system on the bell. Since 3|n we must have 3|m, and by Theorem 26 there are exactly two nonequivalent (Z_m,1) triple systems on the bell. There are thus exactly two nonequivalent ways to arrange the odd vertices of T on the bell of γ. Each of these has two dual extensions: $T = Q \cap W$ and $T' = Q \cap W'$, where W' denotes the set of twins of points in W. □

Now suppose T is a $(Z_n \times Z_n, (0,0))$ triple system with $n \geq 3$. *We will assume that the vertices of T must lie on an irreducible cubic curve* γ. If n is odd, i.e., T has an odd number of vertices, then T cannot be split, and so must have all its vertices on a Type V, Type III, or bell of a Type IV cubic curve γ. The subsets $S = \{(0,x) : x \in Z_n\}$; $R = \{(x,0) : x \in Z_n\}$; $Q = \{(x,x) : x \in Z_n\}$ are each subalgebras of (T,*) equivalent to (Z_n,0) triple systems. These distinct (Z_n,0) triple systems all share the flex point (0,0). But by Theorem 23, only one such triple system exists on γ. We conclude that no such triple system T can exist. Now suppose n is a multiple of 4, say $n = 4m$. Then the points $\{(0,0), (0,2m), (2m,0), (2m,2m)\}$ are all square-roots of (0,0), so γ must be a Type IV cubic curve. But these four points are also squares: $(0,0)^*((0,0) = (0,0); (0,m)^*(0,m) = (0,2m)$; $(m,0)^*(m,0) = (2m,0)$ and $(m,m)^*(m,m) = (2m,2m)$ so they must all lie on the bell of γ, since the oval is anticlosed. But the bell can only support two square-roots of any point, so in this case also we can conclude that T cannot exist. Finally suppose $n = 2m$ where m is odd. Since $n \geq 3$, we must also have $m \geq 3$. In this case T contains a subsystem equivalent to a $(Z_m \times Z_m, (0,0))$ triple system which we have shown not to exist. With the proviso indicated by the second sentence of this paragraph we have proved:

Theorem 32 *If $n \geq 3$, then no $(Z_n \times Z_n, (0,0))$ triple system exists.* □

In [4] the existence of various triple systems with base $Z_2 \times Z_n$ is proved. Every finite abelian group is isomorphic to a direct product of cyclic groups. Any such group which contains a subgroup isomorphic to $Z_n \times Z_n$ for $n \geq 3$, cannot, by Theorem 32, serve as the base of a zero-sum triple system. On the other hand, if n,m have no common nontrivial factors, then $Z_n \times Z_m$ is isomorphic to Z_{nm}. Also, evidently no triple system with base $Z_2 \times Z_2 \times Z_2$ and sum (0, 0, 0) exists. With these considerations, we can deduce that among finite abelian groups, only Z_n and $Z_2 \times Z_n$ can serve as the base for a geometric triple system with sum zero.

9 The Connection with Groups

Let γ be an irreducible cubic curve and let γ_0 denote the set of nonsingular points of γ. Any of the five basic types of cubics, illustrated in Figs. 12, 13, 14, 15, 16 contains at least one flex point, say e, and (γ_0,*) is a hypercommutative groupoid. By a well known construction, see for example [1], we can convert this groupoid into an abelian group (γ_0,+) by defining $a + b = e^*(a^*b)$. Then $a + e = e^*(a^*e) = a$, so e is the identity element of (γ_0,+), and a + (a*e) = e* (a* (a*e)) = e*e = e, so a*e is the inverse of a. Finally the associative law is proved as follows: a + (b + c) = a + (e* (b*c)) = e* [a*(e* (b*c)]= (e*e)* [a* (e* (b*c)] = (e*a)* [e* (e* (b*c))] = (e*a)* (b*c), and (a + b) + c = [e* (a*b)] + c = e* [(e* (a*b))*c] = (e*e)* [(e* (a*b))*c] = (e*c)* [e* (e* (a*b))] = (e*c)* (a*b) = (e*a)* (b*c). A geometric triple system T with its vertices on γ_0 is a subalgebra of (γ_0,*). If T contains the flex point e of γ_0, then (T,+) is a subgroup of (γ_0,+). In fact if S is any subalgebra of (γ_0,*) which contains e, then (S,+) is a subgroup of (γ_0,+). The determination of the group (γ_0,+) is therefore important in order to describe its subalgebras and hence to determine the triple systems on γ_0 which contain a flex point. We can also gain information about the subalgebras of (γ_0,*) which do not contain a flex point of γ, since any such subalgebra can be extended to a subalgebra of (γ_0,*) which does contain a flex point as follows.

Lemma 33 *Let γ be an irreducible cubic curve, and let γ_0 denote the set of nonsingular points of γ. If S, T are subalgebras of (γ_0,*), then $S^*T = \{s^*t : s \in S, t \in T\}$ is also a subalgebra of (γ_0,*). Moreover $S^*S = S$.*

Proof Let $s_1, s_2 \in$ S, and $t_1, t_2 \in$ T. Then $(s_1{}^*t_1)^*\ (s_2{}^*t_2) = (s_1{}^*s_2)^*\ (t_1{}^*t_2) \in$ S*T, so S*T is a subalgebra of (γ_0,*). It is clearly true that S*S $\subseteq$ S. On the other hand, if s $\in$ S, then s = s* (s*s), and thus s $\in$ S*S. □

Theorem 34 *Let γ be an irreducible cubic curve with flex point e, and let γ_0 denote the set of nonsingular points of γ. If S is a subalgebra of (γ_0,*) which does not contain any flex point of γ, then the subalgebra $\langle S \cup e\rangle$ of (γ_0,*) generated by $S\cup\{e\}$ is given by $S \cup e^*S \cup S^*\ (e^*S)$ where this is a disjoint union. If S is finite then $|\langle S \cup e\rangle| = 3|S|$.*

Proof Certainly S $\cup$ e*S $\cup$ S* (e*S) contains S $\cup$ {e}, since e = s* (e*s) $\subseteq$ S* (e*S) for any s $\in$ S. Since e is a flex point we have e*e = e, and so {e} is a one point subalgebra of (γ_0,*). It then follows from Lemma 33 that each of e*S, S* (e*S) is a subalgebra of (γ_0,*). We claim that the multiplication table for S $\cup$ e*S $\cup$ S* (e*S) is given by:

(*)	S	e*S	S* (e*S)
S	S	S* (e*S)	e*S
e*S	S* (e*S)	e*S	S
S* (e*S)	e*S	S	S* (e*S)

The diagonal entries follow from Lemma 33, so we need only prove the following:

(i) S* [S* (e*S)] = e*S. *Proof.* Let q $\in$ S* [S* (e*S)], then q = $s_1{}^*\ [s_2{}^*\ (e^*s_3)]$ for some $s_1, s_2, s_3 \in$ S. Then q = $[s_3{}^*\ (s_1{}^*s_3)]^*\ [s_2{}^*\ (e^*s_3)]$ = $[s_3{}^*\ (e^*s_3)]^*\ [(s_1{}^*s_3)^*s_2]$ = $e^*\ [(s_1{}^*s_3)^*s_2] \subseteq$ e*S. Conversely, suppose q $\in$ e*S, then q = e*s = s* [s* (e*s)] $\subseteq$ S* [S* (e*S)], where s is any element of S.

(ii) $(e^*S)^*[S^*(e^*S)] = S$. *Proof.* Let $q \in (e^*S)^*[S^*(e^*S)]$, then $q = (e^*s_1)^*[s_2{}^*(e^*s_3)]$ for some $s_1, s_2, s_3 \in S$. Then $q = [e^*(e^*s_3)]^*(s_1{}^*s_2) = s_3{}^*(s_1{}^*s_2) \in S$. Conversely, suppose $q \in S$, then $q = (e^*q)^*[q^*(e^*q)] \in (e^*S)^*[S^*(e^*S)]$.

From the completion of the table, we obtain that $S \cup e^*S \cup S^*(e^*S)$ is a subalgebra of $(\gamma_0,*)$. Now to show that the subalgebras $S, e^*S, S^*(e^*S)$ are mutually disjoint, suppose $s \in S \cap e^*S$, then $s = e^*s_1$ for some $s_1 \in S$. But then $e = s^*s_1 \in S$, a contradiction. If $s \in S \cap S^*(e^*S)$, then $s = s_1{}^*(e^*s_2)$ for some $s_1, s_2 \in S$. But then $e = s_2{}^*(s^*s_1) \in S$, a contradiction. Finally suppose $q \in (e^*S) \cap [S^*(e^*S)]$, then $q = e^*s_1 = s_2{}^*(e^*s_3)$ for some $s_1, s_2, s_3 \in S$. Then $s_2 = (e^*s_1)^*(e^*s_3) = (e^*e)^*(s_1{}^*s_3) = e^*(s_1{}^*s_3)$, and consequently $e = s_2{}^*(s_1{}^*s_3) \in S$, again contradicting the idempotent free character of S.

(1) The mapping $\psi : S \to e^*S$ defined by $\psi(s) = e^*s$ is easily seen to be a bijection, so $|S| = |e^*S|$. Now define a mapping $\phi : S \to S^*(e^*S)$ by $\phi(s) = s_0{}^*(e^*s)$, where s_0 is any fixed element of S. Again, it is easily seen that ϕ is injective. To show ϕ is onto, let $s_1{}^*(e^*s_2)$ be a typical element of $S^*(e^*S)$, and let $s = (s_2{}^*s_0)^*s_1$, then $\phi(s) = s_0{}^*\left(e^*\left[(s_2{}^*s_0)^*s_1\right]\right) = [s_2{}^*(s_0{}^*s_2)]^*\left(e^*\left[(s_2{}^*s_0)^*s_1\right]\right) = s_1{}^*(e^*s_2)$. So we have $|S| = |S^*(e^*S)|$ as well. If S is finite then $|\langle S \cup e\rangle| = |S| + |e^*S| + |S^*(e^*S)| = 3|S|$. □

In Theorem 34, if we let $W = S^*(e^*S)$, we saw that $W = s_0{}^*(e^*S)$, where s_0 is any element of S. It follows immediately that $e^*S = s_0{}^*W$, and from this it follows immediately that $S = e^*(s_0{}^*W) = (e^*e)^*(s_0{}^*W) = (e^*s_0)^*(e^*W) = (e^*s_0)^*W$. The last equality follows from the fact that $e \in W$. Thus we can say:

Corollary 35 *Every idempotent-free subalgebra of* $(\gamma_0,*)$ *is a translate of a subalgebra of* $(\gamma_0,*)$ *which does contain an idempotent element.* □

The results of Theorem 34 and Corollary 35 should be compared with Theorem 25 which describes the extension of a $(Z_n,1)$ triple system, possibly with no flex point, to a $(Z_{3n},0)$ triple system.

If $(S,*)$ is any hypercommutative groupoid which contains an idempotent element e, and we define an addition on the elements of S, as above, we obtain an abelian group which we denote by $\Phi_e(S)$. If $(G,+)$ is an abelian group and we define a binary operation $*$ on the elements of G by $a^*b = -a - b$, then we obtain a hypercommutative groupoid with 0 as idempotent element. In case the abelian group G is presented multiplicatively, then we define $a^*b = a^{-1}b^{-1}$. In either case we denote the resulting hypercommutative groupoid by $\Psi(G)$. It is then easily seen that $\Psi\Phi_e(S) = S$ for every hypercommutative groupoid S which contains an idempotent element e. Conversely, if G is any abelian group with identity e, then $\Phi_e(\Psi(G) = G$. The mappings Φ_e, Ψ are thus inverses, and provide a one to one correspondence between members of the variety of abelian groups and members of the variety of hypercommutative groupoids with idempotent e. This correspondence easily extends to subalgebras: If W is a subalgebra of a hypercommutative groupoid $(S,*,e)$, and W contains the idempotent element e, then $\Phi_e(W)$ is a subgroup of

Φ_e (S), and if H is a subgroup of an abelian group G, then by Ψ (H) is a subalgebra of the hypercommutative groupoid Ψ (G). In this context, Theorem 34 and Corollary 35 have the following generalizations:

Theorem 36 *Let (W,*) be a hypercommutative groupoid with idempotent element e, and let S be an idempotent-free subalgebra of W. Then* $\langle S \cup e\rangle = S \cup e^*S \cup S^*\,(e^*S)$ *where this is a disjoint union of subalgebras of W. Moreover S, e*S are isomorphic translates of the subalgebra* $Q = S^*(e^*S)$ *which contains e, and they are cosets of the subgroup* $\Phi_e(Q)$ *of* $\Phi_e(\langle S \cup e\rangle)$*.* □

Note that Theorem 36 is a purely algebraic result with no reference to cubic curves. Further general properties of hypercommutative groupoids can be found in [2]. Now we determine the abelian groups which correspond to the hypercommutative groupoids (γ_0,*) where γ is an irreducible cubic curve. We consider these in order according to the Type of γ.

Theorem 37 *Let* γ *be a Type I cubic, (see Fig. 12), and let* $W = (\gamma_0, ^*)$ *denote the hypercommutative groupoid defined on the nonsingular points of* γ*. Then* $\Phi_\infty(W)$ *is isomorphic to the group (R,+) of real numbers under addition.*

Proof Here, ∞ denotes the point at infinity on vertical lines, which is the unique flex point of γ. The proofs of this Theorem and Theorems 38, 39 are outlined in exercises in [1]. Here we will just fill in some of the details. Define $g : R \to \gamma$ by $g(t) = (t^2, t^3)$, and then define $f : R \to \gamma_0$ by $f(p) = g(1/p)$ if $p \neq 0$, and $f(0) = \infty$. It is easy to show that g, f are bijections. Now regard f as a mapping from the group (R,+) to the group Φ_∞ (W). We will show that f is an isomorphism. First we show that $f(-p) = -f(p)$. If $p \neq 0$, then $f(-p) = g(1/-p) = (1/p^2, -1/p^3)$, and $-f(p) = \infty^* f(p) = \infty^*\, g(1/p) = \infty^*\,(1/p^2, 1/p^3) = (1/p^2, -1/p^3)$ due to the symmetry of γ with respect to the x-axis. If $p = 0$, $f(-p) = f(0) = \infty$, and $-f(p) = \infty^* f(p) = \infty^* \infty = \infty$ since infinity is a flex point hence idempotent. Now suppose $p, q \in R$ and $f(p), f(q), f(-p-q)$ are distinct. We will show that these three points are collinear. If $p = 0$, then $f(p) = \infty$, and we have $\infty, f(q), f(-q)$. We have just shown that $f(-q) = -f(q) = \infty^* f(q)$, so the three points are collinear. We may thus suppose that p, and q, are nonzero, and thus that $f(p) = g(1/p), f(q) = g(1/q)$ and $f(-p-q) = g(1/(-p-q))$. If L denotes the line through f(p) and $f(-p-q)$, and K denotes the line through f(q) and $f(-p-q)$, then we obtain that both lines have slope $\frac{p^2+pq+q^2}{pq(p+q)}$, and so the three points in question are collinear. Next we show that $f(-2p)$ lies on the tangent to γ at f(p). If $p = 0$ then $f(-2p) = f(p) = \infty$, and since $\infty^*\infty = \infty$ we can say that $f(-2p) = \infty$ lies on the tangent to γ at $f(p) = \infty$. So suppose $p \neq 0$, then $f(-2p) = (1/4p^2, -1/8p^3)$. By implicit differentiation of $y^2 = x^3$ we obtain that the slope m_T of the tangent to γ at f(p) is given by $m_T = 3/2p$. Then the equation for T is given by:

$$y - \frac{1}{p^3} = \frac{3}{2p}\left(x - \frac{1}{p^2}\right) \tag{24}$$

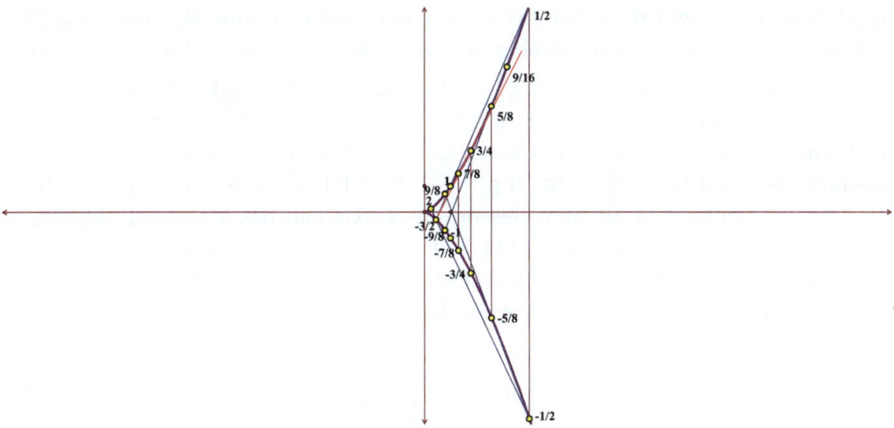

Fig. 28 (R,0) triple system on the nonsingular points of the cubic $y^2 = x^3$

the coordinates of f(2p) satisfy (24) so our contention is proved. Now we are prepared to show that f: $(R, +) \rightarrow \Phi_\infty$ (W) is an isomorphism. The mapping is a bijection, so it remains only to show that it is a group homomorphism. Consider: $f(p+q) = \infty^*(\infty^* f(p+q)) = \infty^* f(-p-q) = \infty^*\big(f(p)^* f(q)\big)$. The last equality holds if $p \neq q$, since we have shown that f(p), f(q) and $f(-p-q)$ are collinear points of W, and it holds if $p = q$ since we have shown that $f(-2p)$ lies on the tangent to γ at f(p), and thus $f(p)^* f(p) = f(-2p)$. Finally, $\infty^*\big(f(p)^* f(q)\big) = f(p) + f(q)$ in accordance with the way addition is defined in $\Phi_\infty(W)$. □

Theorem 37 allows us to construct an (R,0) triple system on a Type I cubic by identifying the point f(p) with p. This construction is illustrated in Fig. 28. The only finite subgroup of (R,+) is the trivial subgroup {0}. This corresponds to $\{\infty\}$ which is thus the only finite subalgebra of $(\gamma_0,^*)$. This agrees with the result mentioned in Theorem 10 which was arrived at by other considerations. Next we do a similar analysis for a Type II cubic, (Fig. 13).

Theorem 38 *Let γ be a Type II cubic with equation $y^2 = x^3 + x^2$ as illustrated in Fig. 13 and let $W = (\gamma_0{}^*)$ denote the hypercommutative groupoid defined on the nonsingular points of γ. Then $\Phi_\infty(W)$ is isomorphic to the group (R,•) of nonzero real numbers under multiplication.*

Proof The Type II curve γ has one singularity at the origin and one flex point (∞) at infinity on vertical lines. The mapping g: $R\backslash\{\pm 1\} \rightarrow \gamma_0\backslash\infty$ defined by $g(t) = (t^2 - 1, t^3 - t)$ is easily seen to be bijective. Then the mapping f: $R\backslash\{0\} \rightarrow \gamma_0$ defined by:

$$f(p) = \begin{Bmatrix} g\left(\frac{p+1}{p-1}\right) \text{ if } p \neq 1 \\ \infty \text{ if } p = 1 \end{Bmatrix} \tag{25}$$

is a bijection. We will show that f is a group isomorphism from (R,•) onto Φ_∞ (W). Suppose p is a nonzero real number, and let $q = \frac{p+1}{p-1}$. Then $f(1/p) = g(-q) = (q^2-1, -q^3+q) = \infty^*(q^2-1, q^3-q) = \infty^* f(p) = -f(p)$. Now we will show that $f(1/p^2)$ lies on the tangent to γ at f(p). If p = 1 then $f(1/p^2) = f(p) = f(1) = \infty$, and since ∞ is a flex point we have $\infty^*\infty = \infty$. So the tangent at $f(p) = \infty$ contains the point $f(1/p^2) = \infty$. If $p = -1$, then $f(1/p^2) = \infty$, and $f(p) = g(0) = (-1, 0)$. The point $(-1, 0)$ of γ lies on the x-axis and has a vertical tangent, so $(-1,0)^*(-1,0) = \infty$. So, again, $f(1/p^2) = \infty$ lies on the tangent to γ at $f(p) = (-1, 0)$. If $p \neq \pm 1$, then f(p) = g(q). Using implicit differentiation we obtain that the slope of the tangent T to γ at f(p) is given by:

$$m_T = \frac{3q^2-1}{2q}. \tag{26}$$

The equation of T is then:

$$T : y - q(q^2-1) = \frac{3q^2-1}{2q}\left(x - q^2 + 1\right). \tag{27}$$

We need to show that $f(1/p^2)$ satisfies (27). Let $w = \frac{1+p^2}{1-p^2}$, then $f(1/p^2) = g(w) = (w^2-1, w^3-w)$, so we must show:

$$w^3 - w - q(q^2-1) = \frac{3q^2-1}{2q}\left(w^2-q^2\right). \tag{28}$$

This can be shown after some considerable algebraic manipulation which we will spare the reader. Next, let a, b $\in$ R and suppose f(a), f(b) and f(1/ab) are distinct. We will show that these three points are collinear. Let $q = \frac{a+1}{a-1}$; $s = \frac{b+1}{b-1}$; $r = \frac{1+ab}{1-ab}$, then $f(a) = (q^2-1, q^3-q)$; $f(b) = (s^2-1, s^3-s)$, and $f(1/ab) = (r^2-1, r^3-r)$. Let m_L denote the slope of the line L through f(a) and f(1/ab), and let m_K denote the slope of the line K through f(b) and f(1/ab). We then obtain:

$$m_L = \frac{r^2+q^2+qr-1}{r+q}; \quad m_K = \frac{r^2+s^2+sr-1}{r+s}. \tag{29}$$

With a little algebra it is easily shown that $m_L = m_K$, and thus the points f(a), f(b), f(1/ab) are collinear. Now to complete the proof that $f : (R, \bullet) \square \Phi_\infty (W)$ is an isomorphism, suppose $a, b \in R\setminus\{0\}$. If a = b, then $f(ab) = f(a^2)$ which lies on the tangent to γ at f(1/a). Thus $f(a^2) = f(1/a)^* f(1/a) = -f(1/a) - f(1/a) =$

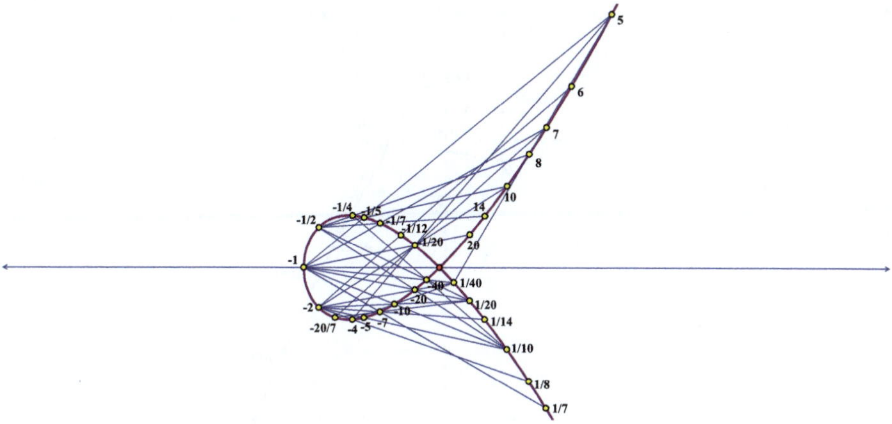

Fig. 29 (R,1) triple system on cubic: $y^2 = x^3 + x^2$

f(a) + f(a). If a $\neq$ b, then f(ab) = f(1/a)*f(1/b) since the points f(ab), f(1/a) and f(1/b) are collinear. Then $f(ab) = -f(1/a) - f(1/b) = f(a) + f(b)$. □

Theorem 38 allows us to construct an (R,1) triple system on a Type II cubic by identifying the point f(p) with p. This construction is illustrated in Fig. 29. The only finite subgroups of (R,1) are $\{1\}$ and $\{1, -1\}$, and thus the only finite subalgebras of (γ_0,*) which contain the idempotent element ∞ are $\{\infty\}$, and $\{\infty, (-1, 0)\}$. By Theorem 34 a subalgebra S of (γ_0,*) which does not contain ∞ would have an extension to a subalgebra Q of (γ_0,*) which does contain ∞. If S is finite than $|Q| = 3|S|$. But there are no such finite subalgebras Q, so we can conclude that $\{\infty\}$ and $\{\infty, (-1, 0)\}$ are the only finite subalgebras of (γ_0,*). This is also in agreement with the result mentioned in Theorem 10.

Theorem 39 *Let γ be a Type III cubic with equation $y^2 = x^3 - x^2$. Let γ_0 denote the set of nonsingular points of γ; let $W = (\gamma_0, *)$ denote the corresponding hypercommutative groupoid and let U denote the group of complex numbers with modulus 1. Then $\Phi_\infty(W)$ is isomorphic to U.*

Proof Define $g : R \to \gamma_0 \backslash \infty$ by $g(t) = \left(t^2 + 1, t^3 + t\right)$ and $f : U \to \gamma_0$ by:

$$f(z) = \left\{ \begin{array}{c} g\left(-cot\frac{\theta}{2}\right) \; if \; z = \cos\theta + i\sin\theta \neq 1 \\ \infty \; if \; z = 1 \end{array} \right\}. \quad (30)$$

It is easily seen that f is a bijection. Let P = (cos θ, sin θ) be any point of U different from (1,0) and let L denote the line through P and (1,0). Let K denote the line through (0,0) and parallel to L, as in Fig. 30. We will show that f(P) = $K \cap \gamma$.The equation of K is given by

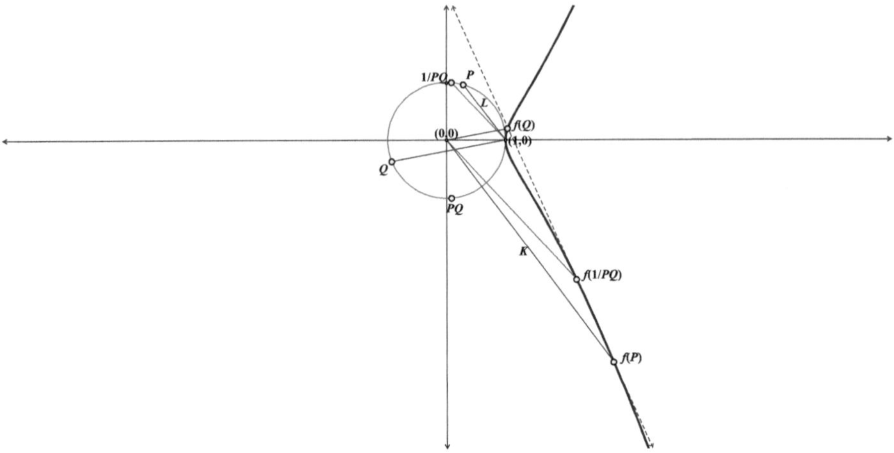

Fig. 30 (U,1) triple system on Type III cubic

$y = \left(\frac{\sin\theta}{\cos\theta - 1}\right)$. Solving together the equations of K and γ, we find that the coordinates of the point $K \cup \gamma$ are given by $\left(\frac{2}{1-\cos\theta}, \frac{-2\sin\theta}{(\cos\theta-1)^2}\right)$. On the other hand:

$$f(P) = \left(1 + cot^2\frac{\theta}{2}, -cot^3\frac{\theta}{2} - cot\frac{\theta}{2}\right). \tag{31}$$

Now using the identity $\cot\frac{\theta}{2} = \frac{1+\cos\theta}{\sin\theta}$ and simplifying, we find indeed that $f(P) = K \cap \gamma$. Next we show that if $z \in U$, then $f(1/z) = -f(z)$ where $-f(z)$ is the inverse of f(z) in the group $\Phi_\infty(W)$. If $z = 1$ then $f(z) = f(1/z) = \infty$, and $-f(z) = \infty^* f(z) = \infty^* \infty = \infty$. If $z \neq 1$, then f(z) is given by (31) and $f(1/z) = f\left(\bar{z}\right) = f(\cos(-\theta) + i\sin(-\theta)) = g(-\cot(-\theta/2)) = g(\cot(\theta/2)) = \left(1 + cot^2\frac{\theta}{2}, cot^3\frac{\theta}{2} + cot\frac{\theta}{2}\right) = \infty^* f(z) = -f(z)$. Now we show that if $z \in U$, then $f(1/z^2)$ lies on the tangent to γ at f(z). If $z = 1$ then $f\left(1/z^2\right) = f(z) = f(1) = \infty$, and since $\infty^*\infty = \infty$ we can say that $f(1/z^2)$ lies on the tangent to γ at f(z). If $z = -1$ then $f\left(1/z^2\right) = \infty$ and $f(z) = f(-1) = f(\cos\pi + I\sin\pi) = g\left(-\cot\frac{\pi}{2}\right) = g(0) = (1, 0)$. Since $(1, 0)^* (1, 0) = \infty$, we have again that $f(1/z^2)$ lies on the tangent to γ at f(z). Now suppose $z \neq \pm 1$. Differentiating implicitly the equation for γ yields: $\frac{dy}{dx} = \frac{3x^2-2x}{2y}$, and by substituting the coordinates $\left(\frac{2}{1-\cos\theta}, \frac{-2\sin\theta}{(\cos\theta-1)^2}\right)$ for f(z) we obtain $m_T = \frac{2+\cos\theta}{-\sin\theta}$ for the slope of the tangent line T to γ at f(z). The equation of T is then given by:

$$y + \frac{2\sin\theta}{(\cos\theta - 1)^2} = \left(\frac{2 + \cos\theta}{-\sin\theta}\right)\left(x - \frac{2}{1 - \cos\theta}\right). \tag{32}$$

Now $1/z^2 = \cos(-2\theta) + i \sin(-2\theta)$, so $f(1/z^2) = g\left(-\cot\left(\frac{-2\theta}{2}\right)\right) = g(\cot\theta) = \left(\cot^2\theta + 1, \cot^3\theta + \cot\theta\right) = \left(\frac{1}{sin^2\theta}, \frac{\cos\theta}{sin^3\theta}\right)$. We leave it to the reader to determine that these coordinates do satisfy (32). In the next step we show that if $w, z \in U$ and $f(w), f(z), f(1/wz)$ are distinct, then these three points are collinear. Suppose first that one of w, z equals 1. Without loss of generality suppose $w = 1$. Then $f(w) = \infty$, so we have the points ∞, f(z), f(1/z). We have shown that $f(1/z) = -f(z) = \infty^* f(z)$, so the three points are collinear. If $wz = 1$ we obtain the same collinear points f(1/z), f(z), ∞, so we can suppose that w, z, wz are not equal to 1. Let $z = \cos\alpha + i \sin\alpha$, and let $w = \cos\beta + i\sin\beta$, then $wz = \cos(\alpha+\beta) + i\sin(\alpha+\beta)$, and $\frac{1}{wz} = \cos(-\alpha-\beta) + i\sin(-\alpha-\beta)$, and:

$$f(z) = \left(\frac{2}{1-\cos\alpha}, \frac{-2\sin\alpha}{(\cos\alpha - 1)^2}\right) \tag{33}$$

$$f(w) = \left(\frac{2}{1-\cos\beta}, \frac{-2\sin\beta}{(\cos\beta - 1)^2}\right) \text{ and} \tag{34}$$

$$f(1/wz) = \left(\frac{2}{1-\cos(\alpha+\beta)}, \frac{2\sin(\alpha+\beta)}{(\cos(\alpha+\beta) - 1)^2}\right). \tag{35}$$

Let L denote the line through f(z) and f(1/wz), and let K denote the line through f(w) and f(1/wz). For ease of notation we let $a = \cos\alpha$; $b = \sin\alpha$; $c = \cos\beta$ and $d = \sin\beta$, then $a^2 + b^2 = 1 = c^2 + d^2$. Further we let $M = 1 - \cos\alpha$; $N = 1 - \cos\beta$; $Q = 1 - \cos(\alpha+\beta)$ and $R = \sin(\alpha+\beta)$. Then the slopes of K, L are then given by:

$$m_L = \frac{RM^2 + bQ^2}{M^2Q - MQ^2} \quad m_K = \frac{RN^2 + dQ^2}{N^2Q - NQ^2} \tag{36}$$

We need to show $m_L = m_K$. This is equivalent to showing:

$$M^2Qd - MQ^2d - MRN^2 = N^2Qb - NQ^2b - NRM^2. \tag{37}$$

The left hand side of (37) when expanded is given by:

$$3acd - bd^2 - 2ad - 2a^2cd + 2abcd^2 - b^2d^3 - bc + 2bc^2 - bc^3 - adc^2 + 2a^2d - a^3cd + a^2bd^2$$

$$2a^2bcd^2 + a^3c^2d + ab^2d^3 + abc - 2abc^2 + abc^3, \tag{38}$$

And the right hand side of (37) is given by:

$$3abc-b^2d-2bc+2ab^2cd-b^3d-ad+2a^2d-a^3d-bca^2-2abc^2+2bc^2-abc^3+b^2c^2d$$

$$2ab^2c^2d + a^2c^3b + cb^3d^2 + acd - 2a^2dc + ca^3d. \tag{39}$$

After making the substitutions $d^2=1-c^2$ and $b^2 = 1-a^2$, the expressions in (38), (39) both simplify to:

$$3abc-bc-d+3a^2d+3acd-a^3cd-b+3bc^2+ba^2-ba^2c^2-ad-a^3d-2bca^2-2abc^2$$

$$abc^3 + c^2d - a^2c^2d - 2ac^2d + 2a^3c^2d + 2a^2c^3b - bc^3 - 2a^2cd. \tag{40}$$

Thus $m_L = m_K$, and so the points f(z), f(w), f(1/wz) are collinear. Now suppose $w, z \in U$ and f(w), f(z), f(1/wz) are distinct. Then $f(wz) = \infty^* (\infty^* f(wz)) = \infty^* (-f(wz)) = \infty^* f(1/wz) = \infty^* \left(f(w)^* f(z)\right) = f(w) + f(z)$. If $w = z$ then $f(wz) = f\left(z^2\right) = \infty^* \left(\infty^* f\left(z^2\right)\right) = \infty^* (-f\left(z^2\right) = \infty^* f\left(1/z^2\right) = \infty^* \left(f(z)^* f(z)\right) = f(z) + f(z)$. If $z = (1/wz)$, then $w = (1/z^2)$ so $f(w) = f\left(1/z^2\right) = f(z)^* f(z)$, and $f(wz) = f(1/z) = -f(z) = \infty^* f(z) = \infty^* (f(z)^* \left(f(z)^* f(z)\right) = \infty^* \left(f(z)^* f(w)\right) = f(z) + f(w)$. In all cases the mapping $f : U \to \Phi_\infty (W)$ is a homomorphism, and thus an isomorphism. □

The finite subgroups of U are all cyclic, so if γ is a Type III cubic, the finite subalgebras of $(\gamma_0{}^*)$ which contain ∞ correspond to finite cyclic groups and thus form $(Z_n,0)$ triple systems with vertex 0 at ∞. Suppose S is a finite subalgebra of $(\gamma_0{}^*)$ which does not contain ∞. By Theorem 34, S has an extension to a subalgebra Q of $(\gamma_0,{}^*)$ which does contain ∞. Q contains a subalgebra M which contains ∞ and Φ_∞ (M) is a subgroup of the cyclic group Φ_∞ (Q) with index 3. The points in S form a coset of Φ_∞ (M) and thus (S,*) is a $(Z_n,1)$ triple system where $3n = |Q|$. We summarize these observations in:

Theorem 40 *If γ is a Type III cubic curve, then every subalgebra of $(\gamma_0{}^*)$ is a $(Z_n,0)$ or a $(Z_n,1)$ triple system.* □

Note that the existence of $(Z_n,0)$ and $(Z_n,1)$ triple systems proved in Theorems 11,12,13 is also proved by Theorem 38. It is evidently known, see [3], that if γ is a Type V cubic, and $W = (\gamma,{}^*)$ then Φ_∞ (W) is isomorphic to U, although in this case we have no explicit formula for the isomorphism. Theorem 40 then applies also if γ is Type V. In case γ is Type IV, then Φ_∞ (W) is isomorphic to $Z_2 \times U$, and besides the $(Z_n,0)$ and $(Z_n,1)$ triple systems, triple systems with base $Z_2 \times Z_n$ can be constructed on γ. The discussion of this last kind of triple system can be found in [4].

Figure Appendix

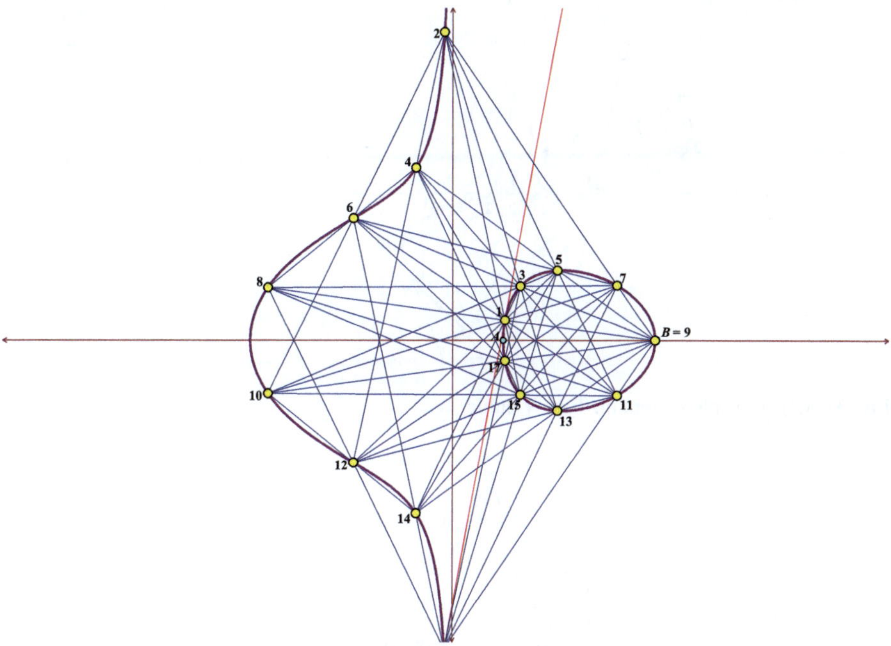

Fig. 31 (Z_{18},0) triple system on Type IV cubic curve with increment=2

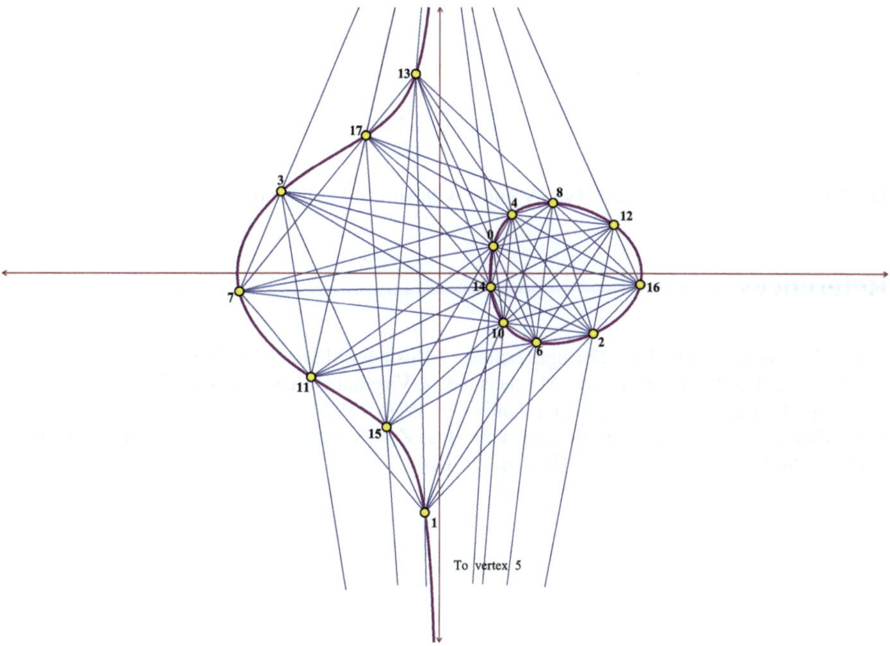

Fig. 32 (Z_{18},1) triple system on Type IV cubic curve with increment 4

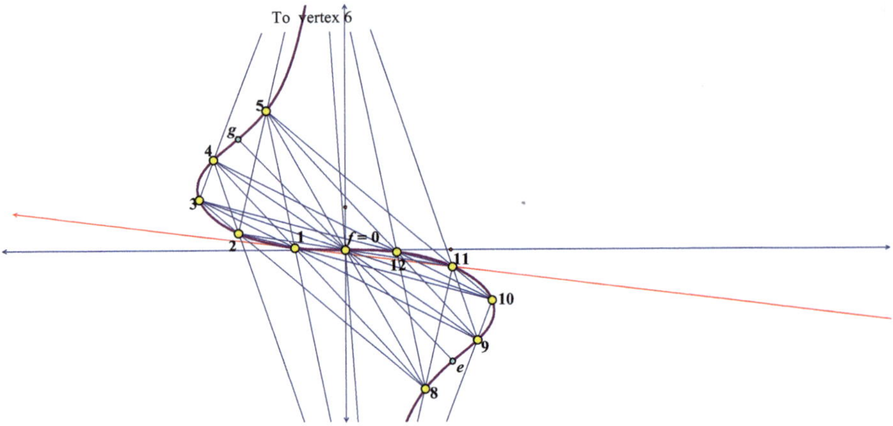

Fig. 33 (Z_{13},0) triple system on Type V cubic

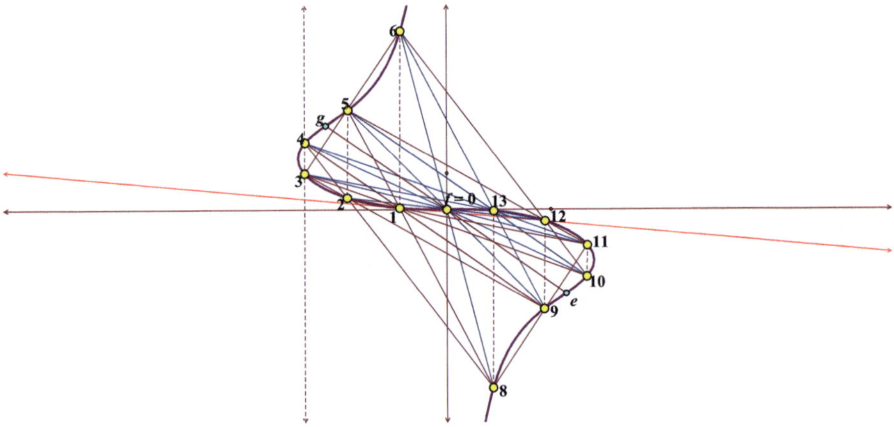

Fig. 34 (Z_{14},0) triple system on Type V cubic with vertex 7 at infinity

References

1. Bix, R.: Conics and Cubics. Springer Science+Business Media, LLC, New York (2006)
2. Harris, M: Thirdpoint groupoids. Master's thesis, Virginia State University, (2008)
3. Knapp, A.: Eliptic Curves. Princeton University Press, (1992)
4. Fletcher, R: Geometric Triple Systems with Base $Z_2 \times Z_n$, submitted for publication in Springer Proceedings in Mathematics and Statistics, (2016)

Geometric Ramifications of Invariant Expressions in the Binary Hypercommutative Variety

Raymond R. Fletcher III

Abstract The *binary hypercommutative (BH) variety*, also referred to as the variety of *thirdpoint groupoids* in Harris (Thirdpoint groupoids, Masters thesis, Virginia State University, 2008), is a collection of algebras associated naturally with the nonsingular points of an irreducible cubic curve. The identities which define these algebras have, as consequences, expressions whose value is invariant under any permutation of the variables comprising the expression. We characterize such expressions and Conjecture that they can be used to determine points of intersection between a given cubic curve and an arbitrary algebraic curve. We prove the Conjecture in the case of the intersection of a cubic γ and a conic β: If γ, β meet in the five points a,b,c,d,e in the real projective plane, then they meet also in the point e*{(a*b)*(c*d)} where a*b denotes the point of intersection besides a,b where the line joining a,b meets γ. The expression e*{(a*b)*(c*d)} is invariant in the BH variety. The Conjecture is also proved for two types of singular cubics. We provide illustrations as strong evidence, but cannot prove similar statements pertaining to the intersection of a nonsingular cubic with higher degree algebraic curves.

Keywords Cubic curve • Algebraic curve • Hypercommutative • Groupoid • Equational variety • Conic

1 Introduction

The *binary hypercommutative (BH) variety* is the collection of algebras which satisfy the identities:

$$(x^*y) = (y^*x) \tag{1}$$

$$x^*\left(x^*y\right) = y \tag{2}$$

R.R. Fletcher III (✉)
Department of Mathematics and Economics, Virginia State University, Petersburg, VA 23806, USA
e-mail: rfletcher210@gmail.com

B. Toni (ed.), *New Trends and Advanced Methods in Interdisciplinary Mathematical Sciences*, STEAM-H: Science, Technology, Engineering, Agriculture, Mathematics & Health, DOI 10.1007/978-3-319-55612-3_3

$$\left(x^*y\right)^* \left(u^*v\right) = \left(x^*u\right)^* \left(y^*v\right) \tag{3}$$

A *word* in this variety is a properly parenthesized expression consisting of variables and the * operation. If T is a BH algebra, and elements from T are substituted for the variables in a word, then the resulting element of T is the *value* of the word with respect to T. A word is *monotone* if no variable in the word is repeated, and a monotone word is *invariant* if the value of the word with respect to any BH algebra remains unchanged under any permutation of the variables. For example, the words a*b, (a*b)*(c*d) are invariant by virtue of identities (1) and (3). We also regard the word consisting of a single variable as invariant. Consider the word e*{(a*b)*(c*d)}. Using identities (1), (2), (3) we have e*{(a*b)*(c*d)} = {a*(e*a)}* {(a*b)*(c*d)} = {a*(a*b)}* {(e*a)*(c*d)} = b*{(e*a)*(c*d)}. Now it is clear that e*{(a*b)*(c*d)} is invariant; we have switched the elements e and b and since (a*b)*(c*d) is invariant, there is nothing special about b. In general, if w is a word which is expressed in the form u*v where u,v are invariant, then w is itself invariant in case it is possible to interchange two variables, one from u and one from v, without changing the value of w. The words u, v are the *principle subwords* of w.

The motivation for the study of the BH variety lies in the theory of cubic curves. If γ is an irreducible cubic curve, then we can define a binary operation on the nonsingular points of γ by setting a*b equal to the third point besides a,b which lies at the intersection of the line [a,b] and γ. This operation is well defined according to the following theorem which can be found in [1].

Theorem 1 *Let L be a line that intersects an irreducible cubic γ at least twice, counting multiplicities. Then L intersects γ exactly three times, counting multiplicities.* □

The product a*a denotes the point of contact besides a, of the tangent to γ at a with the cubic γ. In case a*a = a, the point a is called a *flex* of γ, and the tangent at a intersects γ at a with multiplicity 3. It is known that an irreducible cubic curve has three collinear flexes. Identities (1), (2) are clearly satisfied by this operation. It is a rather remarkable property of irreducible cubics, that identity (3) is also satisfied.

Theorem 2 *If γ is an irreducible cubic curve, then the algebra (γ, *) defined on the nonsingular points of γ lies in the BH variety.* □

The proof of Theorem 2 can also be found in [1]. In Fig. 1 we illustrate identity (3) on a typical irreducible cubic curve. The invariant word e*{(a*b)*(c*d)} has an interesting geometric significance in connection with cubic curves, which we discuss in the next section. In [2] members of the BH variety are referred to as *thirdpoint groupoids* and a quite different set of properties is determined.

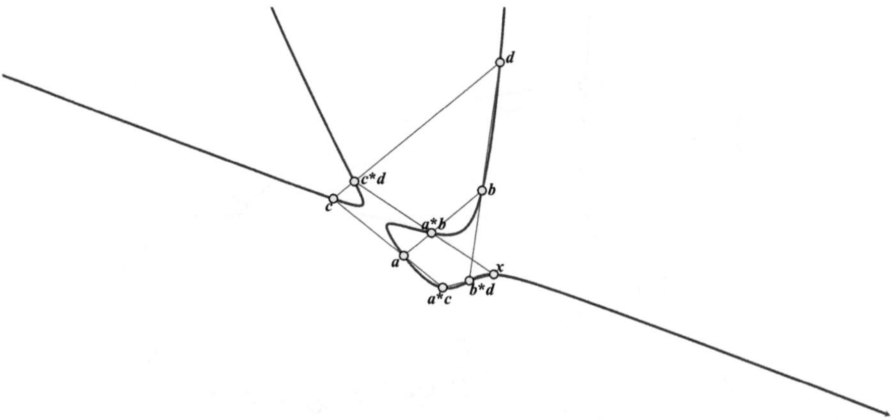

Fig. 1 The hypercommutative axiom: (a*b)*(c*d) = (a*c)*(b*d) = x

2 Intersection of Cubic and Conic

We shall find the following theorem useful.

Theorem 3 *Let G = 0 and H = 0 be distinct algebraic curves of degree n. Assume that there is a conic K = 0 such that I_P(G,K) = I_P(H,K) for every point P in the projective plane and such that K intersects G or H a total of 2n times, counting multiplicities. Then there is a curve W = 0 of degree n-2 such that I_P(G,H) = I_P(G,K) + I_P(G,W) = I_P(H,K) + I_P(H,W).* □

The proof of Theorem 3 can be found in [1]. The notation I_P(G,K) refers to the intersection multiplicity of the curves G,K at the point P. Now we consider a conic which intersects a cubic in five points in the real projective plane.

Theorem 4 *Let K be a conic which meets an irreducible cubic curve γ in points {a,b,c,d,e} in the real projective plane. Then K also meets γ in the point f = a*{(b*c)*(d*e)}.*

Proof Let f denote the sixth point, counting multiplicities, where the conic K meets γ, and let ξ denote the cubic consisting of the three lines [a,f], [e,d], and [b,c]. The conic K intersects each of the two cubics γ, ξ in the same six points, namely a,b,c,d,e,f. By Theorem 3 the intersections of γ, ξ consist of the intersections of either curve with K together with its intersections with a curve W of degree 1, i.e., a line. Thus the points e*d, b*c, q (see Fig. 2), which are the remaining points of intersection of γ, ξ when the intersections with K are removed, must be collinear. Then we must have q = (b*c)*(e*d) and f = a*q = a*{(b*c)*(d*e)}. □

In Theorem 4 we see the geometric significance of the invariant expression a*{(b*c)*(d,e)}: it yields the sixth point of intersection of a conic and a cubic which meet in five points, a,b,c,d,e in the real projective plane. The natural

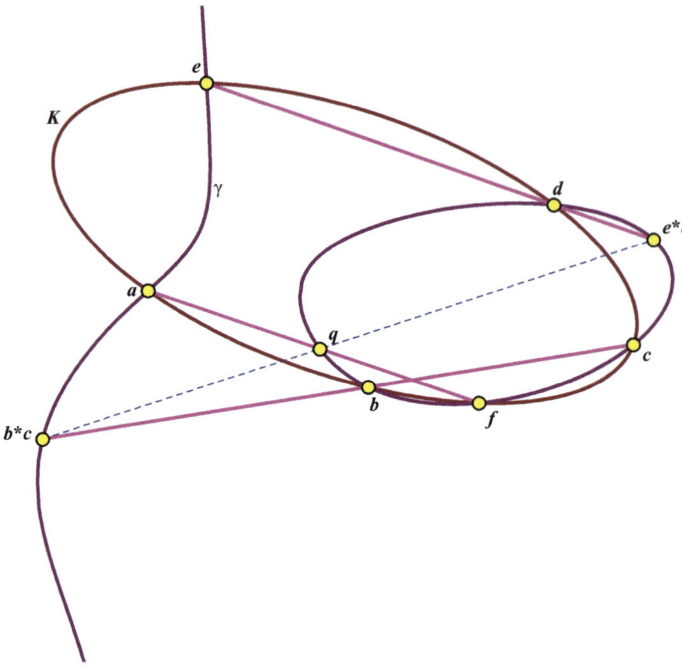

Fig. 2 Intersection of conic and cubic

question then is can we find an eight variable invariant expression which yields the ninth point of intersection of two cubics ? Consider the seven variable monotone expression (c*d)*{b *{(e*f)*(g*h)}} whose principal subwords are invariant, and note that (c*d)*{b * {(e*f)*(g*h)}} = (c*b)*{d*{(e*f)*(g*h)}} by axiom (3). We have interchanged two variables, one from each invariant principal subword, and obtained an equivalent expression. We can immediately conclude that (c*d)*{b * {(e*f)*(g*h)}} is invariant. Now consider the eight variable expression a * {(c*d)*{b * ((e*f)*(g*h))}} whose principal subwords are invariant, and note that a*{(c*d)*{b*((e*f)*(g*h))}}={c*(c*a)}* {(c*d)*{b*((e*f)*(g*h))}}={c*(c*d)}*{(c*a) * ((e*f)*(g*h))}} = d* {(c*a) * ((e*f)*(g*h))}}. We have interchanged the a with the d from the second principal subword to obtain an equivalent expression, and so we can conclude that a * {(c*d)*{b * ((e*f)*(g*h))}} is invariant. Observe also that a *{(c*d)*{b * ((e*f)*(g*h)} = {b*(a*b)}* {(c*d)* {b * ((e*f)*(g*h))}}= {(a*b)*(c*d)}* {(e*f)*(g*h)} so we can conclude that this latter expression is also invariant. In Fig. 3 we illustrate two cubic curves which meet in nine points, a,b,c,d,e,f,g,h,x where h is the point at infinity on vertical lines. The figure indicates that x = {(a*b)*(c*d)}* {(e*f)*(g*h)}, so we have evidence that the invariant expression {(a*b)*(c*d)}* {(e*f)*(g*h)} yields the ninth point of intersection of two irreducible cubics which meet in the eight points a,b,c,d,e,f,g,h.

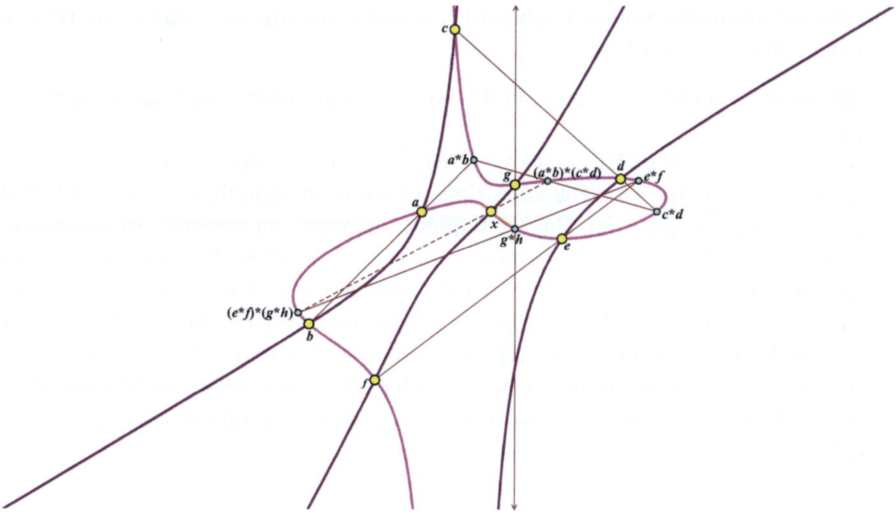

Fig. 3 Intersecting cubics

3 Invariant Expressions in the BH Variety

The observations of the previous section suggest that invariant expressions in the BH variety may have significant geometric ramifications regarding the intersections of cubic curves with algebraic curves of various degrees. It is important then to answer the following questions: (1) Do invariant expressions exist of any given length ? (2) Are invariant expressions of the same length equivalent ? The constructions in Section 2 provide a clue to a general procedure for constructing invariant expressions.

Theorem 5 *Let w be a word in the BH variety and suppose w = (b*c)*v and v = a*u where u,v are words, v is invariant, and u does not contain the variables b,c. Then w is invariant.*

Proof Consider: w = (b*c)*(a*u) = (b*a)*(c*u). We have interchanged two variables of w, one from the invariant principal subword b*c and the other from the invariant principle subword a*u, without changing the value of w. We can conclude that w is invariant. □

Theorem 6 *Let w be a word in the BH variety and suppose w = a*v and v = (b*c)*u where u,v are words, v is invariant, and u does not contain the variable a. Then w is invariant.*

Proof Consider: w = a*{(b*c)*u} = {b*(a*b)}* {(b*c)*u}= {b*(b*c)}*{(a*b)*u} = c* {(a*b)*u}. We have interchanged the variable a from the first principal invariant subword (which is just the single variable a) with the variable c from the second invariant principal subword, without changing the value of w. We can conclude that w is invariant. □

Theorems 5 and 6 allow us to build the following infinite sequence of invariant expressions:

E_1 = a; E_2 = a*b; E_4 = (a*b)*(c*d); E_5 = e*{(a*b)*(c*d)}= e*E_4; E_7 = (f*g)*E_5; E_8 = h*E_7; etc. The subscripts represent the length, i.e., the number of variables, in the expression. So our sequence contains an invariant expression of length p precisely when $p \equiv 1 \pmod 3$ or $p \equiv 2 \pmod 3$. We shall see that no invariant word can have a length which is a multiple of 3. When we write the word E_p*E_q we really mean for E_p and E_q to be represented with disjoint variable sets. So, for example, the notation E_4*E_4 represents the word {(a*b)*(c*d)}*{(e*f)*(g*h)} which we have seen to be invariant, and which we have shown to be equivalent to a*{(c*d)*{b*((e*f)*(g*h))}} = E_1*E_7 = E_8. This observation leads to our next result.

Theorem 7 *If p, q are both congruent to 1 (mod 3) or both congruent to 2 (mod 3), then* E_p*E_q= E_{p+q}, *and* E_{p+q} *is invariant.*

Proof Note that when $p \equiv 2 \pmod 3$ we have E_p= E_1*E_{p-1}, and when $p \equiv 1 \pmod 3$ we have E_p= E_2*E_{p-2}. Clearly the theorem holds if $p+q < 5$. Let $k \geq 5$ and assume inductively that the theorem holds whenever $p+q < k$. Suppose p+q = k, and $p\equiv 2 (\mathrm{mod}\ 3)$ and $q\equiv 2 (\mathrm{mod}\ 3)$. Then E_p*E_q = (E_1*E_{p-1})*(E_1*E_{q-1}) = (E_1*E_1)*(E_{p-1}*E_{q-1}) = E_2*E_{p+q-2} = E_{p+q}. In the last equality we use the fact that $p+q \equiv 1 (\mathrm{mod}\ 3)$, and in the second to last equality we use the inductive hypothesis. Now suppose p+q = k, and $p \equiv 1 (\mathrm{mod}\ 3)$ and $q \equiv 1 (\mathrm{mod}\ 3)$. Without loss of generality we can assume that $q \geq 4$. Then E_p*E_q = E_p*(E_2*E_{q-2}) = E_p*(E_2*(E_1*E_{q-3}) since $q-2 \equiv 2 (\mathrm{mod}\ 3)$. Then E_p*(E_2*(E_1*E_{q-3}) = E_p*{(x*y)*(w*E_{q-3})}={y*(x*y)}*{(y*E_p)*(w*E_{q-3})} = x * {(y*E_p)*(w*E_{q-3})} = x*(E_{p+1}*E_{q-2}) = x*E_{p+q-1}= E_{p+q}. Since $p+q \equiv 2 \pmod 3$ or $p+q \equiv 1 \pmod 3$, the word E_{p+q} is invariant. □

Let (G,+) be an abelian group with identity 0, and let g be a fixed element of G. If we define a binary operation * on the elements of G by a*b = g – a – b, then it is readily seen that (G,*) is a BH algebra. If we take the group of integers (Z,+) for the abelian group and set g = 0 we obtain a*b = -a – b; (a*b)*(c*d) = a+b+c+d; e*{(a*b)*(c*d)} = -e – a – b – c – d, etc., and we find that if $p \equiv 1 \pmod 3$, then the valuation of E_p in this model is a sum of the variables which comprise the word E_p, and if $p\equiv 2 (\mathrm{mod}\ 3)$, then the valuation of E_p is a sum of the inverses of the variables in E_p. In the former case we say that E_p is *positive,* and in the latter case we say E_p is *negative.* The product of two positive expressions is negative, and the product of two negative expressions is positive. If a word w = u*v is the product of a positive expression u and a negative expression v, then w cannot be invariant since it is not invariant in the integer model. For example, if we take u = (a*b)*(c*d) and v = (e*f), then u*v = {(a*b)*(c*d)}*(e*f) = (a+b+c+d)*(-e –f) = -a –b – c – d +e +f. This last expression is not invariant in the integer model, for example, if a = e = f = 1 and b = c = d = 0, then -a – b – c – d + e + f = 0, but -e – f – c – d + a + b = -1, and thus u*v cannot be invariant. We can conclude that the word E_p*E_q is not invariant if $p \equiv 1 (\text{mod } 3)$ and $q \equiv 2$ (mod 3). In our final theorem of this section we give a complete determination of invariant expressions in the BH variety.

Theorem 8 *A word w is invariant in the BH variety iff* $w = E_p$ *for some positive integer p which is not a multiple of 3.*

Proof We have shown in Theorems 5 and 6 that E_p is invariant whenever p is not a multiple of 3. So suppose w is an invariant BH word, and let |w| denote its length. We proceed by induction on |w|. If $|w| = 1$ or 2, then $w = E_1$ or $w = E_2$. If $|w| = 3$, then $w = E_1 * E_2$ is not invariant since E_1 is positive and E_2 is negative. Assume inductively that if an invariant BH word y has $3 \leq |y| < p$ and |y| not a multiple of 3, then $y = E_{|y|}$. Suppose $|w| = p$ and let u, v denote the principal subwords of w, so $w = u*v$. If 3 divides |w|, then it must be that either 3 divides |u| or 3 divides |v|. Suppose without loss of generality that 3 divides |u|, but then u is not invariant by our inductive hypothesis, and as u is a subword of w, the word w also cannot be invariant. Now suppose $p \equiv 1$ (mod 3), then it must be that $|u| \equiv 2$ (mod 3) and $|v| \equiv 2$ (mod 3). Since w is invariant, the subwords u, v must also be invariant, and since $|u|<p$ and $|v|<p$, we obtain by the inductive hypothesis that $u = E_{|u|}$ and $v = E_{|v|}$. Then by Theorem 7, $w = E_{|u|} * E_{|v|} = E_{|u|+|v|} = E_{|w|} = E_p$. If $p \equiv 2$ (mod 3), then $|u| \equiv 1$(mod 3) and $|v| \equiv 1$(mod 3), and by the same argument we obtain $w = E_p$. □

4 Conclusion

We seek a proof for the following Conjecture:

Conjecture 9 Let γ be an irreducible cubic curve and let ξ be an algebraic curve of positive degree n which meets γ in $p = 3n$ nonsingular points in the real projective plane. If the word $E_{p\text{-}1}$ is evaluated at any 3n-1 of the intersection points, then the remaining point of intersection is obtained.

As further evidence for Conjecture 9 we illustrate in Fig. 4, the intersection of a *moon quartic* Q with a *three-pronged* cubic curve L. The moon-shaped quartic was obtained by inverting an ellipse. In this Figure E_{11} is given by: h*{(a*b)*{f*{(k*e)*{d*((j*g)*(c*m)}}}}, and its value is the 12th point of intersection x of γ and ξ.

We can prove Conjecture 9 in a few cases by using parameterizations. In accordance with [1], irreducible cubics can be put into five Types. Type I consists of cubics which can be transformed into $zy^2 = x^3$, these are known as *cuspidal cubics*; Type II consists of cubics which can be transformed into $x^3 + y^3 = 3xyz$, these are called *crunodal cubics*; Type III consists of cubics that can be put in the form $z(x^2 + y^2) = y(3x^2 - y^2)$, these are called *acnodal cubics.* These are the three types of singular irreducible cubics, each named after the type of singularity they contain. A cuspidal cubic is the inverse of a parabola by a point on the parabola;

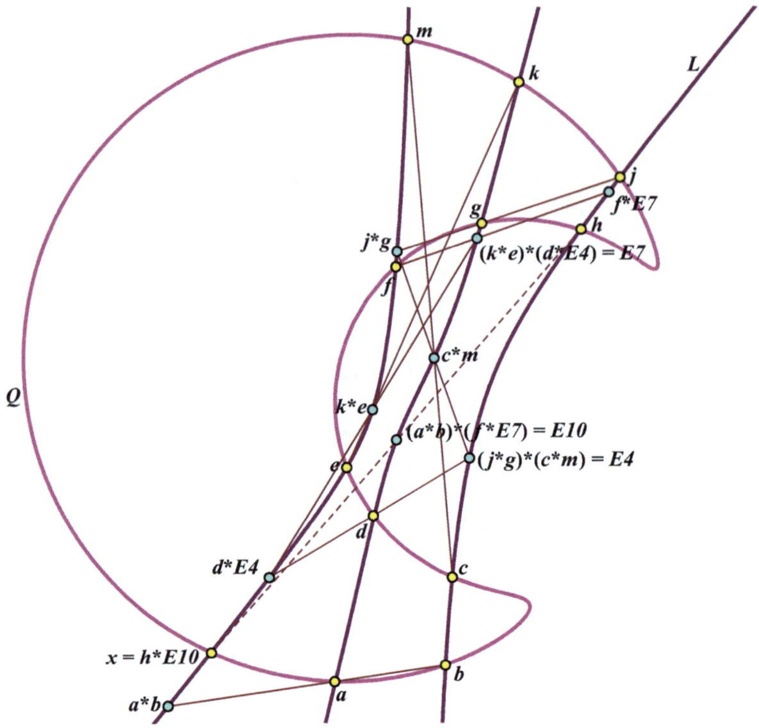

Fig. 4 Cubic meets moon quartic

a crunodal cubic is the inverse of a hyperbola by a point on the hyperbola, and an acnodal cubic is the inverse of an ellipse by a point on the ellipse. We have given homogenized forms of the equations for these cubics all of which can be found in [3]. The remaining two Types encompass the nonsingular cubics: Type IV consists of those cubics which can be put into the form $z^2x = y(y\text{-}x)(y - k^2x)$, where $0 < k^2 < 1$. These are cubics which consist of two *circuits* which we have referred to as the *bell* and the *oval.* Type V cubics consist of a single circuit and can be put in the form $z^2(y - x) = (y + x)(a^2x^2 + b^2y^2)$, where $a^2 + b^2 = 1$. Again these equation forms are due to Hilton [3]. We can prove Conjecture 9 for cuspidal cubics as follows.

Theorem 10 *Conjecture 9 holds true in case γ is a cuspidal cubic.*

Proof We may transform γ so that its equation becomes $zy^2 = x^3$, and then parameterize by setting $x = t$, $y = 1$, and $z = t^3$. Let P_1, P_2, P_3 be three collinear points on γ with parameters t_1, t_2, t_2, respectively. If the points lie on the line $\lambda x + \mu y + \nu z = 0$, then the parameters are roots of the equation

$$f(t) = \lambda t + \mu + \nu t^3, \tag{4}$$

and thus f(t) must also be given by

$$f(t) = \nu\,(t-t_1)\,(t-t_2)\,(t-t_3)\,. \tag{5}$$

The coefficient of t^2 in (4) is 0, and the coefficient of t^2 in (5) is $-\nu(t_1 + t_2 + t_3)$, so we must have $t_1 + t_2 + t_3 = 0$. Thus we have

If 3 points of $zy^2 = x^3$ are collinear, then the sum of their parameters is 0, and conversely.

Up to this point is the argument given by Hilton in [3]. We will just extend the argument from a line to an algebraic curve ξ of degree n. Suppose $\lambda \cap \xi$ contains the 3n points $P_1, \ldots, P_{3n}$ with parameters $t_1, \ldots, t_{3n}$, respectively. If ξ is given by F(x,y,z), then the parameters must be roots of the polynomial

$$f(t) = F\left(t, 1, t^3\right). \tag{6}$$

The high power term in f(t) arises from the term νz^n in F(x,y,z), and is given by νt^{3n}. The term in f(t) with the next highest power arises from the term $\mu x z^{n-1}$ in F(x,y,z), and is given by μt^{3n-2}, thus f(t) has no term of degree 3n-1. Since the parameters $t_1, \ldots, t_{3n}$ are roots of f(t), f(t) must also be given by

$$f(t) = \nu\,(t-t_1)\,(t-t_2)\,(t-t_3)\ldots.(t-t_{3n})\,. \tag{7}$$

The coefficient of t^{3n-1} in (7) is $-\nu(t_1 + t_2 + \ldots + t_{3n})$, which must be 0 by comparison with (6). So we conclude

If λ is a cuspidal cubic and ξ is an algebraic curve of degree n which meets λ in 3n points with parameters $t_1, \ldots, t_{3n}$, then $t_1 + \ldots + t_{3n} = 0$.

The invariant expression $E_2 = P_1{*}P_2$ has parameter $-t_1 - t_2$. (Since P_1, P_2, and $P_1{*}P_2$ are collinear, the sum of their parameters is 0 as we have shown above.) We can then determine that $E_4 = (P_1{*}P_2){*}(P_3{*}P_4)$ has parameter $t_1 + t_2 + t_3 + t_4$ and $E_5 = P_5{*}\{(P_1{*}P_2){*}(P_3{*}P_4)\}$ has parameter $-t_1 - t_2 - t_3 - t_4 - t_5$. In general, if $s \equiv 1 \bmod 3$, then E_s has parameter $t_1 + t_2 + \ldots. + t_s$, and if $s \equiv 2 \bmod 3$, then E_s has parameter $-(t_1 + t_2 + \ldots. + t_s)$. Since $3n-1 \equiv 2 \bmod 3$, E_{3n-1} has parameter $-(t_1 + t_2 + \ldots. + t_{3n-1}) = t_{3n}$. Since t_{3n} is the parameter of P_{3n}, we have $E_{3n-1} = P_{3n}$ in accordance with Conjecture 9. □

A similar approach can be used to prove Conjecture 9 in case γ is a crunodal cubic.

Theorem 11 *Conjecture 9 holds in case γ is a crunodal cubic.*

Proof In [3] it is given that a crunodal cubic can be put in the form $x^3 + y^3 = 3xyz$ when we obtain a parameterization: $x = 3t$, $y = 3t^2$, $z = 1 + t^3$. Suppose points with parameters t_1, t_2, t_3 lie in the line $\lambda x + \mu y + \nu z = 0$. Then the parameters are roots of the equations

$$\lambda\,(3t) + \mu\left(3t^2\right) + \nu\left(1 + t^3\right) = 0, \tag{8}$$

$$\nu\,(t-t_1)\,(t-t_2)\,(t-t_3) = 0. \tag{9}$$

Since (8) and (9) must be identical, their constant terms must be identical and thus $\nu = -\nu(t_1\, t_2\, t_3)$.

We conclude that the product of the parameters of three collinear points on γ *is* -1.

This much is given as an exercise in [3]. We will now generalize this result by replacing the line with an arbitrary algebraic curve ξ of degree n. Let $F(x,y,z) = 0$ be the equation of ξ and let $P_1, \ldots, P_{3n}$ be the points of $\gamma \cap \xi$ with parameters $t_1, \ldots, t_{3n}$, respectively. These parameters must be roots of the polynomial

$$g(t) = F\left(3t, 3t^2, 1 + t^3\right) = 0. \tag{10}$$

The term cz^n in the equation of ξ gives rise to the high power term of g(t), namely ct^{3n}. The parameters must then also be roots of the equation

$$c\,(t-t_1)\,(t-t_2)\ldots(t-t_{3n}) = 0. \tag{11}$$

The term $cz^n = c(1+t^3)^n$ also gives rise to the constant term of (10), namely c. The constant terms in (10), (11) must be identical so we have $c = c(-1)^{3n}(t_1 t_2 \ldots t_{3n})$ which yields

$$t_1 t_2 \ldots t_{3n} = (-1)^{3n}. \tag{12}$$

Using the above shown fact that the product of parameters of three collinear points on γ is -1, we obtain the following results: if $s \equiv 1 \bmod 6$, then the parameter of E_s is $t_1 t_2 \ldots t_s$; if $s \equiv 2 \bmod 6$, then the parameter of E_s is $(-1)/t_1 t_2 \ldots t_s$; if $s \equiv 4 \bmod 6$, then the parameter of E_s is $-t_1 t_2 \ldots t_s$; if $s \equiv 5 \bmod 6$, then the parameter of E_s is $1/t_1 t_2 \ldots t_s$. If n is odd, then $3n-1 \equiv 2 \bmod 6$ and E_{3n-1} has parameter $(-1)/t_1 t_2 \ldots t_{3n-1} = (-1)^{3n}/t_1 t_2 \ldots t_{3n-1} = t_{3n}$ in accordance with (12). As a consequence $E_{3n-1} = P_{3n}$ as stipulated by Conjecture 9. If n is even, then $3n-1 \equiv 5 \bmod 6$ and E_{3n-1} has parameter $1/t_1 t_2 \ldots t_{3n-1} = (-1)^{3n}/t_1 t_2 \ldots t_{3n-1} = t_{3n}$, and again we obtain $E_{3n-1} = P_{3n}$. □

According to [3] an acnodal cubic can be put in the form $z(x^2 + y^2) = y(3x^2 - y^2)$ and can then be parameterized by $x = \cos\phi$; $y = \sin\phi$, and $z = \sin 3\phi$. It is then shown that three points on this cubic with parameters ϕ_1, ϕ_2, ϕ_3 are collinear iff $\phi_1 + \phi_2 + \phi_3 \equiv 0 \bmod \pi$, and then it is claimed, without proof, that if six points with parameters $\phi_1, \phi_2, \phi_3, \phi_4, \phi_5, \phi_6$ lie on a conic, then $\phi_1 + \phi_2 + \phi_3 + \phi_4 + \phi_5 + \phi_6 \equiv 0 \bmod \pi$, and conversely. This last result as well as similar results for nonsingular cubics we are unable to obtain.

References

1. Bix, R.: Conics and Cubics, a Concrete Introduction to Algebraic Curves, 2nd edn. Springer Sci. and Business Media, LLC, New York (2006)
2. Harris, M.: Thirdpoint groupoids, Masters thesis, Virginia State University, (2008)
3. Hilton, H.: Plane Algebraic Curves. Oxford University Press, Oxford (1920)

Geometric Ramifications of Invariant Expressions in the Ternary Hypercommutative Variety

Raymond R. Fletcher III

Abstract The *Ternary Hypercommutative (TH)* variety is an equational class of algebras determined by the identities:

(i) [a,b,c] is invariant
(ii) [a,b,[a,b,c]] = c
(iii) [[a,b,c],[d,e,f],[g,h,j]] is invariant.

A *word* in this variety is a properly parenthetized expression consisting of variables and the ternary operation [, ,]. If T is a TH algebra, and elements from T are substituted for the variables in a word, then the resulting element of T is the *value* of the word with respect to T. A word is *monotone* if no variable is repeated, and a monotone word is *invariant* if its value with respect to any TH algebra remains unchanged under any permutation of its variables. Thus to say [a,b,c] is invariant and is a shorthand way to express the identities [a,b,c] = [a,c,b] = [b,a,c] = [b,c,a] = [c,a,b] = [c,b,a]. The geometric significance of the TH variety is found in the study of intersection properties of circles with certain types of algebraic curves. For example, if α is an ellipse and a, b, c are any three points of α, then the circle (a,b,c) meets α in a unique fourth point, counting multiplicities, which we denote by our ternary operation [a,b,c]. Identities (i), (ii) are clearly satisfied by this operation, and in Fletcher (Group Circle Systems on Conics, New Frontiers of Multidisciplinary Research in STEAM-H, Springer International Publishing, Cham, 2014) it is shown that axiom (iii) also holds. We call algebraic curves whose intersection with circles satisfies the TH axioms *supercyclic*. We characterize supercyclic curves and determine the invariant words in the TH variety. Evidence is also given for a Conjecture regarding the significance of these words with respect to intersection properties of supercyclic curves and general algebraic curves.

Keywords Plane algebraic curve • Equational variety • Intersections of algebraic curves • Supercyclic curves • Invariant expressions • Inversion

---

R.R. Fletcher III (✉)
Department of Mathematics and Economics, Virginia State University, Petersburg, VA 23806, USA
e-mail: rfletcher210@gmail.com

B. Toni (ed.), *New Trends and Advanced Methods in Interdisciplinary Mathematical Sciences*, STEAM-H: Science, Technology, Engineering, Agriculture, Mathematics & Health, DOI 10.1007/978-3-319-55612-3_4

1 Introduction

This article is a sequel to and follows the same format as [2] where geometric ramifications of invariant expressions in the *binary hypercommutative (BH)* variety are examined. The BH variety is an equational class of algebras which satisfies the identities:

$$a^*b = b^*a \tag{1}$$

$$a^*\left(a^*b\right) = b \tag{2}$$

$$\left(a^*b\right)^*\left(c^*d\right) = \left(a^*c\right)^*\left(b^*d\right) \tag{3}$$

This variety can also be defined by saying a*b and (a*b)*(c*d) are invariant and keeping (2) as stated. It can thus be seen that the TH algebra is a *ternary version* of the BH variety. As explained in [2], the BH variety has its application in the study of cubic curves. If γ is an irreducible cubic curve and a,b are nonsingular points on γ, we define a*b to be the third point, counting multiplicities which lies on the line ab and on γ. In this interpretation identities (1), (2) are clear. Identity (3) requires some proof which can be found in [3]. In [2] it is shown that the expression a*((b*c)*(d*e)) is invariant in the BH variety and in fact yields the 6th point of intersection of a conic which meets γ in the 5 points a,b,c,d,e. Also in [2] all the invariant BH words are determined and conjectured to yield points of intersection of a cubic curve with an arbitrary algebraic curve. We follow a similar procedure here: we will determine all invariant words in the TH variety with the hope that similar geometric results may obtain.

2 Supercyclic Curves: Conics and Conic Inverses

In [1] it is shown that if α is a parabola, ellipse, or hyperbola, then a ternary operation [, ,] can be defined on the points of α by setting [a,b,c] equal to the fourth point, counting multiplicities which lies on circle (a,b,c) and on α. The expression [a,a,b] refers to the point besides a, b which lies on the circle tangent to α at point a, and which contains point b. The expression [a,a,a] refers to the point on α besides a which lies on the unique circle which intersects α at point a with multiplicity 3. It is also possible that [a,a,a] = a, in which case there exists a circle which meets α at a with multiplicity 4. In [1] it is shown that this ternary operation determines a TH algebra on the points of α. Identity (iii) in the TH variety can be replaced by the seven variable identity:

$$[[a,b,c],[d,e,f],q] = [[a,b,d],[c,e,f],q], \tag{4}$$

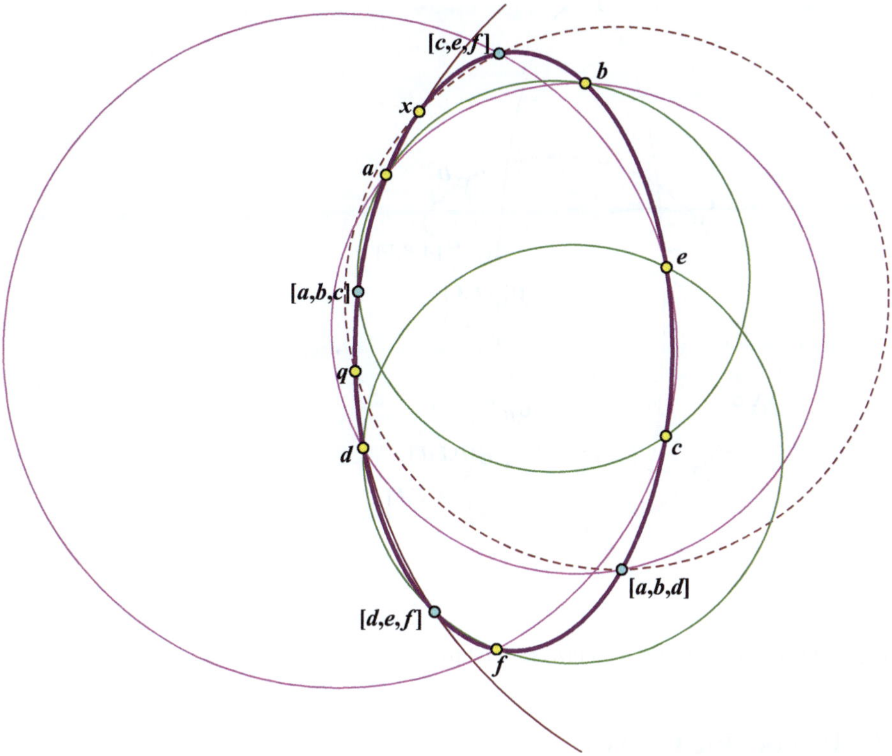

Fig. 1 [[a,b,c],[d,e,f],q] = x = [[a,b,d],[c,e,f],q]

so to show (α, [, ,]) is a TH algebra, it suffices to verify this identity. In Fig. 1 we illustrate identity (4) on an ellipse.

We call an algebraic curve upon which a TH algebra can be defined *supercyclic.* Thus for an algebraic curve γ to be supercyclic it must be the case that every circle or line which meets γ in three points a,b,c also meets γ in a unique fourth point, counting multiplicities, which we designate as [a,b,c]. Then identity (3) or (4) must be satisfied by this ternary operation. Thus every noncircular conic is supercyclic. Since inversion maps circles to circles, and supercyclicity is an intersection property involving circles, it follows that the inverse of any supercyclic curve is again supercyclic. Actually, a circle which passes through the center of inversion is inverted onto a line, so we must also consider lines which meet γ in four points. (Properties of inversion can be found in [4].) If ω is the circle of inversion and the center of ω lies on a noncircular conic, then inversion through ω yields a supercyclic cubic curve. On the other hand, if the center of ω does not lie on the conic, then a supercyclic quartic curve is obtained. In Fig. 2 we illustrate (4) on the inverse of an ellipse with the center of inversion off the ellipse.

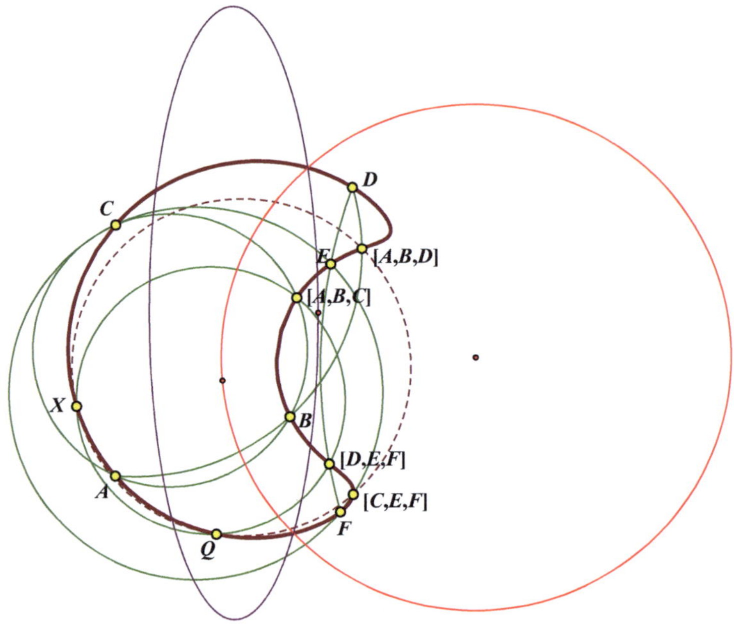

Fig. 2 Identity (4) illustrated on inverse of ellipse

3 Protocubic Curves

We shall call *protocubic* an irreducible algebraic curve which satisfies the BH identities (1), (2), (3). Thus if ξ is protocubic, then it must be the case that every line which meets ξ in two points a, b meets ξ in a unique third point a*b. We will show that the inverse of a supercyclic curve γ by a circle with center on γ is protocubic and that a protocubic curve is nothing more than an irreducible cubic. We shall make use of a construct found in [5]. If G is an abelian group, g ∈ G and ϕ:G→∏ an injective mapping from G into the projective plane such that whenever a,b,c ∈ G and a+b+c = g, the corresponding points ϕ(a), ϕ(b), ϕ(c) are collinear, then we call the set of points ϕ(G) and the associated collinear triples, a *(G,g) geometric triple system.* If we let G = Z the group of integers under addition, and if we take g = 1, we obtain a (Z,1) triple system.

Theorem 1 *Let γ be supercyclic and let q be a point on γ. Then the inverse of γ by a circle with center q is protocubic.*

Proof Let ξ denote the inverse of γ by a circle ω with center q, and define a binary operation ∘ on the points of γ by setting a ∘b = [q,a,b]. If p is any point on γ, let p′ denote the inverse of p with respect to ω. Since circle (q,a,b) passes through the center of inversion, the points a′, b′, (a ∘b)′, are collinear on ξ. Define a binary operation * on ξ by setting a′* b′= (a ∘b)′, then certainly identities (1),

(2) are satisfied by *. To check identity (3), let a′, b′, c′, d′ be any four points on ξ. Then (a′* b′)* (c′*d′) = (a ∘b)′ * (c ∘d)′ = ((a ∘b) ∘ (c ∘d))′ = ([q,a,b]∘[q,c,d])′ = [[q, [q,a,b],[q,c,d]]′ = [[q, [q,a,c],[q,b,d]]′ = ([q,a,c]∘[q,b,d])′ = (a ∘ c)′ * (b ∘ d)′ = (a′* c′)* (b′ * d′) = ((a ∘ c) ∘ (b ∘ d))′ = (a′ * c′)* (b′ * d′). □

Theorem 2 *Every protocubic curve is an irreducible cubic curve.*

Proof Let ξ be protocubic and let p be a nonsingular nonflex point of ξ. We construct a (Z,1) triple system on ξ as follows. Choose point p as vertex 0 and let vertex 1 be the point 0*0. Let vertex -1 = 1*1; let 2 = 0*-1 and -2 = 1*2. If k ≥ 3 and vertices {0, ± 1, ± 2, ..., ± (k-1)} have been assigned to points on ξ, let (i) k = 0*(-k+1), and (ii) -k = 1*k. We assume that vertex 0 (the generating vertex) is chosen so that no point on ξ is assigned more than one integer label. In [5] it is shown that this construction produces a (Z,1) triple system on ξ in case ξ is an irreducible cubic curve, but the proof only employs the identity (3) which characterizes a BH algebra. Thus the proof given in [5] is equally valid if we assume only that ξ is protocubic. It is also shown in [5] that the vertices of a (Z,1) triple system must lie on an irreducible cubic curve. Let η denote the irreducible cubic which contains the vertices of the (Z,1) triple system we have constructed on ξ. Then the irreducible algebraic curve ξ intersects the cubic η in infinitely many points. This can only happen if η is a factor of ξ, but since ξ is irreducible we must have that the two curves are identical, and thus ξ is an irreducible cubic. □

Now putting Theorems 1 and 2 together we have:

Corollary 3 *Every supercyclic curve is the inverse of an irreducible cubic curve.* □

4 Self-Inversive Character of Supercyclic Curves

Irreducible cubic curves belong to five Types in accordance with the classification given in [3]. Type I are those cubics which can be transformed into $y^2 = x^3$; Type II are cubics which can be transformed into $y^2 = x^2(x+1)$; Type III are those which can be transformed into $y^2 = x^2(x-1)$. These three types are singular, each with a single singularity; Type I curves have a *cusp*; Type II curves have a self-crossing point known as a *crunode,* and Type III curves contain an isolated point called an *acnode*. Type IV curves are those which can be transformed into $y^2 = x(x-1)(x-w)$; w>1, and Type V are those which can be transformed into $y^2 = x(x^2 + kx + 1)$; $-2 < k < 2$. Types IV and V are nonsingular; Type V cubics consist of one connected component when regarded in the projective plane, and Type IV cubics have two connected components: the *bell* and the *oval*. If ξ is Type IV, then the bell of ξ is a subalgebra of (ξ,*). Now suppose ξ is the inverse of a supercyclic curve γ by a circle ω with center q on γ. By Corollary 3, ξ is an irreducible cubic, but also as the inverse of a supercyclic curve, it is supercyclic. It is shown in [6] that ξ contains a single point at infinity which we denote by ∞. From the collinear triple {a,b,a*b} of ξ we obtain [a,b,a*b] = ∞, and thus a*b = [a,b,∞], a useful expression relating the two operations which can be defined on the points of ξ. Let u be a point of ξ in $\sqrt{\infty}$, i.e., u*u = ∞, and let p ∈ $\sqrt{u}$ so (p*p)*(p*p) = ∞. From the equation (p*p)*(p*p) = ∞ we obtain ∞ = [p,p,∞]* [p,p,∞] = [[p,p,∞] ,[p,p,∞], ∞]] = [[p,p,p],[p,∞,∞],∞] , which implies [p,p,p] = [[p,∞ , ∞], ∞ , ∞] = p. From all this we obtain that ξ contains a *[, ,]—idempotent element.* Inversion maps [, ,]—idempotent elements to [, ,]—idempotent elements, so the inverse of p with respect to ω is a [, ,]—idempotent element of the supercyclic curve γ. We label this point with 0, so [0,0,0] = 0 , and will build a *(Z,0) circle system* on γ.

Let G be an abelian group and g a fixed element of G, and let ψ:G→∏ be an injective mapping from G into the projective plane so that no five points in ψ(G) are cocyclic. If for every four element subset {a,b,c,d} with a+b+c+d = g, the corresponding points {ψ(a), ψ(b), ψ(c), ψ(d)}, are either cocyclic or collinear, then we call the set of points ψ(G) and the associated cocyclic, (or collinear), four element sets, a *(G,g) circle system.* Properties of circle systems can be found in [6, 7]. We take G to be the additive group of integers (Z,+) and g to be the identity 0, to obtain our (Z,0) circle system. Now to construct a (Z,0) circle system on γ we have already found a point which we have labeled 0 and which satisfies [0,0,0]=0 as is required in a (Z,0) circle system. The remaining vertices of our intended circle system are defined as follows. Choose vertex 1 on γ so that [1]≠ 0,1 and let -1 = [0,0,1]; let vertex 2 = [0,-1,-1] and vertex -2 = [0,0,2]; let vertex 3 = [0,-1,-2] and vertex -3 = [0,0,3]. In general, for k≥0 we let (i) k = [0,-1,-k+1], and (ii) -k = [0,0,k]. First we show that formulas (i), (ii) also hold when k<0. If k<0, then -k>0 and by (i) we have -k = [0,-1,k+1]. Replacing k by k-1 yields -(k-1) = [0,-1,k] or k = [0,-1, -k+1] which is (i). On the other hand, (ii) immediately implies k = [0.0.-k] which is (ii) with k replaced by -k. We assume that vertex 1 (the generating vertex) is chosen so that no two points of γ receive the same integer label. To show that the points so labeled form a (Z,0) circle system we must show that [a,b,c] = -a-b-c for any integers a,b,c. First we show that [0,a,b] = -a-b ∀a,b ∈ Z. Consider: [0,1,k] = [0, [0,0,-1], [0,0,-k]] = [0, [0,0,0], [0,-1,-k]] = [0,0,k+1] = -1-k. Now let m >1 and assume inductively that [0,m,k] = -k-m ∀k∈Z. Consider: [0,m+1,k] = [0, [0,-1,-m], [0,0,-k]] = [0,[0,0,-m],[0,-1,-k]] = [0,m,k+1] = -k-1-m. It now follows by induction that [0,m,k] = -m-k ∀m≥0. This equation also holds if m<0. We have shown that [0,a,b] = -a-b ∀a,b ∈ Z. To finish the argument consider: [c,a,b] = [c, [0,0,-a], [0,0,-b]] = [c, [0,0,0], [0,-a,-b]] = [c,0,a+b] = [0,c,a+b] = -c-a-b.

We have shown that a (Z,0) circle system can be constructed on a supercyclic curve γ. It is shown in [6] that the vertices of a (Z,0) circle system lie on an algebraic curve which is the inverse of an irreducible self-inversive cubic curve, so we can conclude with the following strengthening of Corollary 3.

Theorem 4 *Every supercyclic curve is the inverse of an irreducible self-inversive cubic curve.* □

5 Self-Inversive Cubic Curves

In light of Theorem 4, it becomes important to obtain a description of irreducible self-inversive cubic curves. Suppose ξ is such a curve and that ω is a circle which inverts ξ onto itself. The center X of ω must lie on ξ since X must be the inverse of the point at infinity on ξ. We may by translation suppose that X lies at the origin, and by dilation we may also suppose that the radius of ω is 1. In [6] the equation of a cubic which inverts onto itself via the unit circle is given as:

$$F(x,y) = (ax + by)\left(x^2 + y^2 + 1\right) + ex^2 + fxy + gy^2 = 0. \tag{5}$$

In (5) we must have at least one of a, b nonzero, else we no longer have a cubic. Also we assume that the coefficients are chosen so that (5) remains irreducible. In this section we would like to show that every irreducible self-inversive cubic curve is supercyclic. In case (5) represents a nonsingular cubic ξ, it is shown in [6] that ξ is supercyclic, so we need only consider here the singular case. Suppose ξ is an irreducible singular cubic which inverts onto itself via the unit circle ω. A singular cubic has a unique singularity. For the curve ξ, this singularity must invert onto itself and so must lie on ω. By rotating ξ about the origin, we may assume that this singularity occurs at the point (1,0). Starting with (5) as the equation for ξ, and assuming ξ has its unique singularity at (1,0), we will develop a formula for ξ. Since (1,0) lies on (5), we immediately obtain $2a + e = 0$. Since (1,0) is a singular point we must have $I_{(1,0)}(L, F(x,y)) \geq 2$ for all lines L through (1,0). (The notation $I_{(1,0)}(L, F(x,y))$ denotes the intersection multiplicity of L and F at (1,0).) If L has slope m, then the equation of L is given by $G(x,y) = y - m(x\text{-}1) = 0$. Translating so that the singularity lies at the origin, we obtain:

$$I_{(1,0)}(L, F(x,y)) = I_{(0,0)}(G(x+1,y), F(x+1,y)) =$$

$$I_{(0,0)}\left(y-mx, \underline{(a(x+1)+by)\left(x^2+2x+2+y^2\right)-2a(x+1)^2+f(x+1)y+gy^2}\right). \tag{6}$$

This, in accordance with [3], is given by the smallest positive degree of H(x,mx), where H(x,y) is the underlined portion of (6), Consider:

$$H(x,mx) = (ax+a+bmx)\left(x^2+2x+2+m^2x^2\right)-2a\left(x^2+2x+1\right)+fmx(x+1) \\ +g\,m^2x^2 \tag{7}$$

For the smallest positive degree of (7) to be larger than or equal to 2, we must arrange for the coefficient of x in (7) to be 0. This yields $m(2b + f) = 0$, so either $f = \text{-}2b$ or $m = 0$. In case $m = 0$, (7) becomes $H(x,0) = ax^3 + ax^2$ which already has a zero coefficient for x. Making the substitutions $e = \text{-}2a$ and $f = \text{-}2b$ in (5) we obtain:

$$F(x,y) = (ax + by)\left(x^2 + y^2 + 1\right) - 2ax^2 - 2bxy + gy^2 = 0 \text{ point at } (1,0) \tag{8}$$

As the equation of an irreducible cubic curve which inverts onto itself through the unit circle, and which contains a singular point at (1,0). To be complete we must also consider the possibility that L is vertical. For this we need to insure that $I_{(0,0)}(x, H(x,y)) \geq 2$. The following Properties given in [3] will be needed to compute the desired intersection multiplicity.

P1: $I_{(0,0)}(x,y) = 1$,
P2: $I_{(0,0)}(F,G) = I_{(0,0)}(F, G + FH)$
P3: $I_{(0,0)}(F,GH) = I_{(0,0)}(F,G) + I_{(0,0)}(F,H)$
P4: If the curve g does not contain the origin, then $I_{(0,0)}(F,G) = 0$.

Now consider: $I_{(0,0)}(x, H(x,y)) = I_{(0,0)}(x, (a(x+1)+by)(x^2+ 2x + 2 + y^2) - 2a(x+1)^2+ f(x+1)y + gy^2)= I_{(0,0)}(x, (a+by)(2 + y^2) - 2a +fy + gy^2+ xK(x)$ where K(x) is a polynomial in x only. In Property P2 we let F= x; $G= (a+by)(2 + y^2) - 2a +fy + gy^2+ xK(x)$, and $H = -K(x)$ to obtain:

$$I_{(0,0)}\left(x, H(x,y)\right) = I_{(0,0)}\left(x, (a+by)\left(2+y^2\right) - 2a + fy + gy^2\right) = I_{(0,0)}\left(x, (a+g)y^2 + (2b+f)y + by^3\right) = I_{(0,0)}\left(x, y\left(by^2 + (a+g)y + (2b+f)\right)\right). \tag{9}$$

Now using Properties P1, P3 we obtain: $I_{(0,0)}(x, H(x,y))= I_{(0,0)}(x, y) + I_{(0,0)}(x, by^2+ (a+g)y + (2b+f)) = 1 + I_{(0,0)}(x, by^2+ (a+g)y + (2b+f))$. If $by^2+ (a+g)y + (2b+f))$ does not contain the origin, then by Property P4, $I_{(0,0)}(x, by^2+ (a+g)y + (2b+f)) = 0$ and thus $I_{(0,0)}(x, H(x,y))= 1$. Since we require $I_{(0,0)}(x, H(x,y)) \geq 2$, it must be that $by^2+ (a+g)y + (2b+f))$ contains the origin, and thus $2b + f = 0$. We again obtain $f = -2b$ and thus confirm (8) as the general equation of an irreducible cubic curve ξ which inverts onto itself via the unit circle, and which contains its unique singularity at (1,0).

We will invert (8) by a unit circle with center at the singular point (0,1). To do this we will translate (8) so that the singularity lies at the origin, and then invert through the unit circle. The desired translation is given by:

$$F(x+1, y) = ax^3 + by^3 + axy^2 + bx^2y + ax^2 + (a+g)y^2 = 0. \tag{10}$$

To invert through the unit circle we replace x by $x/(x^2+ y^2)$ and y by $y/(x^2+ y^2)$ in (10) and then clear fractions to obtain:

$$ax^3 + by^3 + axy^2 + bx^2y + \left(x^2+y^2\right)\left(ax^2 + (a+g)y^2\right) = \left(x^2+y^2\right)\left(ax + by + ax^2 + (a+g)y^2\right) = 0. \tag{11}$$

Ignoring the double point at the origin, we obtain the conic $ax + by + ax^2+(a+g)y^2) = 0$. Our singular cubic ξ is the inverse of a conic and thus is supercyclic. We have proved:

Theorem 5 *Every irreducible self-inversive cubic curve is supercyclic.* □

In the equation:

$$ax + by + ax^2 + (a + g)\, y^2 = 0, \tag{12}$$

we note that $g \neq 0$ since otherwise we would obtain a circle whose inverse is again a circle, (or a line), and not our cubic ξ. Also $a \neq 0$ since otherwise (12) is reducible. If g = -a, then (12) reduces to a parabola and the singularity of ξ is a cusp. We can put (12) in the form:

$$x^2 + x + \frac{a+g}{a} y^2 + \frac{b}{a} y = 0. \tag{13}$$

We see that if $\frac{a+g}{a} > 0$, then (13) is an ellipse and the singularity of ξ is an acnode, i.e., an isolated point, and if $\frac{a+g}{a} < 0$, then (13) is a hyperbola and the singularity of ξ is a crunode, i.e., a point where the curve crosses itself to form a loop. This same information can also be gained by using partial derivatives. If M(x,y) = 0 is a curve, then according to [8], a *double point,* i.e., a singularity is a crunode, acnode, or cusp depending on whether:

$$\frac{\delta^2 M}{\delta x \delta y}^2 >, < or = \frac{\delta^2 M}{\delta x^2} \frac{\delta^2 M}{\delta y^2}. \tag{14}$$

By taking (10) for M(x,y) in (14) the reader may wish to verify the above-mentioned results. Since $x^2 + y^2$ is a factor of the high power terms in (5), this is an example of a *circular cubic.* In [8] it is shown that a circular cubic ξ is self-inversive with respect to four mutually orthogonal circles. If ξ is Type IV, then three of the circles with respect to which it is self-inversive are real. These real circles are also described in [6]. The fourth circle is unreal but has a real center on ξ at the radical center of the three real circles. In [6] this point is described as the center of a circle which *antiinverts* ξ onto itself. (The *antiinverse* of a point P with respect to a real circle ω with center O is the reflection through O of the inverse of P with respect to ω.) If ξ is Type V, then only two of the four self-inverting circles are real and these are orthogonal. If ξ is Type I, then it is the inverse of a parabola by a point on the parabola. The axis of symmetry of the parabola inverts into a circle which inverts ξ onto itself. If the center of inversion is taken as the intersection of the axis of the parabola with the parabola, then ξ reflects onto itself via the same axis. Thus ξ either inverts onto itself via one real circle, or has a single axis of symmetry. If ξ is Type II, then ξ is the inverse of a hyperbola. The two axes of symmetry of the hyperbola invert either to two orthogonal circles which invert ξ onto itself or possibly to two perpendicular axes of symmetry of ξ, or even to one axis of symmetry and to one self-inverting circle with center on this axis. Similar conditions occur in case ξ is Type III and thus the inverse of an ellipse. In light of Theorem 4, these same self-inversive or reflective properties also apply to supercyclic curves. So we have:

Theorem 6 *Let γ be a supercyclic curve. Then either (i) γ is self-inversive through three mutually orthogonal real circles or (ii) γ has two perpendicular axes of symmetry and is also self-inversive through a real circle with center at the crossing point of its axes or (iii) γ has one axis of symmetry and is also self-inversive through two orthogonal circles each with centers on the axis or (iv) γ is self-inversive through exactly two orthogonal circles or (v) γ has exactly two perpendicular axes of symmetry and is not self-inversive through any circle or (vi) γ has one axis of symmetry and exactly one real self-inverting circle with its center on the axis or (vii) γ inverts onto itself through only one circle or (viii) γ has one axis of symmetry and is not self-inversive through any circle.* □

6 Bicircular Quartics

Our Eq. (5) represents a *circular cubic*. It is given in [8] that the inverse of a circular cubic with respect to a circle with center on the cubic is again a circular cubic, and the inverse with respect to a circle with center not on the cubic is a *bicircular quartic*. A *bicircular quartic* is an algebraic curve of degree 4 which can be put into the form

$$\left(x^2 + y^2\right)^2 + ax^2 + by^2 + 2gx + 2fy + c = 0. \tag{15}$$

We have shown that a circular cubic is supercyclic, and since supercyclicity is preserved by inversion, it follows that a bicircular quartic is also supercyclic. It is also given in [8] that the inverse of a bicircular quartic with respect to a circle with center not on the curve is a bicircular quartic and the inverse with respect to a circle with center on the curve is a circular cubic. Since inversion maps self-inverting circles, (or axes of symmetry) to self-inverting circles, (or axes of symmetry), a bicircular quartic is self-inversive with respect to one, two orthogonal, or three mutually orthogonal real circles. In Fig. 3 we illustrate a circular cubic δ of Type V which inverts onto itself via the orthogonal circles α, β. Also depicted is the bicircular quartic δ' and the circles α' and β' which are inverses of δ, α, β, respectively, with respect to the circle w. We chose w with center not on the cubic δ and also so that w is orthogonal to α, and thus $\alpha = \alpha'$. The bicircular quartic δ' inverts onto itself via the orthogonal circles α' and β'. In Fig. 4 we illustrate a circular cubic δ of Type IV and its bicircular quartic inverse δ' via circle w. The cubic δ inverts onto itself via the three mutually orthogonal real circles α, β, γ, and its inverse δ' inverts onto itself via the mutually orthogonal circles α', β', γ' which are the inverses of α, β, γ, respectfully with respect to w. Again we have chosen circle w to be orthogonal to α, and thus $\alpha = \alpha'$. We can now state our final description of supercyclic curves.

Theorem 7 *An algebraic curve γ is supercyclic iff it is a noncircular conic, a circular cubic, or a bicircular quartic.* □

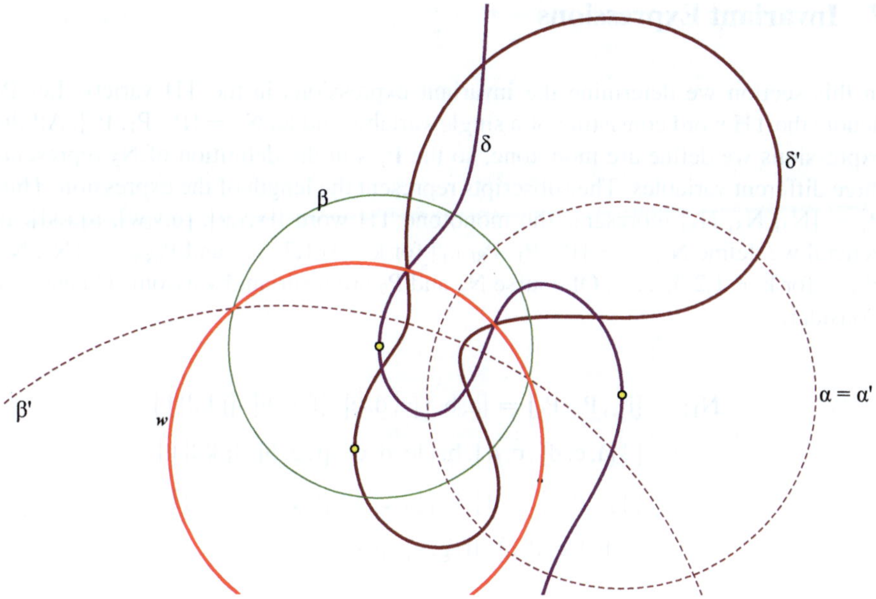

Fig. 3 Bicircular quartic obtained as inverse of circular cubic

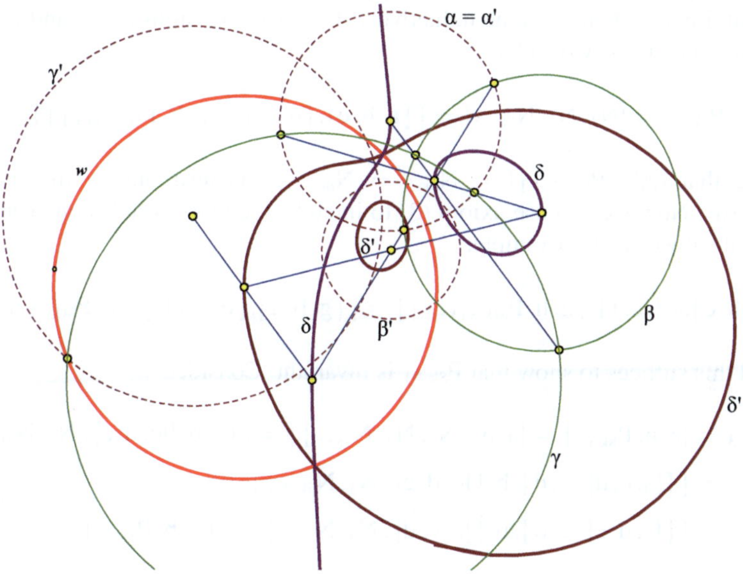

Fig. 4 Type IV self-inversive cubic and bicircular quartic inverse

7 Invariant Expressions

In this section we determine the invariant expressions in the TH variety. Let P_1 denote the TH word consisting of a single variable and let $N_3 = [P_1, P_1, P_1]$. All the expressions we define are monotone, so the P_1's in the definition of N_3 represents three different variables. The subscripts represent the length of the expression. Thus $P_9 = [N_3, N_3, N_3]$ represents the monotone TH word [[x,y,z], [u,v,w], [q,r,s]]. In general we define $N_{8k+3} = [P_1, P_1, P_{8k+1}]$ for k = 0,1,2, ... and $P_{8k+1} = [N_3, N_3, N_{8k-5}]$ for k = 1,2,3, ... Of course N_3 and P_9 are invariant by axioms (1) and (3). Consider:

$$\begin{aligned} N_{11} &= [P_1, P_1, P_9] = [a, b, [\,[c, d, e]\,, [f, g, h]\,, [j.k.l]\,]\,] \\ &= [\,[\,[a, c, d]\,, c, d\,], b, [\,[c, d, e]\,, [f, g, h]\,, [j.k.l]\,]\,] \\ &= [\,[\,[c, d, e]\,, c, d\,], b, [\,[a, c, d]\,, [f, g, h]\,, [j.k.l]\,]\,] \\ &= [e, b, [\,[c, d, a]\,, [f, g, h]\,, [j.k.l]\,]\,]\,. \end{aligned}$$

This is all that is needed to show that N_{11} is invariant. It suffices, in general, to show that two variables can be interchanged between each pair among the three principal subwords of a TH word, provided each principal subword is itself invariant. Let $k \geq 2$ and assume inductively that the expressions N_{8t+3} and P_{8t+1} are invariant when $t < k$. Consider:

$$P_{8k+1} = [N_3, N_3, N_{8k-5}] = [\,[\,a, b, c\,], [\,d, e, f\,], [\,g, h, P_{8(k-1)+1}\,]\,]\,.$$

Note that $[g,h, P_{8(k-1)+1}] = N_{8k-5} = N_{8(k-1)+3}$ is invariant by the inductive hypothesis. Now we can use axiom (3) to interchange two variables one each from principal subwords, for example,

$$[\,[\,a, b, c\,], [\,d, e, f\,], [\,g, h, P_{8(k-1)+1}\,]\,] = [\,[\,g, b, c\,], [\,d, e, f\,], [\,a, h, P_{8(k-1)+1}\,]\,],$$

And this suffices to show that P_{8k+1} is invariant. Consider:

$$\begin{aligned} N_{8k+3} &= [a, b, P_{8k+1}] = [a, b, [\,N_3, N_3, N_{8k-5}\,]\,] = [\,[\,a, b, [[c, d, e]\,, N_3, N_{8k-5}]\,] \\ &= [\,[\,[a, c, d]\,, c, d\,], b, [\,[c, d, e]\,, N_3, N_{8k-5}\,]\,] \\ &= [\,[\,[c, d, e]\,, c, d\,], b, [\,[a, c, d]\,, N_3, N_{8k-5}\,]\,] = [\,e, b, P_{8k+1}] \end{aligned}$$

We have interchanged the variable a with the variable e from the invariant word P_{8k+1}, which suffices to show N_{8k+3} is invariant. By induction we have proved

Theorem 8 *The expressions $N_{8k+3} = [P_1, P_1, P_{8k+1}]$ and $P_{8k+1} = [N_3, N_3, N_{8k-5}]$ are invariant in the TH variety for each positive integer k.* □

Theorem 9 *If q,r,s are positive integers each congruent to 1 (mod 8), then* $[P_q, P_r, P_s] = N_{q+r+s}$ *and if t,u,v are positive integers each congruent to 3 (mod 8), then* $[N_t, N_u, N_v] = P_{t+u+v}$.

Proof We can suppose q≥r≥9 with q,r ≡ 1 (mod 8). The theorem is clearly true if q=r=s=1 or if t = u = v = 3. It is also true by definition in case two from among {q,r,s} equal 1, or if two from among {t,u,v} equal 3. Assume inductively that $[P_x, P_y, P_z] = N_{x+y+z}$ and $[N_x, N_y, N_z] = P_{x+y+z}$ if $x+y+z < q+r+s$. We can suppose q≥ r ≥9. Consider:

$$\begin{aligned}[P_q, P_r, P_1] &= [P_q, P_r, y] = [P_q, [N_3, N_3, N_{r-6}], y] \\ &= [[[P_q, w, z], w, z], [[x, w, z], N_3, N_{r-6}], y] \\ &= [[[x, w, z], w, z], [[P_q, w, z], N_3, N_{r-6}], \\ &\quad y] = [x, y, [N_{q+2}, N_3, N_{r-6}]] = [P_1, P_1, P_{q+r-1}] = N_{q+r+1}.\end{aligned}$$

Note that in the last step we have used the inductive hypothesis since (q+2) + 3 + (r-6) = r+ q − 1 < q+r+s.

We have shown

$$[P_q, P_r, P_1] = N_{q+r+1} \text{ if } q, r \text{ are positive integers congruent to 1 (mod 8)}. \quad (16)$$

Now suppose s, q, r are positive integers, each congruent to 1 (mod 8), and with q≥r≥9. Consider:

$$\begin{aligned}&[P_q, P_r, P_s] = [P_q, [N_3, N_3, N_{r-6}], P_s] = [[[P_q, w, z], w, z], [[x, w, z], N_3, N_{r-6}], P_s] \\ &= [[[x, w, z], w, z], [[P_q, w, z], N_3, N_{r-6}], P_s] = [x, [N_{q+2}, N_3, N_{r-6}], P_s] \\ &= [P_{q+r-1}, P_s, P_1] = N_{q+r+s}.\end{aligned}$$

The last equality is due to (16). Finally suppose t,u,v are positive integers each congruent to 3 (mod 8). Consider:

$$\begin{aligned}[N_t, N_u, N_3] &= [[x, y, P_{t-2}], [a, b, P_{u-2}], N_3] = [[x, y, a], [b, P_{t-2}, P_{u-2}], N_3] \\ &= [N_3, N_3, N_{t+u-3}] = P_{t+u+3}.\end{aligned}$$

We have shown

$$[N_t, N_u, N_3] = P_{t+u+3} \text{ if } t, u \text{ are positive integers each congruent to 3 (mod 8)}. \quad (17)$$

Now suppose t,u,v are positive integers each congruent to 3 (mod 8). Consider:

$$[N_t, N_u, N_v] = [[x, y, P_{t-2}], [a, b, P_{u-2}], N_v] = [[x, y, a], [b, P_{t-2}, P_{u-2}], N_v]$$
$$= [N_v, N_{t+u-3}, N_3] = P_{t+u+v}.$$

The last equality is due to (17). □

Theorem 10 *Every invariant word in the TH variety is equivalent to P_{8k+1} or N_{8k+3}, $k = 0,1,2, \ldots$.*

Proof If a,b,c ∈ Z, and we define a ternary operation on Z by setting [a,b,c] = −a − b − c, we obtain a TH algebra we call the *integer model*. A TH word is *negative* if its valuation in the integer model is a sum of the (additive) inverses of the variables in the word. For example, N_3 = [a,b,c] is negative as is N_{11} = [x. y, [[a,b,c], [d,e,f], [g,h,j]]] = -x −y −a −b −c −d −e −f −g −h −j. A TH word is *positive* if its valuation in the integer model is a sum of its variables. For example, P_9 = [[a,b,c], [d,e,f], [g,h,j]] = a+b+c+d+e+f+g+h+j is positive. In fact P_{8k+1} is positive and N_{8k+3} is negative for every nonnegative integer k. Clearly, for a TH word to be invariant it must be either positive or negative. It is easily seen that P_1 and P_9 are the only positive words with length ≤ 9, and N_3 and N_{11} are the only negative words with length ≤ 11. Let Z = [U,V,W] be an invariant TH word with principal subwords U,V,W and with length k. Suppose inductively that every invariant TH word with length less than k is equivalent to P_{8t+1} or N_{8t+3} for some positive integer t. For Z to be invariant, it must be either positive or negative. If Z is positive, then each of its principal subwords U,V,W must be negative. Since these subwords have length less than k, we can invoke the inductive hypothesis to conclude that U = N_p , V = N_q, and W = N_r for positive integers p,q,r each congruent to 3 (mod 8). Then by Theorem 9, Z = [N_p , N_q , N_r] = P_{p+q+r}. Similarly, if Z is negative, then its principal subwords must each be positive, say U = P_s , V = P_t and W = P_u. Then Z = [P_s , P_t, P_u] = N_{s+t+u} by Theorem 9. □

8 Conclusion

Our main purpose is to present the following Conjecture involving the intersection of certain supercyclic curves with certain algebraic curves.

Conjecture Let γ be a bicircular quartic curve and let ξ be an arbitrary algebraic curve with odd degree n = 2t+1. Suppose γ and ξ meet in 4n real points A_1 , A_2 , ... , A_{4n} , then the valuation of the word E_{4n-1} = E_{8t+3} at A_1 , A_2 , ... , A_{4n-1} equals A_{4n}.

In Figs. 5 and 6 we present evidence for the Conjecture. In Fig. 5 the quartic γ meets the cubic L in the 12 points A,B,C,D,E,F,G,H,J,K,L. The invariant expression E_{11} is evaluated in the form

$$E_{11} = [E, H, [[A, I, K], [B, D, J], [C, F, L]]]. \tag{18}$$

Fig. 5 Bicircular quartic meets cubic

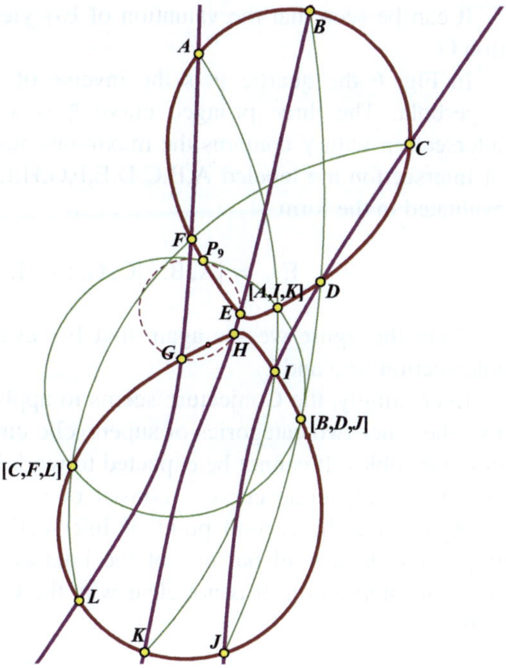

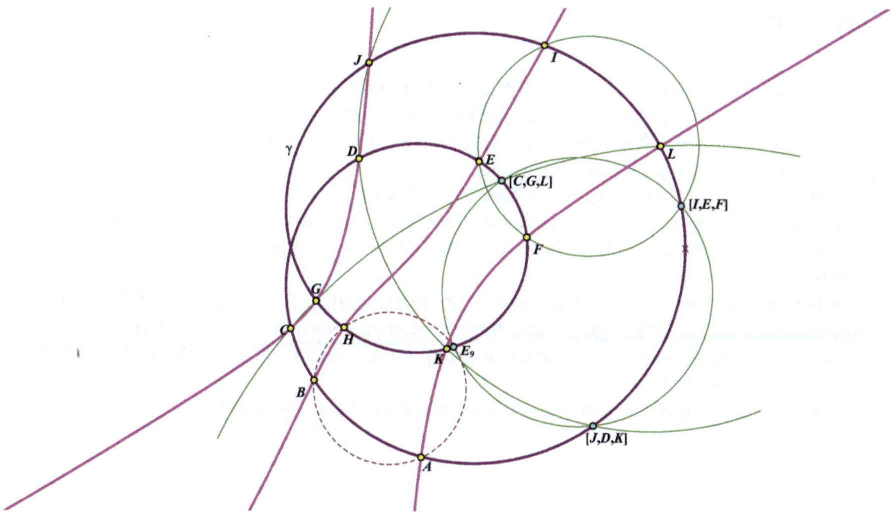

Fig. 6 Inverse of hyperbola meets cubic

It can be seen that the valuation of E_{11} yields the remaining point of intersection G.

In Fig. 6 the quartic γ is the inverse of a hyperbola by a point not on the hyperbola. The three pronged curve ξ is a typical cubic designed so that its intersection with γ contains the maximum number (12) of real points. The points of intersection are labeled A,B,C,D,E,F,G,H,I,J,K,L and the invariant word E_{11} is evaluated in the form

$$E_{11} = [A, B, [[C, G, L], [I, E, F], [J, D, K]]]. \qquad (19)$$

From the figure we see again that E_{11} evaluates to H, the remaining point of intersection of γ and ξ.

Interestingly, the Conjecture seems to apply only to bicircular quartics and not two the other two categories of supercyclic curves, namely noncircular conics and circular cubics. It cannot be expected to work in the case of conics if, for example, we let our algebraic curve ξ consist of an odd number of lines. Since we can freely change the second point of intersection of one of the lines, it is clearly impossible to use all but one of the intersection points to predict the remaining point of intersection. It is not clear why the Conjecture fails in case γ is a circular cubic.

References

1. Fletcher, R.: Group Circle Systems on Conics, New Frontiers of Multidisciplinary Research in STEAM-H, vol. 90. Springer International Publishing (2014)
2. Fletcher, R.: Geometric Ramifications of Invariant Expressions in the Binary Hypercommutative Variety, submitted for publication, Springer (2016)
3. Bix, R.: Conics and Cubics. Springer, New York (2006)
4. Pedoe, D.: Geometry, a Comprehensive Course. Cambridge University Press, Cambridge (1970)
5. Fletcher, R.: Geometric Triple Systems with Base Z and Z_n, submitted for publication, Springer (2016)
6. Fletcher, R.: Self-Inversive Cubic Curves. In: Mathematical Sciences with Multidisciplinary Applications, vol. 157, Springer Proceedings in Mathematics and Statistics (2016)
7. Robinson, D.: Self-inversive properties of group circle systems, Masters thesis, Virginia State University Library (2015)
8. Hilton, H.: Plane Algebraic Curves. Oxford University Press, Oxford (1920)

The Psychoneuroimmunological Influences of Recreational Marijuana

Larry Keen II, Arlener D. Turner, Deidre Pereira, Clive Callender, and Alfonso Campbell

Abstract *Background*: Marijuana is the most widely used illicit substance in the USA and self-reported use has remained steady over the past decade. Numerous publications examine the influence of marijuana use on various facets of human physiology including neurocognitive function, immune function, and illness symptom control, each discussing marijuana's influence in a narrow or compartmentalized fashion. However, there is a scant literature discussing the empirical and clinical implications of the intersection of these constructs. The primary objective of this review is to review and synthesize this disparate literature and propose future research directions. Thus, this review examines the literature that relates the influence of marijuana to: (1) neurocognitive function; (2) immune function; (3) treatment uses; and (4) propose future directions.

Methods: Clinical and nonclinical empirical studies were collected and utilized to inform this review. The authors used PubMed search engine as the primary mechanism used to identify relevant articles.

L. Keen II (✉)
Department of Psychology, Virginia State University, 1 Hayden Drive, PO Box 9079, Petersburg, VA, 23806, USA
e-mail: LKeen@vsu.edu

A.D. Turner
Department of Behavioral Sciences, Rush University Medical Center, 1653 W Congress Pkwy, Chicago, IL, USA
e-mail: Arlener_Turner@rush.edu

D. Pereira
Department of Clinical and Health Psychology, University of Florida, 1225 Center Drive, Gainesville, FL, USA
e-mail: DPereira@phhp.ufl.edu

C. Callender
College of Medicine, Howard University Hospital, 520 W St NW, Washington, DC, USA
e-mail: CCallender@howard.edu

A. Campbell
Department of Psychology, Howard University, 2400 Sixth St NW, Washington, DC, USA
e-mail: ACampbell@howard.edu

B. Toni (ed.), *New Trends and Advanced Methods in Interdisciplinary Mathematical Sciences*, STEAM-H: Science, Technology, Engineering, Agriculture, Mathematics & Health, DOI 10.1007/978-3-319-55612-3_5

Conclusion: Given the legalization efforts of recreational marijuana use, there is a need to discuss health and treatment effects of marijuana use from a more comprehensive, psychoneuroimmunological or biopsychosocial framework. We will discuss the need for an interdisciplinary research and future steps regarding the examination of marijuana use.

Keywords Marijuana • Immune function • Cytokine • Cognitive function • Imaging • Cannabinoids

1 Introduction

1.1 Foundation and Statistics

Marijuana is the most widely used illicit drug in the USA, with approximately 5.7 million people above the age of 12 reporting daily marijuana use over the past year [1]. Approximately 19 million individuals report using marijuana in the past 30 days [1] and nearly 9% of individuals who use marijuana meet the criteria for dependence or abuse [2]. Marijuana has been used for various recreational and medicinal reasons for over 200 years [3]. In fact, it is possible that the cannabis plant has been used for medical purposes since 2737 BC [4] and may be the oldest psychotropic drug [5].

Given the recent movement for the legalization of marijuana in the USA over the last decade, 23 states and the District of Columbia have legalized marijuana in some form, which portends an increase in marijuana users in the coming years. This forecast is based on the idea that marijuana use is directly related to changes in the risk of legal punishment [6]. With decreases in legal ramifications for marijuana use or possession, more individuals may report using. As the routes of administration expand from smoking marijuana to utilizing edibles and vaporizing butane hash oil, marijuana use is becoming widely more acceptable than in previous decades. Additionally, the National Institute on Drug Abuse has suggested that given the evolution of marijuana growth, marijuana may actually be increasing in Delta 9-tetrahydrocannabinol (THC), the main psychoactive compound in marijuana [2]. This idea has not been well examined in the epidemiological or psychopharmacological literature, as many studies are focusing on any use of marijuana and not examining the specified types used recreationally. This suggests that individuals who are newly initiated into marijuana use may be at higher risk for any negative effects of the drug [2].

In general, marijuana is seen as a "gateway" drug, leading to subjectively more harmful drugs [7]. However, this position has become more controversial, given competing theories [8, 9]. Empirically, there have been many studies examining marijuana's influence on various physiological states [10], pathological conditions [11], and neurocognitive outcomes [12]. However, to date, researchers have not examined the intersection among these constructs and their subsequent potential influences on health outcomes. This notion is substantiated by a PubMed search using keywords representing each field, such as "marijuana," "immune function," and "cognition" (and their variants), yielding less than eight results. None of which examine the intersection of these constructs relative to epidemiological constructs or health outcomes. This is imperative, as fields become more interdisciplinary and treatment or care programs are looking to generate more comprehensive forms of care.

1.2 Theoretical Framework

Overall, there is a large amount of research that examines marijuana use with decision-making processes [13], a cellular literature that examines marijuana use and disease susceptibility [14], but the authors are unaware of any current studies that incorporate each of these components in the same statistical or theoretical model. Historically, previous research have attempted to provide interdisciplinary models, such as the psychoimmunological model, which calls for an integration of psychological factors in addition to immune factors to elucidate the mechanisms involved in disease conditions [15]. There are models that include the physiological systems needed for a comprehensive view of marijuana's influence on the human body, such as Maier and Watkins [16] posited bidirectional pathway that connects the immune system to the brain. The authors suggest that through this bidirectional pathway, much of the complex interplay between psychological phenomena and adaptive immune response can be examined. Unfortunately, these authors did not include an environmental or behavioral component that could account for substance use. There are conceptual models that would fit this paradigm, such as the biopsychosocial model, which attempts to integrate various mechanisms in order to mitigate disease or identify primary vectors [17]. The current review looks to present different components in the empirical literature in hopes of substantiating a psychoneuroimmunological model of recreational marijuana use and infectious disease susceptibility.

2 Marijuana and Neurocognitive Function

2.1 Marijuana Constituents and Function in CB1 Saturated Cortical Regions

Marijuana contains over 60 different cannabinoids, but only two constituents are currently the main focus of research: THC and cannabidiol (CBD). Previous research suggests that this is due to the role of THC as the main psychoactive constituent and CBD's role as the most abundant naturally occurring constituent [18, 19]. While CBD does not appear to impair cognition [20, 21], THC is thought to be the primary constituent that influences neurocognitive performance [22, 23]. THC binds to the Cannabinoid-1 (CB1) receptors, which are distributed in various areas of the central nervous system. Generally, CB1 receptors are found presynaptically, in cortical areas that are responsible for the behavioral and pharmacological effects of marijuana use. These areas include the dorsolateral prefrontal and orbitofrontal regions, motor areas, hippocampus, amygdala, the striatum, and the cerebellum [19, 24]. The preponderance of CB1 receptors in the central nervous system suggests that the introduction of exogenous cannabinoids in the human body would interfere with normal neurocognitive processes governed by these brain regions when levels of this substance exceeded values required for normal functioning of the cannabinergic system [22, 25]. For example, due to the high densities of the CB1 in the cerebral cortex, the hippocampus, and the basal ganglia, previous research suggests that cannabinoids are involved in attentional and memory processes, as well as in executive functions as these areas are the neural substrates for these specific neurocognitive domains [12, 26–28]. Much of the early research into the neurocognitive consequences of marijuana use focused on acute intoxication with the inference that it leads to impairments in memory and attention (see [29, 30] *for review*). However, later examinations, as with this review, focus more on the non-acute effects, which are less clear cut but show more diffuse deficits.

2.2 *Marijuana and Neuroimaging Studies Findings*

Neuroimaging studies have begun to examine the neural substrates underlying the neurocognitive consequences of marijuana use, with varying results. Specifically, resting state examinations using positron emission tomography (PET) during acute intoxication have shown an increase in global metabolism and regional metabolic increases in the orbitofrontal cortex, prefrontal cortex (PFC), and basal ganglia in long-term marijuana users which was not seen in nonusers [31–35]. Moreover, while there were minimal to no differences between groups on task performance, PET and functional magnetic resonance imaging (fMRI) studies involving neurocognitive challenge have reported differential activation during executive functioning and memory tasks between recently abstinent marijuana users and nondrug using controls. For example, PET examinations have indicated that chronic marijuana users show decreased blood flow in the PFC, increased blood flow in the cerebellum, and altered lateralization of the hippocampus for memory tasks when compared to nondrug using controls [36]. Previous research utilizing PET also suggests hypoactivity in the anterior cingulate and prefrontal cortices, and hyperactivity in the hippocampus during executive functioning tasks [37]. Interestingly, neither Block et al. [36] nor Eldreth et al. [37] found differences in the actual performance on the neurocognitive tasks, despite the differential activation in different cortical regions. Examinations of brain activation in recently abstinent chronic marijuana users using fMRI indicate increased activation in the PFC, anterior cingulate, and basal ganglia when compared to controls [38–40]. This line of research posits that both resting

Table 1 Marijuana's influence on cortical regions and structures

Type of study	Duration	Outcome	Source
Resting state PET	Acute	Increases in global brain glucose metabolism with marked increases in the specific regions of the orbitofrontal cortex, prefrontal cortex, and basal ganglia	[31–35]
Resting state PET	Acute	Global increases in cerebral blood flow with marked increases in the regions of the prefrontal cortex, anterior cingulate, and the insular cortex	[32–34]
PET with memory task	Non-acute	Decreased cerebral blood flow in the prefrontal cortex, increased cerebral blood flow in the cerebellum, and altered lateralization of the hippocampus	[36]
PET with executive function task	Non-acute	Hypoactivity in the anterior cingulate and prefrontal cortices, and hyperactivity in the hippocampus	[37]
fMRI with memory task	Non-acute	Increased global brain activation with marked increases in the prefrontal cortex, anterior cingulate, and basal ganglia	[38, 39]
fMRI with executive function task	Non-acute	Increased activation in the prefrontal, insular, and parietal cortices	[40]

PET positron emission tomography scan, *fMRI* functional magnetic resonance imaging

state and activated neuroimaging paradigms indicate changes in metabolism and activation in cortical areas responsible for neurocognitive processes in marijuana users [12]. Similarly, heavy, long-term recently abstinent marijuana users show more widespread brain activation than controls when attempting a spatial working memory task [39]. These results suggest that even in the absence of performance differences, marijuana users display persistent metabolic and activation alterations in the brain. The results of neuroimaging research that demonstrate the effects of marijuana on the brain are presented in Table 1.

Previous research suggests that the enhanced brain activity demonstrated by marijuana users indicates a need for increased neural effort or a change in strategy to meet task demands and maintain good task performance [41]. This would account for the lack of differences reported between marijuana users and controls to be subtle. Neurocognitive examinations of marijuana use's consequences posit deficits in attention [42–45], memory [21, 39, 46], processing speed [43], verbal learning [43, 47–52], and executive function [46, 49, 53–59].

2.3 Marijuana, Neurocognition, and Study Discrepancies

Most researchers agree that the neurocognitive dysfunction related to marijuana use extends beyond the period of intoxication, but how far beyond that period has less of a consensus. Some findings suggest that the negative impact of marijuana use on cognition is temporary and reversible [60–62], while others indicate that impairments are still seen 28 days to 3 months after last marijuana use and that deficits may increase as years of regular marijuana use increase [37, 45, 63]. These inconsistencies may stem from differences in methodology. For example, it has been indicated that examinations of non-acute effects of marijuana on cognition suffer from possible methodological issues surrounding the parameters of marijuana use. The specific parameters mentioned were frequency of marijuana use, duration of marijuana use, and age of onset of marijuana use [64]. Each of these parameters is associated with neurocognitive dysfunction when examined separately; therefore, examining a population that is heterogeneous on any of these parameters could confound results. However, the more likely culprit in these inconsistencies may be the subtlety of deficits shown by marijuana users, especially given that neuroimaging research indicates that marijuana users show metabolic and activation differences even in the absence of performance differences. Moreover, adult marijuana users who initiated use before 16 years of age have been associated with increased cortical activation possibly reflecting more effort due to suboptimal cortical efficiency during neurocognitive challenge [65]. These results coupled with findings that memory deficits are more pronounced in younger marijuana samples suggest that the maturing brains of teenagers may be more vulnerable to the harmful effects of marijuana use on the brain and show lasting effects [22]. Overall, the considerable variability among the findings presented in the literature regarding the long-term effects of marijuana use on cognition sets a solid foundation to begin exploring the potential moderators or more proximal factors in this relationship.

3 Marijuana and the Immune System

3.1 Marijuana and Cannabinoids

Numerous researchers have explored the relationship between marijuana (or its' constituents) and immune function over the past 35 years. Most of the literature is based on experimental research in laboratory settings that involve in human and murine vitro studies (for review, see [14, 66]). Previous research points to the inhibitory influence of cannabinoids anti-inflammatory or modulatory properties on the cytokine network. The endocannabinoid system has been suggested to be associated with the modulation of the cytokine network. However, the literature also points to the exogenous cannabinoids as the potential immunosuppressant [67]. Very few studies have examined the acute effects for marijuana smoking on immune cells, but there is also a dearth of literature examining long-term effects of marijuana use on immune system cells or function [68]. This is imperative to examine, given that marijuana is used recreationally by many with age of initiation of use as early as 12 years old [1]. With so many individuals using marijuana recreationally, even inconsistent use or sporadic use, researchers should explore marijuana use in the context of subsequent immunological effects. Differences in patterns of use

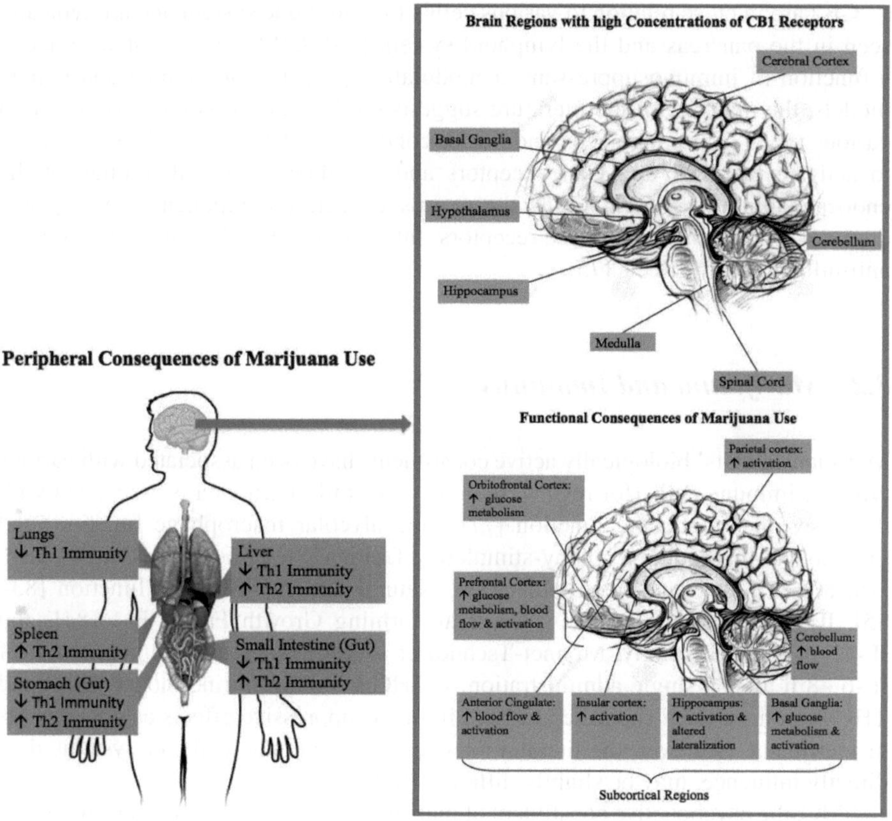

Fig. 1 The influence of marijuana on the peripheral and central nervous systems. This figure details cortical immune functions in each region or area

may predict future infections or conditions comprised of proinflammatory cytokine dysfunction. Overall, this body of research has focused on the marijuana and immune system relationship at the cellular level.

Immune cells express both CB1 (primarily in the central nervous system) and CB2 (primarily in the peripheral nervous system) receptors, suggesting a physiological connection between the endocannabinoid system and the immune system (see Fig. 1). This endocannabinoid system plays a pivotal role in the mitigation of functional immunity do to its direct ties to the hypothalamic pituitary axis and influence on the proliferation of natural killer cells and other adaptive immune cells [69, 70]. However, the exact role or connection between these two systems is still controversial [71]. In exploring the potential physiological overlap between these two systems, researchers have found that the cannabinoid system influences the production and proliferation of some immune cells, specifically cytokines. Comparison among many of the published findings is problematic, given the different types of experimental methodologies and quantification of outcomes [72].

CB2 are in close relation to various cells of the immune system and are typically seen in the pancreas and the lymphoid system [73]. CB2 are suggested to act as a function of immunosuppressant or modulation [74]. In both animal and human models, the majority of the literature suggests CB2 as anti-inflammatory based on various types of immune cells and health conditions [19]. Though THC is known to activate the CB1 and CB2 receptors and use them to signal throughout the endocannabinoid network, CBD, the non-psychoactive component in marijuana, does not typically activate these receptors. Interestingly, CBD is suggested to have anti-inflammatory effects [75].

3.2 *Marijuana and Immunity*

Marijuana and its' biologically active constituents have been associated with various types of immune cells (for review, see [76]). Overall, marijuana is associated with decreases in lymphocyte function [77], lung alveolar macrophage function, and granulocyte-macrophage colony-stimulating factor [78], TNF-alpha [78, 79], IL-6 [78, 80, 81], IL-8 and neutrophils [82], Natural Killer (NK) cells function [83–85], IL-2 [83, 84], and IL-10 [83], Transforming Growth Factor B-1 [84], and IL-17 [81]. Interestingly, Monnet-Tschudi et al. [86] reported an upregulation of IL-6 48 h after a single administration of THC within a murine model. CBD and THC both have been suggested to have immunosuppressive effects as constituents of marijuana. However, the literature suggests that immune cells, or systems they directly influence, may be slightly different [87].

THC, the psychoactive constituent of marijuana, is associated with various agents within the immune system (for review, see [88]). THC is associated with suppressed Natural Killer (NK) cells function [89], CD40 ligand [90], IL-8, TNF, IFN [91], T cells, IFN, and mRNA related to Th1 cytokine production [92], IL-12 [93, 94], TNF [95, 96], T cells, and mRNA related to Th1 cytokine production [92]. In murine models, researchers have reported an inverse relationship between THC and IL-2 [97], natural killer cells [98], increases in Th2 antibodies and decreases in Th1 antibodies [99], NK cells, IFN, IL-2, T and B lymphocytes [100], IL-12 and IFN [101, 102], IFN [103, 104], IL-6 [105], and Th1 related dendritic cells [106].

Verhoeckx et al. [107] discussed the potential damage done by THC through marijuana use and suggested exploring THC-acid as it may be metabolized via a different pathway. THC-acid is a chemical, when burned, converts to THC. These researchers reported that THC-acid not only inhibited TNF levels but did so for a longer period of time than THC.

THC was not associated with IL-2 function [77]. THC and CBD have also been shown to increase the expression of IL-10, an anti-inflammatory cytokine [81]. Moreover, THC was reported to increase mRNA levels for Th2 cytokine levels [92]. In murine models, THC is reported to upregulate IL-4 and IL-10 [108, 109], IL-4 [101], IL-10, and TGF-β [104]. See Table 2 for a listing of the cells influenced by marijuana.

Table 2 Marijuana's influence on Th1 and Th2 immune cells

Cellular structure	Influence	Source
Lymphocyte	Decrease	[77]
Alveolar macrophage (lung)	Decrease	[78]
Granulocyte-macrophage colony-stimulating factor	Decrease	[78]
CRP	Decrease	[110, 111]
CRP	Increase	[112]
TNF-alpha	Decrease	[78, 80, 91, 95, 96]
IL-6	Decrease	[78, 80, 81, 105]
IL-8	Decrease	[82, 91]
Neutrophils	Decrease	[82]
NK cells	Decrease	[85]
NK cells	Decrease	[83, 84]
NK cells	Decrease	[89]
IL-2	Decrease	[83, 84]
IL-10	Decrease	[83]
TGF B-1	Decrease	[84]
IL-17	Decrease	[81]
CD40 ligand	Decrease	[90]
IFN	Decrease	[91, 92]
T cells mRNA	Decrease	[92]
IL-12	Decrease	[93, 94]
IL-2	Increase	[97, 100]
NK cells	Decrease	[98]
Th1 antibodies	Decrease	[99]
Th2 antibodies	Increase	[99]
NK cells, IFN, T & B lymphocytes	Decrease	[100]
IL-12, IFN	Decrease	[101, 102]
IFN	Decrease	[103, 104]
Th1 dendritic cells	Decrease	[106]
IL-10	Increase	[81]
mRNA (Th2)	Increase	[92]
IL-4, IL-10	Increase	[108, 109]
IL-4	Increase	[101]
IL-10 and TGF-B	Increase	[104]

CRP C-reactive protein, *TNF* tumor necrosis factor, *IL-6* interleukin 6, *IL-8* interleukin-8, *NK Cells* natural killer cells, *IL-2* interleukin 2, *IL-10* interleukin 10, *IL-17* interleukin 17, *CD40 ligand* cluster of differentiation 40, *IFN* interferon, *mRNA* messenger ribonucleic acid, *IL-2* interleukin 2, *IL-4* interleukin 4, *IL-12* interleukin 12, *TGF-B* transforming growth factor

To date, very few published studies have explored this relationship at the epidemiological level. Among these studies, self-reported lifetime use of marijuana use was found to be associated with lower levels of IL-6 [80] and TNF [79]. In addition, there are very few epidemiological studies that have explored the relationship between marijuana use and inflammatory markers, specifically [79, 80,

110–112]. Of these, Alshaarawy and Anthony [110] have the largest sample, utilizing a population-based sample. These researchers reported that recent marijuana use was associated with lower C-Reactive Protein (CRP) levels in a nonclinical, predominately White sample. In contrast to Alshaarawy and Anthony [110], Costello et al. [112] reported a positive association between CRP and marijuana use in the past 3 months among children aged 9–21 years in a longitudinal examination. However, these researchers also reported that this association did not survive correction for covariates. Additionally, Rajavashisth et al. [113] found that higher levels of CRP were associated with non-marijuana users in comparison to former and frequent users in a sample of predominately White, middle-aged population-based study. In contrast to these inverse associations between marijuana use and inflammatory markers, Muniyappa et al. [111] reported that chronic marijuana users have higher CRP levels than their non-marijuana using counterparts in a small sample matched on age, sex, and BMI. Taken together, these studies begin a foundation for epidemiological exploration of marijuana and immune function. This line of research will allow for the examination of incidence of infectious disease conditions associated with various patterns of marijuana use. Potential findings could inform clinical trials and subsequent prevention or intervention programs that are based on biobehavioral processes and not just pure epidemiological constructs.

3.3 Marijuana Use and Infection Susceptibility

Marijuana is commonly used in combination with illicit substances. In accordance with this, individuals who inject drugs may leave themselves more susceptible to various diseases due to the immunosuppressive influence of marijuana in conjunction with opiates or stimulants [114]. Specifically, some researchers suggest that marijuana use exacerbates vascular and proinflammatory symptomatology in immunity based conditions, such as HIV [91]. Moreover, marijuana use (either medical or recreational) has been reported to be associated with the development of invasive pulmonary aspergillosis [115–117]. This finding suggests how the immunosuppression produced by marijuana use may function in individuals who are already or at risk for immune dysfunction. Cabral and Pettit [118] discuss this issue, suggesting that cannabinoids increase susceptibility to infectious diseases. Moreover, Huemer et al. [119] reported that the use of marijuana a day before infection augmented the severity of the subsequent cowpox infection in a group of mice. Though these researchers focus on animal based models, this sets a strong foundation for future research to begin exploring this research question in humans, even at the epidemiological or global level.

4 Marijuana Use and Symptom Palliation

In 1964, THC was isolated and utilized to control nausea during chemotherapy [3]. After the discovery of the endocannabinoid system [120], clinicians and researchers began to elucidate the therapeutic nature of marijuana and its constituents [72]. Moreover, with the budding exploration of marijuana as a therapeutic agent, few regions of the USA have begun employing medical marijuana programs to distribute marijuana for medicinal purposes. Medicinal marijuana users who are approved for these programs: (1) are individuals with more severe symptomatology or diseases; (2) may be individuals with a history of substance use; (3) may potentially develop dependence or have negative health outcomes based on the medicinal marijuana use [121]. In addition to the decreases in physiological symptomatology, improving the quality of life for patients with severe or chronic conditions may also play a critical role in the rise of medicinal marijuana usage. In the alleviation of pain, palliative care looks to improve the quality of life for the patient as much as possible. This raises a very interesting discussion, as there are two clear sides to the conversation regarding medicinal marijuana use. One side cites various side effects and the potential harmful effects of marijuana use in any capacity. While the other speaks to the quality of life for the truly suffering patients and weight the alleviation of disease symptomatology more than the short-term effects of marijuana use. Philipsen et al. [122] coin this debate Nonmalfeasance ("First, do no harm") versus Beneficence ("Do all the good you can").

Overall, the literature exploring the therapeutic potential for marijuana and its constituents is growing. Such conditions include but are not limited to neurodegeneration, irritable bowel syndrome, brain injury, arthritis, and vascular inflammation (for review, see [123, 124]). The underlying theme for the relationship between marijuana's therapeutic utility and these conditions is the inflammatory function proximal to each condition.

Endocannabinoids control various central nervous system and peripheral nervous system functions, such as movement, memory, cognitive function, neuroendocrine secretion, and immune system modulation [72]. Endocannabinoids also have the ability to regulate neurotransmission, which may be the key in neurological diseases such as multiple sclerosis or Parkinson's disease where you have overactive neurotransmission. Previous research posits that marijuana does not present harmful or negative outcomes, as long as it is used in moderation [125]. Much of the focus has been on marijuana's mitigating influence on various disease symptoms (nausea, appetite, etc.); however, there is a growing literature examining the antitumoral function of THC within cancer [126].

Given the anti-inflammatory effects of THC, various medical fields have begun exploring the treatment usages. For example, Gaffal et al. [103] reported that topically applied THC may attenuate inflammation related to allergy in an experimental murine sample in the field of dermatology. Previous research has reported that marijuana use may improve Hepatitis C virus treatment adherence and reduce the likelihood of virologic relapse after treatments [125]. Marijuana has been reported to decrease disease activity and even reduce the need for prescribed medication in individuals with Crohn's disease [127]. Further, within the Naftali et al. [127] examination of marijuana's influence on Crohn's disease, researchers posited that it was not only the anti-inflammatory effects of marijuana that aided in disease symptoms amelioration, but the gastrointestinal effects (i.e., reduction in diarrhea) as well. Interestingly, a recent study reveals that an ultralow dose of THC may reduce myocardial damage if administered before or up to 24 h after induced myocardial infarction in mice [128]. This finding suggests a potential cardioprotective component of marijuana use, seemingly in dose–response relationship.

Utilizing smoking marijuana as the primary route of administration may deter various health professionals from promoting its health benefits. In the addiction literature, one commonly sees the use of marijuana paired with other illicit drugs or legal substances such as alcohol. However, smoking may be the preferred route of administration due to the efficiency in receiving the desired effect. Specifically, smoking introduces the exogenous cannabinoids to the bloodstream faster, yielding the sought anti-inflammatory effect. Nevertheless, some researchers suggest that consuming marijuana orally may have the same potent influence as smoking [127]. This may be critical if the goal is to alleviate pain, nausea, or any other symptoms in the gut.

5 Conclusions

Taken together, this literature presents a pattern in the effects of marijuana on immune function. Specifically, marijuana seems to suppress the Th1 branch of the immune system, while potentially augmenting various agents within the Th2 branch of the immune system. Given the abundance of CB1 receptors in the central nervous system and majority of the CB2 receptors are in the periphery, marijuana seems to have a concomitant effect on the human physiology. Marijuana also influences the central nervous system; cerebral and subcortical structures involved in memory, decision-making, processing speed, and autonomic regulation are all affected by marijuana use. The potential bidirectional influence of marijuana can be traced throughout the nervous system with the principles outlined in psychoneuroimmunology [129]. Specifically, the alteration of cortical activation relating to the HPA axis and the autonomic nervous system in addition to the various neurocognitive effects portrays a comprehensive portrayal of marijuana's influence on human physiology. Identifying the psychoneuroimmunological influences of marijuana use will aid in examining and evaluating both deleterious and health promoting effects to the human physiology and behavior. However, the complex interplay between the peripheral and central nervous systems altered by marijuana use may vary based on an individual's health status.

The number of individuals using marijuana will continue to increase with various states choosing to legalize the recreational use. With legalization, comes the use of edibles and other variation in routes of administration. Though these items are safe for consumption, the dosage of THC and other chemicals must be regulated and studied further. As individuals are gaining freedom to access/use marijuana legally in various forms, policy makers and individuals involved in regulating marijuana in the USA may need to examine if there are increases in polysubstance uses. Marijuana use is associated with the constructs reported in this review; however, the literature does not discuss how polysubstance users immune, neurocognitive, and disease susceptibility vary by substance grouping. Recreational use of marijuana has varied effects on the human anatomy. Identifying different mechanisms from various fields of research may elucidate the nexus of influence that surrounds recreational marijuana use. This identification may work through models such as the biopsychosocial model, which incorporates physiological, psychological, and social factors that may be applied to the examination of marijuana and potential disease effects or development. At a cellular level, marijuana has both immunosuppressive and anti-inflammatory effects. However, negative effects of marijuana use include risk of opportunistic infections among individuals with impaired cell-mediated immunity as well as impaired neurocognitive functioning across recreational and chronic users. Although there is a growing body of literature highlighting these cellular, clinical, and population-based effects, there is an extremely scant amount of research, to our knowledge, that examines the psychoneuroimmunological effects of marijuana use.

There is a large literature on the influence of marijuana uniquely on immunosuppression at the cellular level, neuropsychological processes, and potential treatment and disease utilities. However, much of this literature does not empirically intersect. Future research should begin to discuss marijuana use from a psychoneuroimmunological perspective in order to provide comprehensive understanding of data and performance from various samples. Moreover, negative data reporting is a phenomenon that has limited the progression of the literature. Researchers must attempt to report null findings as well as significant findings in order to provide a more complete picture of the influence of recreational marijuana use. Having a comprehensive literature that intersects multiple fields would allow for a clear presentation of marijuana's effects, further informing policy and legalization efforts.

Role of Funding Source This project was supported by grant #P20 MD000512 from National Center on Minority Health and Health Disparities.

Conflict of Interest No conflict declared.

Contributors Drs. Keen II, Turner, Pereira, Callender, and Campbell contributed to the literature search and writing/editing of this article and have approved the final article.

References

1. Substance Abuse and Mental Health Services Administration: Substance Abuse and Mental Health Services Administration, Results from the 2013 National Survey on Drug Use and Health: Summary of National Findings, NSDUH Series H-48, HHS Publication No. (SMA) 14-4863. Rockville, MD (2014)
2. National Institute on Drug Abuse: Marijuana Retrieved from http://www.drugabuse.gov/publications/drugfacts/marijuana on August 6, 2015
3. Vincent, B.J., McQuiston, D.J., Einhorn, L.H., Nagy, C.M., Brames, M.J.: Review of cannabinoids and their antiemetic effectiveness. Drugs. **25**, 52–62 (1983)
4. Li, H.L.: An archaeological and historical account of cannabis in China. Econ. Bot. **28**(4), 437–448 (1973)
5. Ben Amar, M.: Cannabinoids in medicine: a review of their therapeutic potential. J. Ethnopharmacol. **105**(1), 1–25 (2006)
6. Pacula, R.L.: Marijuana use and policy: What we know and have yet to learn. NBER Reporter Online, (Winter 2004/05), 22–24 (2004)
7. Hall, W., Degenhardt, L.: Adverse health effects of non-medical cannabis use. Lancet. **374**(9698), L1383–L1391 (2009)
8. Hall, W.D., Lynskey, M.: Is cannabis a gateway drug? Testing hypotheses about the relationship between cannabis use and the use of other illicit drugs. Drug Alcohol Rev. **24**(1), 39–48 (2005)
9. Morral, A.R., McCaffrey, D.F., Paddock, S.M.: Reassessing the marijuana gateway effect. Addiction. **97**(12), 1493–1504 (2002)
10. Raphael, B., Wooding, S., Stevens, G., Connor, J.: Comorbidity: cannabis and complexity. J. Psychiatr. Pract. **11**(3), 161–176 (2005)
11. Tan, W.C., Lo, C., Jong, A., Xing, L., FitzGerald, M.J., Vollmer, W.M., Buist, S.A., Sin, D.D., Vancouver Burden of Obstructive Lung Disease (BOLD) Research Group: Marijuana and chronic obstructive lung disease: a population-based study. Can. Med. Assoc. J. **180**(8), 814–820 (2009)
12. Lundvquist, T.: Cognitive consequences of cannabis use: comparison with abuse of stimulants and heroin with regard to attention, memory and executive functions. Pharmacol. Biochem. Behav. **81**, 319–330 (2005)
13. Cousijn, J., Wiers, R.W., Ridderinkhof, K.R., Brink, W., Veltman, D.J., Porrino, L.J., Goudriaan, A.E.: Individual differences in decision making and reward processing predict changes in cannabis use: a prospective functional magnetic resonance imaging study. Addict. Biol. **18**(6), 1013–1023 (2013)
14. Klein, T.W., Friedman, H., Specter, S.: Marijuana, immunity and infection. J. Neuroimmunol. **83**(1), 102–115 (1998)
15. Jemmott, J.B., Locke, S.E.: Psychosocial factors, immunologic mediation, and human susceptibility to infectious diseases: how much do we know? Psychol. Bull. **95**(1), 78 (1984)
16. Maier, S.F., Watkins, L.R.: Cytokines for psychologists: implications of bidirectional immune-to-brain communication for understanding behavior, mood, and cognition. Psychol. Rev. **105**(1), 83 (1998)
17. Suls, J., Rothman, A.: Evolution of the biopsychosocial model: prospects and challenges for health psychology. Health Psychol. **23**(2), 119 (2004)
18. ElSohly, M.A.: Marijuana and The Cannabinoids. Humana Press, Totowa, New Jersey (2007)
19. Rom, S., Persidsky, Y.: Cannabinoid receptor 2: potential role in immunomodulation and neuroinflammation. J. NeuroImmune Pharmacol. **8**(3), 608–620 (2013)
20. Bhattacharyya, S., Fusar-Poli, P., Borgwardt, S., et al.: Modulation of mediotemporal and ventrostriatal function in humans by Delta9-tetrahydrocannabinol: a neural basis for the effects of Cannabis sativa on learning and psychosis. Arch. Gen. Psychiatry. **66**(4), 442–451 (2009)

21. Schoeler, T., Bhattacharyya, S.: The effect of cannabis use on memory function: an update. Subst. abuse Rehabil. **4**, 11–27 (2013)
22. Crane, N.A., Schuster, R.M., Fusar-Poli, P., Gonzalez, R.: Effects of cannabis on neurocognitive functioning: recent advances, neurodevelopmental influences, and sex differences. Neuropsychol. Rev. **23**(2), 117–137 (2013)
23. Fernández-Serrano, M.J., Pérez-García, M., Verdejo-García, A.: What are the specific vs. generalized effects of drugs of abuse on neuropsychological performance? Neurosci. Biobehav. Rev. **35**(3), 377–406 (2011)
24. Wright, M.J., Vandewater, S.A., Taffe, M.A.: Cannabidiol attenuates deficits of visuospatial associative memory induced by Δ9tetrahydrocannabinol. Br. J. Pharmacol. **170**(7), 1365–1373 (2013)
25. Thames, A.D., Arbid, N., Sayegh, P.: Cannabis use and neurocognitive functioning in a non-clinical sample of users. Addict. Behav. **39**(5), 994–999 (2014)
26. Behan, B., Connolly, C.G., Datwani, S., Doucet, M., Ivanovic, J., Morioka, R., Stone, A., Watts, R., Smyth, B., Garavan, H.: Response inhibition and elevated parietal-cerrebellar correlations in chronic adolescent cannabis users. Neuropharmacology. **84**, 131–137 (2014)
27. Herkenham, M., Lynn, A.B., Little, M.D., Johnson, M.R., Melvin, L.S., de Costa, B.R., Rice, K.C.: Cannabinoid receptor localization in brain. Proc. Natl. Acad. Sci. U. S. A. **87**(5), 1932–1936 (1990)
28. Vik, P.W., Cellucci, T., Jarchow, A., Hedt, J.: Cognitive impairment in substance abuse. Psychiatr. Clin. N. Am. **27**(1), 97–109 (2004)
29. Gonzalez, R.: Acute and non-acute effects of cannabis on brain functioning and neuropsychological performance. Neuropsychol. Rev. **17**, 347–361 (2007)
30. Ranganathan, M., D'Souza, D.C.: The acute effects of cannabinoids on memory in humans: a review. Psychopharmacology. **188**, 425–444 (2006)
31. Mathew, R.J., Wilson, W.H.: Substance abuse and cerebral blood flow. Am. J. Psychiatry. **148**, 292–305 (1991)
32. Mathew, R.J., Wilson, W.H., Coleman, R.E., Turkington, T.G., DeGrado, T.R.: Marijuana intoxication and brain activation in marijuana smokers. Life Sci. **60**(23), 2075–2089 (1997)
33. Mathew, R.J., Wilson, W.H., Chiu, N.Y., Turkington, T.G., Coleman, R.E.: Regional cerebral blood flow and depersonalization after tetrahydrocannabinol adminstration. Acta Psychiatr. Scand. **100**, 67–75 (1999a)
34. Mathew, R.J., Wilson, W.H., Turkington, T.G., Coleman, R.E.: Cerebellar activity and disturbed time sense after THC. Brain Res. **797**, 183–189 (1999b)
35. Volkow, N.D., Gillespie, H., Mullani, N., Tancredi, L., Grant, C., Valentine, A., Holister, L.: Brain glucose metabolism in chronic marijuana users at baseline and during intoxication. Psychiatr. Res. Neuroimaging. **67**, 29–38 (1996)
36. Block, R.I., O'Leary, D.S., Hichwa, R.D., Augustinack, J.C., Boles Ponto, L.L., Ghoneim, M.M., Arndt, S., Hurtig, R.R., Watkins, G.L., Hall, J.A., Nathan, P.E.: Effects of frequent marijuana use on memory-related regional cerebral blood flow. Pharmacol. Biochem. Behav. **72**, 237–250 (2002)
37. Eldreth, D.A., Matochik, J.A., Cadet, J.L., Bolla, K.I.: Abnormal brain activity in prefrontal brain regions in abstinent marijuana users. NeuroImage. **23**(3), 914–920 (2004)
38. Jager, G., Block, R.I., Luijten, M., Ramsey, N.F.: Cannabis use and memory brain function in adolescent boys: a cross-sectional multicenter functional magnetic resonance imaging study. J. Am. Acad. Child Adolesc. Psychiatry. **49**, 561–572 (2010)
39. Kanayama, G., Rogowska, J., Pope, H.G., Gruber, S.A., Yurgelynn-Todd, D.A.: Spatial working memory in heavy cannabis users: a functional magnetic resonance imaging study. Psychopharmacology. **176**, 239–247 (2004)
40. Tapert, S.F., Schweinsburg, A.D., Drummond, S.P., Paulus, M.P., Brown, S.A., Yang, T.T., Frank, L.R.: Functional MRI of inhibitory processing in abstinent adolescent marijuana users. Psychopharmacology. **194**, 173–183 (2007)

41. Bossong, M.G., Jager, G., Bhattacharyya, S., Allen, P.: Acute and non-acute effects of Cannabis on human memory function: a critical review of neuroimaging studies. Curr. Pharm. Des. **20**(13), 2114–2125 (2014)
42. Abdullaez, Y., Posner, M.I., Nunnally, R., Dishion, T.J.: Functional MRI evidence for inefficient attentional control in adolescent chronic cannabis abuse. Behav. Brain Res. **215**, 45–47 (2010)
43. Medina, K.L., Schweinsburg, A.D., Cohen-Zion, M., Nagel, B.J., Tapert, S.F.: Effects of alcohol and combined marijuana and alcohol use during adolescence on hippocampal volume and asymmetry. Neurotoxicol. Teratol. **29**, 141–152 (2007)
44. Solowij, N., Michie, P.T., Fox, A.M.: Differential impairments of selective attention due to frequency and duration of cannabis use. Biol. Psychiatry. **15**, 731–739 (1995)
45. Solowij, N., Stephens, R., Roffman, R., Babor, T., Kadden, R., Miller, M., Christiansen, K., McRee, B., Vendetti, J., The Marijuana Project Research Group: Cognitive functioning of long-term heavy cannabis users seeking treatment. J. Am. Med. Assoc. **287**, 1123–1131 (2002)
46. Dougherty, D.M., Mathias, C.W., Dawes, M.A., Furr, R.M., Charles, N.E., Liguori, A., Shannon, E.E., Acheson, A.: Impulsivity, attention, memory, and decision-making among adolescent marijuana users. Psychopharmacology. **226**(2), 307–319 (2013)
47. Hanson, K.L., Winward, J.L., Schweinsburg, A.D., Medina, K.L., Brown, S.A., Tapert, S.F.: Longitudinal study of cognition among adolescent marijuana users over three weeks of abstinence. Addict. Behav. **35**, 970–976 (2010)
48. Harvey, M.A., Sellman, J.D., Porter, R.J., Frampton, C.M.: The relationship between non-acute adolescent cannabis use and cognition. Drug Alcohol Rev. **26**, 309–319 (2007)
49. Lisdahl, K.M., Price, J.S.: Increased marijuana use and gender predict poorer cognitive functioning in adolescents and emerging adults. J. Int. Neuropsychol. Soc. **18**, 678–688 (2012)
50. Mathias, C.W., Blumenthal, T.D., Dawes, M.A., Liguori, A., Richard, D.M., Bray, B., Tong, W., Dougherty, D.M.: Failure to sustain prepulse inhibition in adolescent marijuana users. Drug Alcohol Depend. **116**, 110–116 (2011)
51. McHale, S., Hunt, N.: Executive function deficits in short-term abstinent cannabis users. Hum. Psychopharmacol. Clin. Exp. **23**, 409–415 (2008)
52. Tapert, S.F., Granholm, E., Leedy, N.G., Brown, S.A.: Substance use and withdrawal: neuropsychological functioning over 8 years in youth. J. Int. Neuropsychol. Soc. **8**, 873–883 (2002)
53. Battisti, R.A., Roodenrys, S., Johnstone, S.J., Pesa, N., Hermens, D.F., Solowij, N.: Chronic cannabis user show altered neurophysiological functioning on Stroop task conflict resolution. Psychopharmacology. **212**, 613–624 (2010)
54. Day, A.M., Metrik, J., Spillane, N.S., Kahler, C.W.: Working memory and impulsivity predict marijuana-related problems among frequent users. Drug Alcohol Depend. **131**(1), 171–174 (2013)
55. Gonzalez, R., Schuster, R.M., Mermelstein, R.J., Vassileva, J., Martin, E.M., Diviak, K.R.: Performance of young adult cannabis users on neurocognitive measures of impulsive behavior and their relationship to symptoms of cannabis use disorders. J. Clin. Exp. Neuropsychol. **34**, 962–976 (2012)
56. Grant, J.E., Chamberlain, S.R., Schreiber, L., Odlaug, B.L.: Neuropsychological deficits associated with cannabis use in young adults. Drug Alcohol Depend. **121**, 159–162 (2012)
57. Pope, H.G., Yurgelun-Todd, D.: The residual cognitive effects of heavy marijuana use in college students. J. Am. Med. Assoc. **275**, 521–527 (1996)
58. Schuster, R.M., Crane, N.A., Mermelstein, R., Gonzalez, R.: The influence of inhibitory control and episodic memory on the risky sexual behavior of young adult cannabis users. J. Int. Neuropsychol. Soc. **18**, 827–833 (2012)
59. Solowij, N., Jones, K.A., Rozman, M.E., Davis, S.M., Ciarrochi, J., Heaven, P.C., Pesa, N., Lubman, D.I., Yücel, M.: Reflection impulsivity in adolescent cannabis users: a comparison with alcohol-using and non-substance-using adolescents. Psychopharmacology. **219**, 575–586 (2012)

60. Kalant, H.: Adverse effects of cannabis on health: an update of the literature since 1996. Prog. Neuro-Psychopharmacol. Biol. Psychiatry. **28**(5), 849–863 (2004)
61. Lyketsos, C.G., Garrett, E., Liang, K.Y., Anthony, J.C.: Cannabis use and cognitive decline in persons under 65 years of age. Am. J. Epidemiol. **149**(9), 749–800 (1999)
62. Pope Jr., H.G., Gruber, A.J., Hudson, J.I., Huestis, M.A., Yurgelun-Todd, D.: Neurpsychological performance in long-term cannabis users. Arch. Gen. Psychiatry. **58**(10), 909–915 (2001)
63. Bolla, K.I., Brown, K., Eldreth, D., Tate, K., Cadet, J.L.: Dose-related neurocognitive effects of marijuana use. Neurology. **59**(9), 1337–1343 (2002)
64. Wagner, D., Becker, B., Gouzoulis-Mayfrank, E., Daumann, J.: Interactions between specific parameters of cannabis use and verbal memory. Prog. Neuro-Psychopharmacol. Biol. Psychiatry. **34**, 871–876 (2010)
65. Becker, B., Wagner, D., Gouzoulis-Mayfrank, E.S., Daumann, J.: The impact of early-onset cannabis use on functional brain correlates of working memory. Prog. Neuro-Psychopharmacol. Biol. Psychiatry. **34**, 837–845 (2010)
66. Rieder, S.A., Chauhan, A., Singh, U., Nagarkatti, M., Nagarkatti, P.: Cannabinoid-induced apoptosis in immune cells as a pathway to immunosuppression. Immunobiology. **215**(8), 598–605 (2010)
67. Suárez-Pinilla, P., López-Gil, J., Crespo-Facorro, B.: Immune system: a possible nexus between cannabinoids and psychosis. Brain Behav. Immun. **40**, 269–282 (2014)
68. Klein, T.W., Newton, C., Larsen, K., Lu, L., Perkins, I., Nong, L., Friedman, H.: The cannabinoid system and immune modulation. J. Leukoc. Biol. **74**(4), 486–496 (2003)
69. Crowe, M.S., Nass, S.R., Gabella, K.M., Kinsey, S.G.: The endocannabinoid system modulates stress, emotionality, and inflammation. Brain Behav. Immun. **42**, 1–5 (2014)
70. Herbert, T.B., Cohen, S.: Stress and immunity in humans: a meta-analytic review. Psychosom. Med. **55**(4), 364–379 (1993)
71. Pestonjamasp, V.K., Burstein, S.H.: Anandamide synthesis is induced by arachidonate mobilizing agonists in cells of the immune system. Biochim. Biophys. Acta (BBA) Lipids Lipid Metab. **1394**(2), 249–260 (1998)
72. Croxford, J.L., Yamamura, T.: Cannabinids and the immune system: potential for the treatment of inflammatory diseases? J. Neuroimmunol. **165**, 3–18 (2005)
73. Munro, S., Thomas, K.L., Abu-Shaar, M.: Molecular characterization of a peripheral receptor for cannabinoids. Nature. **365**, 61–65 (1993)
74. Miller, A.M., Stella, N.: CB2 receptor mediated migration of immune cells: it can go either way. Br. J. Pharmacol. **153**(2), 299–308 (2008)
75. Malfait, A.M., Gallily, R., Sumariwalla, P.F., Malik, A.S., Andreakos, E., Mechoulam, R., Feldmann, M.: The nonpsychoactive cannabis constituent cannabidiol is an oral anti-arthritic therapeutic in murine collagen-induced arthritis. Proc. Natl. Acad. Sci. **97**(17), 9561–9566 (2000)
76. Klein, T.W., Newton, C., Friedman, H.: Cannabinoid receptors and immunity. Immunol. Today. **19**(8), 373–381 (1998)
77. Specter, S., Lancz, G., Hazelden, J.: Marijuana and immunity: tetrahydrocannabinol mediated inhibition of lymphocyte blastogenesis. Int. J. Immunopharmacol. **12**(3), 261–267 (1990)
78. Baldwin, G.C., Tashkin, D.P., Buckley, D.M., Park, A.N., Dubinett, S.M., Roth, M.D.: Marijuana and cocaine impair alveolar macrophage function and cytokine production. Am. J. Respir. Crit. Care Med. **156**(5), 1606–1613 (1997)
79. Keen II, L., Turner, A.D.: Differential effects of self-reported lifetime marijuana use on interleukin-1 alpha and tumor necrosis factor in African American adults. J. Behav. Med. **38**, 527–534 (2015)
80. Keen, L., Pereira, D., Latimer, W.: Self-reported lifetime marijuana use and interleukin-6 levels in middle-aged African Americans. Drug Alcohol Depend. **140**, 156–160 (2014)
81. Kozela, E., Juknat, A., Kaushansky, N., Rimmerman, N., Ben-Nun, A., Vogel, Z.: Cannabinoids decrease the 17th inflammatory autoimmune phenotype. J. NeuroImmune Pharmacol. **8**(5), 1265–1276 (2013)

82. Roth, M.D., Arora, A., Barsky, S.H., Kleerup, E.C., Simmons, M., Tashkin, D.P.: Airway inflammation in young marijuana and tobacco smokers. Am. J. Respir. Crit. Care Med. **157**(3), 928–937 (1998)
83. Pacifici, R., Zuccaro, P., Pichini, S., Roset, P.N., Poudevida, S., Farré, M., Segura, J., de la Torre, R.: Modulation of the immune system in cannabis users. JAMA. **289**(15), 1929–1931 (2003)
84. Pacifici, R., Zuccaro, P., Farré, M., Poudevida, S., Abanades, S., Pichini, S., Langohr, K., Segura, J., De La Torre, R.: Combined immunomodulating properties of 3, 4-methylenedioxymethamphetamine (MDMA) and cannabis in humans. Addiction. **102**(6), 931–936 (2007)
85. Roth, M.D., Whittaker, K., Salehi, K., Tashkin, D.P., Baldwin, G.C.: Mechanisms for impaired effector function in alveolar macrophages from marijuana and cocaine smokers. J. Neuroimmunol. **147**(1), 82–86 (2004)
86. Monnet-Tschudi, F., Hazekamp, A., Perret, N., Zurich, M.G., Mangin, P., Giroud, C., Honegger, P.: Delta-9-tetrahydrocannabinol accumulation, metabolism and cell-type-specific adverse effects in aggregating brain cell cultures. Toxicol. Appl. Pharmacol. **228**(1), 8–16 (2008)
87. Kozela, E., Pietr, M., Juknat, A., Rimmerman, N., Levy, R., Vogel, Z.: Cannabinoids Δ9-tetrahydrocannabinol and cannabidiol differentially inhibit the lipopolysaccharide-activated NF-κB and interferon-β/STAT proinflammatory pathways in BV-2 microglial cells. J. Biol. Chem. **285**(3), 1616–1626 (2010)
88. Roth, M.D., Baldwin, G.C., Tashkin, D.P.: Effects of delta-9-tetrahydrocannabinol on human immune function and host defense. Chem. Phys. Lipids. **121**(1), 229–239 (2002)
89. Ongrandi, J., Spector, S., Horvath, A., Friedman, H.: Combined in vitro effect of marijuana and retrovirus on the activity of mouse natural killer cells. Pathol. Oncol. Res. **4**(3), 191–199 (1998)
90. Ngaotepprutaram, T., Kaplan, B.L., Kaminski, N.E.: Impaired NFAT and NFκB activation are involved in suppression of CD40 ligand expression by Δ 9-tetrahydrocannabinol in human CD4+ T cells. Toxicol. Appl. Pharmacol. **273**(1), 209–218 (2013)
91. Srivastava, M.D., Srivastava, B.I.S., Brouhard, B.: Δ 9 Tetrahydrocannabinol and cannabidiol alter cytokine production by human immune cells. Immunopharmacology. **40**(3), 179–185 (1998)
92. Yuan, M., Kiertscher, S.M., Cheng, Q., Zoumalan, R., Tashkin, D.P., Roth, M.D.: Δ 9-Tetrahydrocannabinol regulates Thl/Th2 cytokine balance in activated human T cells. J. Neuroimmunol. **133**(1), 124–131 (2002)
93. Klein, T.W., Newton, C., Larsen, K., Chou, J., Perkins, I., Lu, L., Nong, L., Friedman, H.: Cannabinoid receptors and T helper cells. J. Neuroimmunol. **147**(1), 91–94 (2004)
94. Roth, M.D., Castaneda, J.T., Kiertscher, S.M.: Exposure to Δ9-tetrahydrocannabinol impairs the differentiation of human monocyte-derived dendritic cells and their capacity for T cell activation. J. NeuroImmune Pharmacol. **10**(2), 333–343 (2015)
95. Fischer-Stenger, K., Pettit, D.D., Cabral, G.A.: Delta 9-tetrahydrocannabinol inhibition of tumor necrosis factor-alpha: suppression of post-translational events. J. Pharmacol. Exp. Ther. **267**(3), 1558–1565 (1993)
96. Kusher, D.I., Dawson, L.O., Taylor, A.C., Djeu, J.Y.: Effect of the psychoactive metabolite of marijuana delta^9-tetrahydrocannabinol (thc), on the synthesis of tumor necrosis factor by human large granular lymphocites. Cell. Immunol. **154**, 99–108 (1994)
97. Snella, E., Pross, S., Friedman, H.: Relationship of aging and cytokines to the immunomodulation by delta-9-tetrahydrocannabinol on murine lymphoid cells. Int. J. Immunopharmacol. **17**(12), 1045–1054 (1995)
98. Wang, M., Richards, A.L., Friedman, H., Djeu, J.Y.: Selective inhibition of natural killer but not natural cytotoxic activity in a cloned cell line by delta-9-tetrahydrocannabinol. J. Leukoc. Biol. **50**(2), 192–197 (1991)

99. Newton, C.A., Klein, T.W., Friedman, H.: Secondary immunity to Legionella pneumophila and Th1 activity are suppressed by delta-9-tetrahydrocannabinol injection. Infect. Immun. **62**(9), 4015–4020 (1994)
100. Massi, P., Sacerdote, P., Ponti, W., Fuzio, D., Manfredi, B., Viganó, D., Rubino, T., Bardotti, M., Parolaro, D.: Immune function alterations in mice tolerant to Δ 9-tetrahydrocannabinol: functional and biochemical parameters. J. Neuroimmunol. **92**(1), 60–66 (1998)
101. Klein, T.W., Newton, C.A., Nakachi, N., Friedman, H.: Δ9-tetrahydrocannabinol treatment suppresses immunity and early IFN-γ, IL-12, and IL-12 receptor β2 responses to Legionella pneumophila infection. J. Immunol. **164**(12), 6461–6466 (2000)
102. Newton, C.A., Lu, T., Nazian, S.J., Perkins, I., Friedman, H., Klein, T.W.: The THC-induced suppression of Th1 polarization in response to Legionella pneumophila infection is not mediated by increases in corticosterone and PGE2. J. Leukoc. Biol. **76**(4), 854–861 (2004)
103. Gaffal, E., Cron, M., Glodde, N., Tüting, T.: Anti-inflammatory activity of topical THC in DNFB-mediated mouse allergic contact dermatitis independent of CB1 and CB2 receptors. Allergy. **68**(8), 994–1000 (2013)
104. Zhu, L.X., Sharma, S., Stolina, M., Gardner, B., Roth, M.D., Tashkin, D.P., Dubinett, S.M.: Δ-9-Tetrahydrocannabinol inhibits antitumor immunity by a CB2 receptor-mediated, cytokine-dependent pathway. J. Immunol. **165**(1), 373–380 (2000)
105. Chang, Y.H., Lee, S.T., Lin, W.W.: Effects of cannabinoids on LPS-stimulated inflammatory mediator release from macrophages: involvement of eicosanoids. J. Cell. Biochem. **81**(4), 715–723 (2001)
106. Lu, T., Newton, C., Perkins, I., Friedman, H., Klein, T.W.: Cannabinoid treatment suppresses the T-helper cell-polarizing function of mouse dendritic cells stimulated with Legionella pneumophila infection. J. Pharmacol. Exp. Ther. **319**(1), 269–276 (2006)
107. Verhoeckx, K.C., Korthout, H.A., van Meeteren-Kreikamp, A.P., Ehlert, K.A., Wang, M., van der Greef, J., Rodenburg, R.J., Witkamp, R.F.: Unheated Cannabis sativa extracts and its major compound THC-acid have potential immuno-modulating properties not mediated by CB 1 and CB 2 receptor coupled pathways. Int. Immunopharmacol. **6**(4), 656–665 (2006)
108. McKallip, R.J., Nagarkatti, M., Nagarkatti, P.S.: Δ-9-Tetrahydrocannabinol enhances breast cancer growth and metastasis by suppression of the antitumor immune response. J. Immunol. **174**(6), 3281–3289 (2005)
109. Ramarathinam, L., Pross, S., Plescia, O., Newton, C., Widen, R., Friedman, H.: Differential immunologic modulatory effects of tetrahydrocannabinol as a function of age. Mech. Ageing Dev. **96**(1), 117–126 (1997)
110. Alshaarawy, O., Anthony, J.C.: Cannabis smoking and serum C-reactive protein: a quantile regressions approach based on NHANES 2005–2010. Drug Alcohol Depend. **147**, 203–207 (2015)
111. Muniyappa, R., Sable, S., Ouwerkerk, R., Mari, A., Gharib, A.M., Walter, M., Courville, A., Hall, G., Chen, K.Y., Volkow, N.D., Kunos, G.: Metabolic effects of chronic cannabis smoking. Diabetes Care. **36**(8), 2415–2422 (2013)
112. Costello, E.J., Copeland, W.E., Shanahan, L., Worthman, M., Angold, A.: C-reactive protein and substance use disorders in adolescence and early adulthood: a prospective analysis. Drug Alcohol Depend. **133**(2), 712–717 (2013)
113. Rajavashisth, T.B., Shaheen, M., Norris, K.C., Pan, D., Sinha, S.K., Ortega, J., Friedman, T.C.: Decreased prevalence of diabetes in marijuana users: cross-sectional data from the National Health and Nutrition Examination Survey (NHANES) III. BMJ Open. **2**(1), e000494 (2012)
114. Ugen, K.E., Nyland, S.B.: Injecting drugs of abuse and immunity: implications for HIV vaccine testing and efficacy. Semin. Immunopathol. **28**(3), 281–287 (2006)
115. Cescon, D.W., Page, A.V., Richardson, S., Moore, M.J., Boerner, S., Gold, W.L.: Invasive pulmonary aspergillosis associated with marijuana use in a man with colorectal cancer. J. Clin. Oncol. **26**(13), 2214–2215 (2008)
116. Ruchlemer, R., Amit-Kohn, M., Raveh, D., Hanuš, L.: Inhaled medicinal cannabis and the immunocompromised patient. Support. Care Cancer. **23**(3), 819–822 (2015)

117. Szyper-Kravitz, M., Lang, R., Manor, Y., Lahav, M.: Early invasive pulmonary aspergillosis in a leukemia patient linked to aspergillus contaminated marijuana smoking. Leuk. Lymphoma. **42**(6), 1433–1437 (2001)
118. Cabral, G.A., Pettit, D.A.D.: Drugs and immunity: cannabinoids and their role in decreased resistance to infectious disease. J. Neuroimmunol. **83**(1), 116–123 (1998)
119. Huemer, H.P., Lassnig, C., Bernhard, D., Sturm, S., Nowotny, N., Kitchen, M., Pavlic, M.: Cannabinoids lead to enhanced virulence of the smallpox vaccine (vaccinia) virus. Immunobiology. **216**(6), 670–677 (2011)
120. Lee, M.A.: The Discovery of the Endocannabinoid system. Retrieved from http://www.beyondthc.com/wp-content/uploads/2012/07/eCBSystemLee.pdf (2010)
121. Fischer, B., Murphy, Y., Kurdyak, P., Goldner, E., Rehm, J.: Medical Marijuana Programs-why might they matter for public health and why should we better understand their impacts? Prev. Med. Rep. **2**, 53–56 (2015)
122. Philipsen, N., Butler, R.D., Simon-Waterman, C., Artis, J.: Medical marijuana: a primer on ethics, evidence, and politics. J. Nurse Pract. **10**(9), 633–640 (2014)
123. Klein, T.W.: Cannabinoid-based drugs as anti-inflammatory therapeutics. Nat. Rev. Immunol. **5**(5), 400–411 (2005)
124. Klein, T.W., Cabral, G.A.: Cannabinoid-induced immune suppression and modulation of antigen-presenting cells. J. NeuroImmune Pharmacol. **1**(1), 50–64 (2006)
125. Sylvestre, D.L., Clements, B.J., Malibu, Y.: Cannabis use improves retention and virological outcomes in patients treated for hepatitis C. Eur. J. Gastroenterol. Hepatol. **18**(10), 1057–1063 (2006)
126. Lorente, M., Torres, S., Salazar, M., Carracedo, A., Hernández-Tiedra, S., Rodríguez-Fornés, F., Garcia-Taboada, E., Meléndez, B., Mollejo, M., Campos-Martin, Y., Lakatosh, S.A.: Stimulation of the midkine/ALK axis renders glioma cells resistant to cannabinoid antitumoral action. Cell Death Differ. **18**(6), 959–973 (2011)
127. Naftali, T., Lev, L.B., Yablecovitch, D., Half, E., Konikoff, F.M.: Treatment of Crohn's disease with cannabis: an observational study. Isr. Med. Assoc. J. **13**(8), 455–458 (2011)
128. Waldman, M., Hochhauser, E., Fishbein, M., Aravot, D., Shainberg, A., Sarne, Y.: An ultra-low dose of tetrahydrocannabinol provides cardioprotection. Biochem. Pharmacol. **85**(11), 1626–1633 (2013)
129. Maier, S.F., Watkins, L.R., Fleshner, M.: Psychoneuroimmunology: the interface between behavior, brain, and immunity. Am. Psychol. **49**(12), 1004 (1994)
130. Schreiner, A.M., Dunn, M.E.: Residual effects of cannabis use on neurocognitive performance after prolonged abstinence: a meta-analysis. Exp. Clin. Psychopharmacol. **20**, 420–429 (2012)

Image Segmentation with the Aid of the p-Adic Metrics

Andrei Khrennikov and Nikolay Kotovich

Abstract We present the results of numerical simulation for image segmentation based on the chain distance clustering algorithm. The key issue is the use of the p-adic metric, where $p > 1$ is a prime number, at the scale of levels of brightness (pixel wise). In previous studies the p-adic metric was used mainly in combination with spectral methods. In this paper this metric is explored directly, without preparatory transformations of images. The main distinguishing feature of the p-adic metric is that it reflects the hierarchic structure of information presented in an image. Different classes of images match with in general different prime p (although the choice $p = 2$ works on average). Therefore the presented image segmentation procedure has to be combined with a kind of learning to select the prime p corresponding to the class of images under consideration.

Keywords Image segmentation • p-adic metric • Clustering • Clustering algorithm

1 Introduction

In theory of computer vision segmentation of images is defined as partition of an image I into say N disjoint domains R_i, which are treated as homogeneous with respect to some characteristic (for example, the level of brightness, texture, color) and essentially different from other domains with respect to this characteristic, see, e.g., [1–5]:

$$\bigcup_{i=1}^{N} R_i = I,$$

A. Khrennikov (✉)
International Center for Mathematical Modelling, Physics, Engineering, Economics, and Cognitive Science, Linnaeus University, 35195, Växjö, Sweden
e-mail: Andrei.Khrennikov@lnu.se

N. Kotovich
Institute of System Analysis of Russian Academy of Science, 117218, Moscow, Russia

B. Toni (ed.), *New Trends and Advanced Methods in Interdisciplinary Mathematical Sciences*, STEAM-H: Science, Technology, Engineering, Agriculture, Mathematics & Health, DOI 10.1007/978-3-319-55612-3_6

$$R_i \cap R_j = \varnothing \text{ for } i, \ j = 1, 2, \ldots, N; i \neq j,$$

$$P(R_i) = TRUE \text{ for } i = 1, 2, \ldots N,$$

$$P(R_i \cup R_j) = FALSE \text{ for neighboring domains } R_i \text{ and } R_i,$$

where P(R_i) is the predicate of homogeneity on the set of pixels composing the domain $R_{i.}$

Segmentation is one of the most important stages of the analysis of images characteristics. Permanently increasing interest in this topic is expressed in particular in a huge number of publications devoted to different aspects of segmentation of images, development and improvement of various algorithms of segmentation, see, e.g., [1–5]. And the number of publications is growing rapidly, IEEE statistics [6]: in 1970–1979: 11 publications; in 1980–1989: 314 publications; in 1990–1999: 3066 publications; in 2000–2009: 9938 publications.

In different studies segmentation of images was based on the use of various types of distances, e.g., Euclidean metric or its square, Hamming metric, Chebyshov metric, and so on. In particular, we point to the theoretical and experimental studies based on the use of p-adic metrics and more generally ultrametrics [7–10]. For instance, in works [7, 8] there were presented results of numerical experiments on segmentation of images presented as JPEG, MPEG images with the aid of the p-adic metric in the *spectral domain* (the Fourier transform of the image).

In this paper we present the result of numerical experiments on segmentation of images with the aid of chain clustering algorithms with fixed threshold based on the use of the p-adic metric for comparing the levels of brightness associated with pixels. This is a kind of "direct clustering," i.e., without preparatory transformations of images such as based on the spectral methods (methods based on the use of the Fourier transform).

The main distinguishing feature of the p-adic metric is that it reflects the hierarchic structure of information presented in an image. Different classes of images match with in general different prime p (although the choice $p = 2$ works in average). Therefore the presented image segmentation procedure has to be combined with a kind of learning to select the prime p corresponding to the class of images under consideration. In the numerical experiments from which results are presented here, the "p-learning" was done manually. In future it can be done automatically, either with the aid of the direct selection of various primes, $p = 2,3, \ldots,$ or with the aid of more advanced learning procedures.

Coupling of the p-adic metric and more generally ultrametric to cognition and in particular its role in recognition and processing of images by a human brain was enlightened in a series of works [11–18]. In this approach hierarchy of mental information is encoded in hierarchy on a branch of the p-adic tree: *further from the root is associated with less importance.* P-adics and generally ultrametrics provide a possibility to encode the information into topology. Then processing of mental information can be mathematically modeled by dynamical systems in ultrametric spaces (a good mathematical theory was developed only for p-adic spaces). Continuity of the evolution of the information state of the brain ("mental state") with respect to ultrametric preserves the hierarchic structure of information. Decisions and general outputs of the process of thinking are represented by attractors of the dynamical systems. The latter is the standard feature of the dynamical systems approach to cognition. The novelty is the use of ultrametric (mainly the p-adic metric) and the hierarchic structure of the process of approaching attractors—solutions.

For numerical studies for signals of different kinds, see, e.g., the work of Murtagh on ultrametric structure of astrophysical signal data [19]; he also used p-adic metric for hierarchical clustering for search and retrieval [19, 20]. Recently Bradley elaborated novel algorithms for information processing and signal analysis, in particular, for image segmentation [21–24].

P-adic neural networks were applied to data classification problems and learning as well as structuring video-streams of information [25].

2 Ultrametric and p-Adic Spaces: Brief Introduction

We recall, e.g., [11], that a metric ρ on the *space* X is said to be an *ultrametric*, if it satisfies the *strong triangle inequality*:

$\rho(x,y) \leq max\,[\rho(x,z),\rho(z,y)],\ x,y,z \in X.$

p-adic ultrametric on the space $Z_{p,N}$ of *p*-discrete vectors of the length N with coordinates belonging to the set of digits {0,1,...,p-1}, where $p > 1$ is a prime number, is defined in the following way.

Let $x = (\alpha_0,\ldots,\alpha_{n-1})$, $y = (\beta_0,\ldots,\beta_{n-1}) \in Z_{p,N}$. We set $\rho(x,y) = 1/p^k$ if $\alpha_j = \beta_j$, $j = 0,1,\ldots,k-1$, and $\alpha_k \neq \beta_k$. This is an ultrametric.

In the same way the p-adic ultrametric is introduced on the space Z_p of *p*-discrete vectors of infinite length with coordinates belonging to the set of digits {0,1,..., p-1}. In this paper we shall use only vectors with finite number of coordinates. However, in theoretical studies Z_p plays an important role [11].

Geometrically Z_p can be represented as the homogeneous tree with m branches leaving each of its vertexes; the branches leaving a vertex are numbered as $i = 0,1,\ldots,$ p-1. Therefore the branches of the tree are labelled as infinite vectors of the form $x = (\alpha_0,\ldots,\alpha_{n-1},\ldots)$ belonging to the set {0,1,...,p-1}.

Algebraically the elements of Z_p are represented by sums:

$$x = \alpha_0 + \alpha_1 p + \ldots\ldots + \alpha_{N-1} p^{N-1} + \ldots.$$

with naturally defined arithmetic operations of addition, subtraction, and multiplication [11]. This is a ring with respect to these operations. It is called the ring of p-adic integers (because all "usual integers" belong to it).

One can extend this ring by using sums including finite numbers of terms with negative powers of p. This is the field Q_p of p-adic numbers, i.e., the operation of division is also well defined [11].

3 Chain Algorithm for Clustering

In this section we present the basic definitions which will be used in further considerations on chain segmentation algorithms.

A sequence of points $a = x_0, x_1, \ldots, x_{n-1}, x_n = b$ *in a metric space* (X,ρ) *is called an* ε*-chain joining a and b if* $\rho(x_k, x_{k+1}) \leq \varepsilon$ *for any* $k \leq n$. *If there exists an* ε*-chain joining a and b they are said to be* ε*-linkable.*

We recall that ultrametric spaces are characterized by the following result:

Theorem *A metric space is ultrametric if and only if no two points* $a \neq b$ *in it are* ε*-linkable for any* $\varepsilon < \rho\ (a,b)$.

Thus by this theorem, for arbitrary two points of an ultrametric space, there is no ε-chain joining a and b for $\varepsilon < \rho\ (a,b)$ which is strictly less than the distance between these points. This property also implies that a metric space is in fact an ultrametric space.

Let now (X,ρ) be an arbitrary metric space. We set

$\Delta(x,y) = Inf\{\varepsilon{:}x$ and y are ε-linkable$\}$, $x,y \in X$.

This function has all properties of an ultrametric, besides non-degeneration (it can be that $\Delta(x,y) = 0$ for some $x \neq y$). It is the so-called *pseudo-ultrametric*. It is called the *chain distance* between points x and y. We remark that theorem implies *that in an ultrametric space* (X,ρ) *the original metric* ρ *coincides with the corresponding chain distance.*

We use an algorithm of chain clustering. This algorithm belongs to the group of AH-methods (*agglomerative hierarchical clustering*), see, e.g., [26, 27] for the general references.[1] The crucial specialty of our approach is the use of the special metric, the p-adic one.

Step 0 Any object of the set of objects under study can be chosen as the generator of the process of clustering. This object gets two labels, $n = 1$ (its number) and $\rho = 0$ (distances).

Step 1 Then we consider all other objects and take the object such that the distance ρ' between this object and the object with $n = 1$ is minimal. This new object gets $n = 2$ and $\rho = \rho'$.

$\ldots\ldots$

[1]This is the nearest neighbor chain algorithm, that, with the reciprocal or mutual nearest neighbor; this is the standard efficient implementation for all standard agglomerative hierarchical clustering algorithms.

Step N Then we consider all other objects and take the object such that the distance ρ'' between this object and the set of previously chosen objects is minimal.

It is easy to see that thus defined the distance ρ is the chain distance. At each step we take the object such that the distance between this object and the set of objects that have been already enumerated is minimal. This procedure is repeated until all objects are enumerated. Finally, all objects are ordered (enumerated) in a so-called chain and each object in it has the label ρ—the distance to the set of previous objects. In more detailed presentation we have:

1. There is given nonempty set A and a metric on it.
2. Take an arbitrary x from A; assign to it $n = 1$ and distance $\rho = 0$. We define the set M as containing this point x.
3. If all elements of A have been used, then go to the end—to step 8.
4. Take one of the elements of A, say y, having the minimal distance to M.
5. Assign y this minimal distance.
6. Add y to M and enumerate it, $n = n + 1$.
7. Got to step 3.
8. End: all elements of A have been enumerated and to each of them there were assigned the corresponding distance.

Splitting of the set of objects into clusters is based on the following procedure. Let $r_0 > 0$ be some constant (the parameter of clustering). We would like to build clusters in such a way that the chain distance between objects inside one cluster will be $\rho \leq r_0$ and at the same time the chain distance between objects belonging to different clusters will $\rho > r_0$. To do this, we consider labels of all objects and find objects such that $\rho > r_0$. Suppose that there are objects with numbers $n_1, \ldots, n_K$.

The first cluster consists of all objects with $1 \leq n < n_1$; the second one – $n_1 \leq n < n_2$, and so on.

By varying the parameter r_0 we can analyze different variants of clustering.[2]

4 Chain Clustering with a Fixed Threshold

Here consider the following *fast version of the chain clustering method*, namely chain clustering with a fixed threshold.

Let us fix a distance between clusters, $r_0 > 0$. Any object can be taken as the initial object. It gets the label as belonging to the first cluster. The first cluster is constructed by grouping all objects such that the distance from them to the initial object is less than the threshold r_0. Then, for each of these new objects that were taken into the first cluster, the procedure is repeated (by considering only the objects

[2]This (thresholding chains or relation representations, from, e.g., single link clustering, or the minimal spanning tree) has been used for image segmentation in some previous work. A review that includes some of this work is available, e.g., in [28].

that do not already belong to the cluster). When there are no more objects that could be collected to the first cluster, some object that does not belong to the first cluster is taken as the base of the second cluster and so on. If we have N objects, then in the worst case we need $N\,(N-1)/2$ operations of distance computation, but in the best case only N operations.

5 Results: p-Adic Clustering

In this section we present the results of segmentation of images with the aid of the p-adic metric by chain clustering method with fixed threshold r_0. The neighboring pixels are unified in a cluster if the distance between their levels of brightness is less than the fixed threshold. To be more precise, we remark that, as always for colored pictures, we consider three components: (R) red, (G) green, (B) blue, and each of them is quantized by using 256 levels.[3] For each color, we use the p-adic distance and finally unify colors by using the max-metric with respect to colors. We note that by taking max of a family of ultrametrics one again obtains an ultrametric.

The crucial point of our study was the use of the p-adic distance. Here, see Figs. 1, 2, 3, and 4, we proceeded with the value $p = 3$ and the threshold $r_0 = 1/3$. As was already remarked in the introduction, in average, i.e., without specification of the class of images, the simulation with value $p = 3$ produces reasonably good outputs. Heuristically it is clear why it happens. Roughly speaking the value $p = 2$ is too small, the levels of brightness are too often considered (from the 2-adic viewpoint) as close to each other; the value $p = 5$ is too big, the levels of brightness are too often considered (from the 5-adic viewpoint) as far from each other. Of course, this is pure heuristics. In general some classes of symbols are better adapted to, e.g.,

Fig. 1 Sky and rocks, p = 3, $r_0 = 1/3$

[3]We noted that future work would consider other color encoding beyond RGB (YUV, etc.); also dynamic range, here 256, to be extended.

Fig. 2 Land, river, and sky, p = 3, $r_0 = 1/3$

Fig. 3 Land, penguins, and sky, p = 3, $r_0 = 1/3$

Fig. 4 Walls, door-segment, and books, p = 3, $r_0 = 1/3$

2-adic, or 5-adic processing. Again, as was remarked in the introduction, in future more advanced methods of p-learning, p-adaptivity to a class of images, can be elaborated, e.g., by using the color histograms of images. Moreover, for a moment we proceed with the same prime p for all colors, (R), (G), (B). In principle, for some classes of symbols p can depend on colors, i.e., one can get better segmentation by using three different prime p. Finally, we remark that, as well as in modelling of cognition, one need not restrict considerations to the case of the prime numbers. It is possible to apply our algorithm to the case of nonprime $m = 2,3,4,5,6,\dots,$ i.e., to work in the rings of the so-called m-adic numbers [11]. The use of primes is preferable in the process of determination of the presence of a hierarchic structure in the image. We remark that any natural m can be represented as the product of the powers of its prime factors. Therefore the intrinsic m-adic structure of an image is combination of the p-adic structures corresponding to the prime divisors of m. If we select one of them, then we will be able to extract at least the component corresponding to this divisor. For example, take $m = 6$; its prime factors are 2 and 3. If an image has the 6-adic internal structure, its component will be visible already for 2-adic clustering. Thus the use of primes is justified, since it simplifies finding of the proper basis of processing. Of course, if one has sufficient computational abilities and no time restriction, it is possible to experiment even with composite natural numbers to find a better matching.

We remark that for $p = 3$ the possible thresholds are equal to 1, 1/3, 1/9, $\dots,$ corresponding to 3-adic balls of these radii. Hence, we performed clustering with the aid of balls of the radius r = 1/3. One can see that such a simple clustering algorithm gives a reasonably good output.

The use of the p-adic distance is crucial; see further comparison with clustering with the aid of the same algorithm, but with the Euclidean metric, Section 6.

Finally, we present, see Fig. 5, the output of the clustering algorithm for the image represented at Fig. 2 (the left-hand side image) but with the parameters $p = 2$ and $r_0 = 1/2$. The quality of segmentation is not satisfactory, further play with the parameter r_0 does not improve the quality.

Fig. 5 Land, river, and sky, $p = 2$, $r_0 = 1/2$

6 Comparison with the Euclidean Metric

The following three pictures are outputs of the Euclidean distance clustering with different thresholds for the original picture at the left-hand side of Fig. 2. For $r_0 = 20$, Fig. 6, large segments are absent, for $r_0 = 28$, Fig. 7a, segmentation is reasonable, but the sky started to merge with parts of the land surface, further increasing of the threshold, $r_0 = 30$, Fig. 7b, implies further merging of sky with parts of the land surface. The same happens for the picture represent at the left-hand side of Fig. 3, for relatively small thresholds, large segments are absent, too many small details; however, increasing of the threshold leads to disappearance of segments corresponding to penguins, see Fig. 8, where $r_0 = 28$. Thus clustering based on the Euclidean distance leads to merging of important structures which are clearly visible in images obtained as outputs of p-adic clustering.

Fig. 6 Land, river, and sky, Euclidean distance clustering, relatively low threshold $r_0 = 20$

a b

Fig. 7 (**a**, **b**) Land, river, and sky, Euclidean distance clustering, with increasing threshold, $r_0 = 28$ and $r_0 = 30$

Fig. 8 Land, penguins, and sky, Euclidean distance clustering, relatively high threshold $r_0 = 28$

7 Concluding Remarks

This pilot study demonstrated that the p-adic metric can serve as a useful tool for image segmentation. The hierarchical structure encoded with the aid of this metric matches well with hierarchic presentation of information and, in particular, images of human brain. An important advantage of the algorithm explored in this paper is its simplicity. And in spite of simplicity, it produces reasonably good segmentation, *at least comparing with segmentation produced by the similar algorithm, but based on the Euclidean distance.*

We plan to extend our studies and apply this p-adic method of image segmentation to concrete problems of image analysis. In particular, we plan to use it in geology for recognition of special oil-related structures, cf. [24, 29]. Here oil flows in porous disordered media which has the tree-like structure. We remark that geometrically p-adic numbers are represented by homogeneous trees. Therefore p-adic distance can be explored to model clustering of geological structures. However, the straightforward application of the p-adics is not possible. Geological structures are more complex geometrically than simply the homogeneous p-adic trees. They can be represented as ultrametric space. The computer representation of such general ultrametric trees is a separate problem for study.

Finally, we make a remark on the speed of the presented clustering algorithm based on the use of the p-adic metric comparing with the speed of the similar clustering algorithms based on the use of the Euclidean metric. The presented p-adic distance algorithm does not provide speed up in image clustering. Thus from this viewpoint it is not more effective than ones exploring the Euclidean distance.[4]

[4]However, the following comment can be useful to clarify the situation with computational speed up. Clustering that uses a p-adic distance can be considered by analogy with (1) creating a hierarchical clustering, and then (2) cutting the hierarchy to produce a partition. This analogy is more the case if the hierarchy is a contiguity-constrained one. The computational requirements that are referred to quite negatively as regards this work may not be all that bad, especially if one makes comparison relative to straightforward, Euclidean distance or related, segmentation.

The main issue is that the use of p-adic distances leads to extraction of new clusters which are "not visible" for other distances. For some classes of images, p-adic clustering is more appropriate than the standard clustering.

Ideologically we propose to look at the same image from different viewpoints: from Euclidean and from p-adic. Some classes of images have the intrinsic hierarchic structure which can be adequately encoded by p-adics. In the original Euclidean representation this intrinsic hierarchy is shadowed and it becomes visible only in the p-adic representation.

As was pointed out, p-adic clustering matches well with the tree-like structures in images. Such structures are very common in geology, as tree-like capillary networks in porous random media. Therefore we hope that p-adic clustering would be applied oil industry (and plan to realize these expectations in further works, cf. [29–31]).

This paper was financially supported by the grant "Math Modelling of Complex Hierarchic Systems," Linnaeus University.

References

1. Ilea, D.E., Whelan, P.F.: Image segmentation based on the integration of colour–texture descriptors—a review. Pattern Recogn. **44**(N10–11), 2479–2501 (2011)
2. Hill, J., Corona, E., Ao, J., Mitra, S., Nutter, B.: Information theoretic clustering for medical image segmentation. In: Advanced Computational Approaches to Biomedical Engineering, pp. 47–70. Springer, Berlin-Heidelberg (2014)
3. Arbelaez, P., Maire, M., Fowlkes, C., Malik, J.: Contour detection and hierarchical image segmentation. IEEE Trans. Pattern Anal. Mach. Intell. **33**(5), 898–916 (2011)
4. Vieux, R., Benois-Pineau, J., Domenger, J.-P., Braquelair, A.: Segmentation-based multi-class semantic object detection. Multimedia Tools and Applications, pp. 1–22. Springer Verlag (2011)
5. Rossi, E., Benini, S., Leonardi, R., Mansencal, B., Benois-Pineau, J., Clustering of scene repeats for essential rushes preview. International Workshop on Image Analysis for Multimedia Interactive Services (WIAMIS), pp. 234–237, London, United Kingdom (2009)
6. IEEE Explore Digital Library. http://ieeexplore.ieee.org.
7. Benois-Pineau, J., Khrennikov, A.Y., Kotovich, N.V.: Segmentation of images in p-adic and Euclidean metrics. Dokl. Math. **64**, 450–455 (2001)
8. Benois-Pineau, J., Khrennikov, A. (2002). Image segmentation in compressed domain by clustering methods with Euclidean and p-adic metrics. In: Proceedings in Interational Conference in Management of Uncertainty in Knowledge-based Systems (IPMU-2), pp. 969–976, Annecy, France
9. Khrennikov, A.Y., Kotovich, N.V.: Representation and compression of images with the aid of the m-adic coordinate system. Dokl. Akad. Nauk. **387**, 159–163 (2002)
10. Khrennikov, A., Kotovich, N., Borzistaya, E.: Compression of images with the aid of representation by p-adic maps and approximation by Mahler's polynomials. Dokl. Math. **69**(N3), 373–377 (2004)
11. Khrennikov, A.Y.: Non-Archimedean analysis: quantum paradoxes, dynamical systems and biological models. Kluwer, Dordrecht (1997)

A further comparison in future work is likely to be beneficial too: the use of wavelet transforms—hierarchical and hence ideally represented p-adically for image and signal segmentation.

12. Albeverio, S., Khrennikov, A.Y., Kloeden, P.: Memory retrieval as a p-adic dynamical system. Biosystems. **49**, 105–115 (1997)
13. Dubischar, D., Gundlach, V.M., Steinkamp, O., Khrennikov, A.Y.: A p-adic model for the process of thinking disturbed by physiological and information noise. J. Theor. Biol. **197**, 451–467 (1999)
14. Khrennikov, A.Y.: Toward an adequate mathematical model of mental space: conscious/unconscious dynamics on m-adic trees. Biosystems. **90**, 656–675 (2007)
15. Murtagh, F.: Ultrametric model of mind, I: review. P-adic Numbers Ultrametr. Anal. Appl. **4**, 193–206 (2012)
16. Murtagh, F.: Ultrametric model of mind, II: application to text content analysis. P-adic Numbers Ultrametr. Anal. Appl. **4**, 207–221 (2012)
17. Lauro-Grotto, R.: The unconscious as an ultrametric set. Am. Imago. **64**, 535–543 (2007)
18. Khrennikov, A., Kotovich, N.: Towards ultrametric modeling of unconscious creativity. Int. J. Cogn. Inform. Nat. Intell. **8**(4), 12 (2014)
19. Murtagh, F.: On ultrametricity, data coding, and computation. J. Classif. **21**, 167–184 (2004)
20. Murtagh, F., Contreras, P.: Fast, linear time, m-adic hierarchical clustering for search and retrieval using the baire metric, with linkages to generalized ultrametrics, hashing, formal concept analysis, and precision of data measurement. P-Adic Numbers Ultrametr. Anal. Appl. **4**, 45–56 (2012)
21. Bradley, P.: Degenerating families of dendrograms. J. Classif. **25**, 27–42 (2008)
22. Bradley, P.: From image processing to topological modelling with *p*-adic numbers. P-Adic Numbers Ultrametr. Anal. Appl. **2**, 293–304 (2010)
23. Bradley, P.: Mumford dendrograms. Comput. J. **53**, 393–404 (2010)
24. Bradley, P.: Comparing G-maps with other topological data structures. GeoInformatica. **18**, 595–620 (2014)
25. Benois-Pineau, J., Khrennikov, A.: Significance delta reasoning with p-adic neural networks: application to shot change detection in video. Comput. J. **53**, 417–431 (2010)
26. Kaufman, L., Rousseeuw, P.J.: Finding groups in data: an introduction to cluster analysis, 1st edn. John Wiley, New York (1990)
27. Hastie, T., Tibshirani, R., Friedman, J.: Hierarchical clustering. The elements of statistical learning, 2nd edn, pp. 520–528. Springer, New York (2009)
28. Murtagh, F.: Algorithms for contiguity-constrained clustering. Comput. J. **28**, 82–88 (1985)
29. Khrennikov, A., Kozyrev, S.V., Oleschko, K., Jaramillo, A.G., de Jesus Correa Lopez, M.: Application of p-adic analysis to time series. Infinite Dimens. Anal. Quantum Probab. Relat. Top. **16**, 1350030 (2013)
30. Khrennikov, A., Oleschko, K., de Jesús Correa Lopez, M.: Modeling fluid's dynamics with master equations in ultrametric spaces representing the treelike structure of capillary networks. Entropy. **18**(7), 249 (2016)
31. Khrennikov, A., Oleschko, K., de Jesús Correa Lopez, M.: Application of p-adic wavelets to model reaction-diffusion dynamics in random porous media. J. Fourier Anal. Appl. **22**, 809–822 (2016)

The Primes are Everywhere, but Nowhere . . .

Klaudia Oleschko, Andrei Khrennikov, Beatriz F. Oleshko, and Jean-Francois Parrot

Abstract In this note we present the Prime Number Paradigm: "Nature encodes its laws with the aid of prime numbers." We claim that prime number skeletons of data-sets contain the basic structures represented them. This paradigm is illustrated by a variety of data: from astrophysics to geophysics (including petroleum data), and reading comprehension. We remark that the ideas about the fundamental (and even mystical) role played by primes in mathematics and in physics were discussed by many scientists during the last 2000 years. The main advantage of our approach is its close connection with real experimental data. To present deeper patterns in data, the prime skeletons can be extended to p-adic skeletons. P-adic numbers play the important role number theory and recently they started to be widely used in theoretical physics, form string theory and theory of complex disordered systems to geophysics.

Keywords Primes • P-adic fields • Image analysis

K. Oleschko
Centro de Geociencias, Universidad Nacional Autonoma de Mexico (UNAM), Campus UNAM Juriquilla, Blvd. Juriquilla 3001, C.P. 76230, Queretaro, Qro., México
e-mail: olechko@unam.mx

A. Khrennikov (✉)
International Center for Mathematical Modelling, Physics, Engineering, Economics, and Cognitive Science, Linnaeus University, 35195, Växjö, Sweden
e-mail: Andrei.Khrennikov@lnu.se

B.F. Oleshko
Universidad Autonoma de México (UNAM), Centro de Ciencias de la Complejidad (C3), Cuerpo B, Nivel 1, Circuito Mario de la Cueva 20, Cd. Universitaria, Delegación Coyoacán, C.P. 04510, Cd. Mx., México
e-mail: be.oleshko@unam.mx

J.-F. Parrot (✉)
Instituto de Geografía, Universidad Autonoma de México (UNAM), Cd. Universitaria, Delegación Coyoacán, 04510 Ciudad de México, México
e-mail: parrot@igiris.igeograf.unam.mx

B. Toni (ed.), *New Trends and Advanced Methods in Interdisciplinary Mathematical Sciences*, STEAM-H: Science, Technology, Engineering, Agriculture, Mathematics & Health, DOI 10.1007/978-3-319-55612-3_7

1 Introduction

The aim of this chapter is to enlighten a computational (or algorithmic) perspective of prime and p-adic numbers in number theory with special attention to its application to geophysics and physics in general.

One of the main goals of the modern science is to extract (with minimum loss) the space and time patterns of the complex systems from the huge multiscale and multiphysics information coming day by day to the experimental and applied scientific areas. The data encrypted by Nature or during the experimentation are not only enriched with different kind of noise but also are full of unnecessary details. During the last 15 years, we tried to find the best mode to find the skeleton of information coming from the numerous research projects. In a natural way, working with the huge data banks, where each number plays the role of elemental particle, therefore it cannot be excluded from the results interpretation without some serious argument, our efforts were routed to number theory. From the beginning, *the importance of prime numbers in our data pattern was observed* and taken into account during the information processing. Two crucial points have changed our data perception of research results. First of all, it is related with the results of purely curiosity based daily life experiment, accomplished by one of the authors (B.F. Oleshko: see Figs. 1 and 2). The next step is related with the beginning of our Group collaboration with Prof. Khrennikov who gave *the p-adic numbers* approach to physics and especially to some extremely important geophysics data processing.

We recall that in modern science the construction of a bridge between the number theory, and especially between prime numbers, and computer experiments, has begun in the early 1970s of past XX Century by Merkle and Hellman [1]. In their pioneer paper, these authors have established the strong link between the number theory and cryptography envisioning the computational and especially algorithmic perspective for prime numbers [2]. Nowadays, encryption became crucial for modern life and science, from quantum to cosmic researches ([32] 11. 2016). It seems not surprising that the principles of codification developed by Kerchoffs in 1883 have resulted in the development of a dynamic security protocol for face recognition systems using seismic waves [3]. Notwithstanding, in spite of these examples of close relation existing between encryption and prime numbers, the experimental face of this topic is fairly recent. Of course, there are many available well-known encryption algorithms, but only small part of them is related to prime or p-adic numbers. However, after recent Nature publication [4] about the peculiar pattern found in "random" prime numbers, the new wave of computer researches was appeared in order to document numerically the Hardy–Littlewood k-tuple conjecture.

Our challenges are still different from the above exemplified works. We are looking for the basic rules of how does Nature encrypt the natural complex systems by prime numbers. Let us enlist some historical facts in order to understand from where our main ansatz derived.

Fig. 1 (**a**) The original position of the Observer during the moonrise imaging. Note: the size of (**a**) and (**b**) images is 960 by 720 pixels. The original photograph size is 3072 by 2304 pixels. (**b**) Rings around the sun observed on the selected prime numbers map, extracted from the corresponding grayscale photograph. The primes corresponding to 18% of more than seven millions of numbers on the original 3072 by 2304 matrix. For this example, only some primes are colored, but for each number only one color is used. The moon is corresponding to the small *black* point distinguished with difficulty on the down part of the map between the seventh and eighth rings. (**c**) The rings around the sun have disappeared at the critical point, when the moon can be distinguished. (**d**) The prime numbers rings have appeared around the moon, while the rest of primes corresponding to the sun ring have disappeared step by step

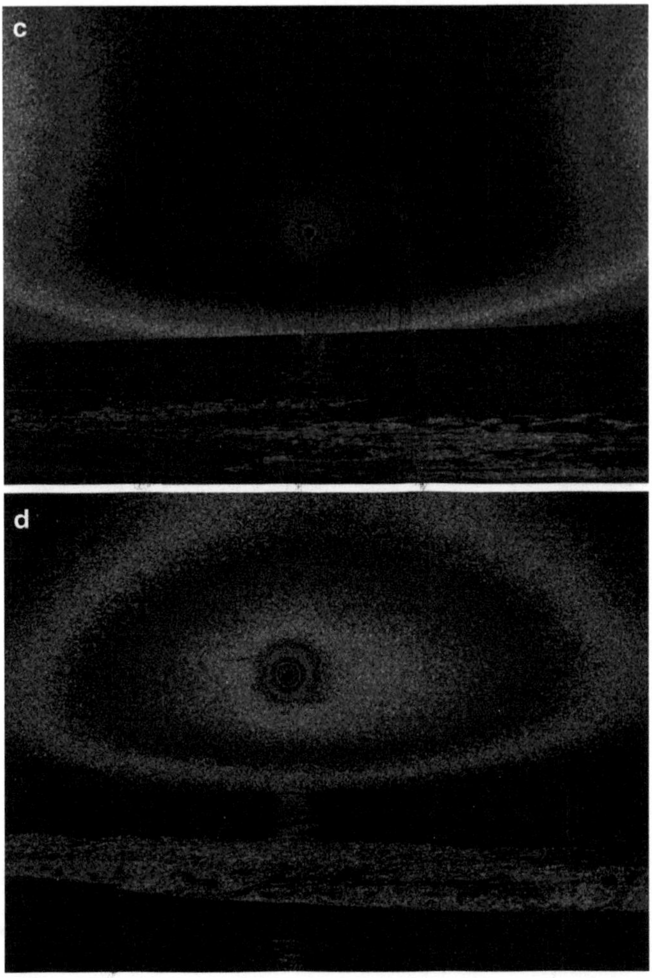

Fig. 1 (continued)

It is well known that prime numbers play the exceptional role not only in mathematics but also in physics, from quantum physics (see above-cited reference) to theory of complex disordered systems and geophysics, information processing, biology, and cognitive science. These numbers are considered as the most fundamental blocks (sometimes called elemental particles or "periodic table of mathematics," 6) of the quantitative representation of natural phenomena. This was recognized already in the Ancient times; for example, in one of the earliest texts mentioning prime numbers, Nicomachus of Gerasa (c.60–c.120 CE), see ([5], [Book 1, chapter XII]), we can find the following statement, p. 222: "the prime and incomposite [. . .] has received this name because it can be measured only by the number which is first and common to all, unity,

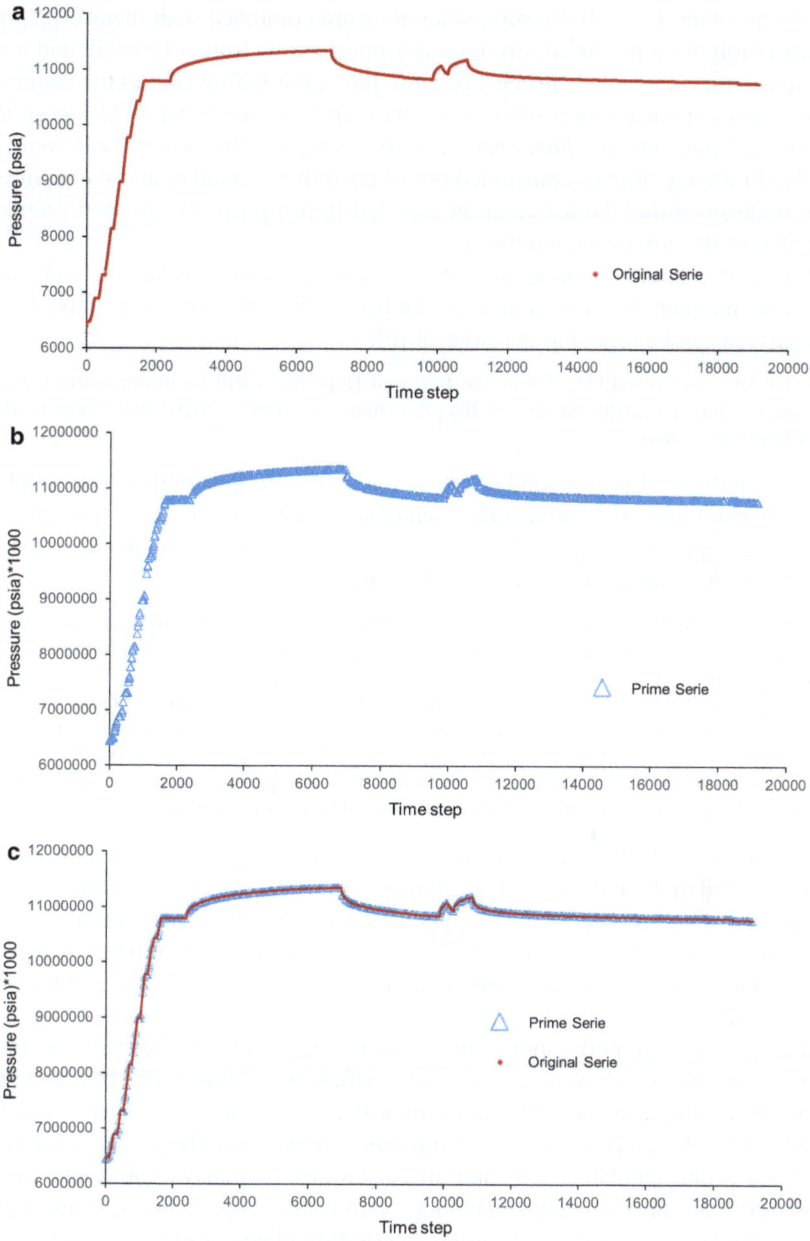

Fig. 2 (**a**) The original graph of the pressure curve. (**b**) The prime number skeleton of the original graph of the pressure curve. (**c**) Shape coincidence of the graphs of the original pressure curve and its prime skeleton

and by no other. [...] To be sure, when they are combined with themselves, other numbers might be produced, originating from them as from a fountain and a root, wherefore they are called "prime", because they exist beforehand as the beginnings of the others. For every origin is *elementary and incomposite, into which everything is resolved and out of which everything is made*, but the origin itself cannot be resolved into anything or constituted out of anything." Crandall and Pomerance [2] have underlined that the information encoded in primes is always more important than that of the composite numbers.

Later, after creation of atomic physics and quantum mechanics, comparison of prime numbers to "the atoms of number theory" became very popular. This comparison can be rooted to the work of [6]:

> It remains unresolved but, if true, the Riemann Hypothesis will go to the heart of what makes so much of mathematics tick: the prime numbers. *These indivisible numbers are the atoms of arithmetic.*

This analogy of prime numbers to atoms and more generally to quanta played the important role in establishing foundations of the information viewpoint on the modern physics, especially quantum mechanics. This viewpoint on physics celebrated in the famous statement of Wheeler [7]:

> *It from bit.* Otherwise put, every 'it'—every particle, every field of force, even the space-time continuum itself—derives its function, its meaning, its very existence entirely even if in some contexts indirectly—from the apparatus-elicited answers to yes-or-no questions, binary choices, bits. 'It from bit' symbolizes the idea that every item of the physical world has at bottom - a very deep bottom, in most instances - an immaterial source and explanation; that which we call reality arises in the last analysis from the posing of yes–no questions and the registering of equipment-evoked responses; in short, that all things physical are information-theoretic in origin and that this is a participatory universe.

The nature is viewed as constructed from elementary information blocks. This viewpoint led to flourishing of the information interpretation of quantum mechanics, *Zeilinger* [8, 9], Brukner and Zeilinger [10], and Plotnitsky and Khrennikov [11] for a variety of approaches. In turn, the information treatment of physics led to the understanding that in physics numbers have to be considered as the fundamental entities [12].

This number-theoretic approach to physics implies recognition of the role of prime numbers, see above citations: " into which everything is resolved and out of which everything is made," "the atoms of arithmetic." Starting with prime numbers, we immediately arrive to the notion of *p-adic numbers*, number fields generated by the fixed prime number $p > 1$, and all its powers. Nowadays, the p-adic number fields serve as a novel mathematical tool for a variety of applications, from quantum physics to theory of complex disordered systems and geology, e.g., [12–18].

2 Primes: From Basics of Number Theory to Fractals

The distribution of prime numbers was one the basic problems which led to creation of number theory, especially analytic number theory. We recall that this is trivially formulated problem of accounting the number of primes $\pi(x)$ in the segment $[0, x]$. One of the first contributions to solution of this problem was due to [33]. He derived the approximate integral formula distribution function $\pi(x)$:

$$Li(N) = \int_2^N \frac{1}{\ln t} dt; \tag{1}$$

So the function $\pi(x)$ is approximately equal to $Li(x)$, $x \to \infty$. This integral representation served as the basis for further study of the asymptotic of $\pi(x)$. Finally, the following celebrating asymptotic relation was obtained:

$$\pi(x) \sim x/\ln x. \tag{2}$$

Nowadays, after development of fractal theory, a lot attention was paid to *the fractal structure of prime numbers and its applications to data analysis*, see, for example, Cattani [19], Cloitre [20], and Batchko [21]. This (random) fractal structure of the set of prime numbers P is encoded not directly in the distribution of primes, but in *the distribution of gaps between the successive primes*:

$$g_n = p_{n+1} - p_n.$$

Theoretical analysis of the structure of these gaps is very complicated and we present just a few basic results. The main efforts were set to estimate asymptotically g_n from below; the best known result is due to Rankin: there exists a constant c>0 such that the equality

$$g_n = \frac{c \ln n \ln \ln n \ln \ln \ln \ln n}{(\ln \ln \ln n)^2}. \tag{3}$$

holds for infinitely many values n. However, for applications, one is interested in the probability distribution of these gaps and its features. This is a very complicated mathematical problem; its theoretical study has not yet led to rigorous mathematical results; the main achievements were approached with the aid of numerical simulation, see papers of Wolf [22, 23] and Scafetta et al. [24].

Denote by $h_N(d)$ *the number of gaps of the length d between successive primes less than N.*

The results of computer simulation obtained [22] shows that $h_N(d)$ is distributed exponentially:

$$h_N(d) \sim \exp\{-d\pi(N)/N\}. \tag{4}$$

We note that this quantity determines the probability of randomly selecting a pair of consecutive primes separated by a gap of size d.

Some authors, e.g., Julia [25] and Wolf [26], proposed statistical physics models in which prime numbers played the role of particles. In particular, in Wolf [26] there was constructed a statistical model of "prime gas," cf. with above discussion on prime atoms and information physics. Here the prime numbers acted as particles; in this model, the nth prime p_n has the energy $g_n = p_{n+1} - p_n$, i.e., the gap between successive primes. By taking into account the asymptotic of the prime counting function, the quantity $h_N(d)$ can be represented as:

$$h_N(d) \sim \exp\{-d/\ln N\}. \tag{5}$$

In Wolf [26], the quantity $T = \ln N$ was interpreted as *temperature of the prime gas*, so $h_N(d)$ was interpreted as the Boltzmann factor.

Recently prime numbers found novel applications to geophysics. Here the prime number structure of data is closely related with the fractal geometry of capillary networks in random porous media [27, 28]. Prime numbers provide compact and informationally rich representation of the underlying fractal structure. By taking into account data which is encoded not only by prime numbers but even their powers, we improve richness of representation. Mathematically the latter is realized in the fields of p-adic numbers. Recently this approach was actively explored to model diffusion of fluids, e.g., oil, oil-in-water, and water-in-oil emulsions in capillary networks [15–18].

3 Searching Primes in Natural Phenomena: Some Representative Examples

As we emphasized in the title of this paper, *prime numbers are everywhere*; they determine the basic number-theoretic structure of various natural phenomena. At the same time, it is not easy to extract the prime number structure from real data. The components of some data-set corresponding to different primes (and their powers generating p-adic fields) "interfere" resulting in the standard real number structure of graphs and images.

Consider, as the important illustration, p-adic physics [12–18]; it is about the p-adic components (for all primes, p>1) of physical quantities; for example, string amplitudes. These components are unified with the aid of *the adele ring*. The integral effect of these components corresponds to the real physical quantity. However, to decompose the latter in its p-adic components, for p=2,3, . . . , 1997, 1999, ..., is the very difficult problem which has not yet been solved. This problem can be compared with the problem of factorization of a natural number n into the prime factors. Up to now, nobody was able to design algorithm that can factor all integers in polynomial time.

In theory of complex disordered systems (spin glasses), the p-adic structure is encoded [29] in the *Parisi matrix*. However, it is merely the exhibition of the ultrametric structure corresponding to the internal hierarchy inside spin glasses and not their number-theoretic features. We also point to a series of works to detect the ultrametric structure in variety of data-sets [30, 31].

We have designed two computer programs in order to extract *the prime numbers and corresponding p-adic structure from multiscale and multiphysics images and time series*. In the following examples, we will show some more interesting examples of contrasting natural systems analysis.

3.1 Image Analysis

3.1.1 Prime Numbers Distribution in Atmosphere

The first example refers to the quantification of some optical effects by primes. The coronae and diffraction were chosen as optical effects which, in words of [34] "teach much about physics and especially optics." The images of moonrise were taken by the Observer (Fig. 1a) during the sunset. The Observer was sitting on the beach taking images of moonrise during around two hours, when the sunset has occurred behind her. This was the Full Moon day in Los Cabos, Mexico, when the moon and sun were perfectly aligned. The primes were extracted directly from these photographic images, showing, first, the clear ring-like patterns of sun light diffraction on the water saturated atmosphere, and, then, the moon rings. The contrasting images of critical phenomena (Fig. 1c) occurred during the change of sun rings pattern (Fig. 1b) to moon pattern (Fig. 1d), which were documented not only visually but also by prime numbers statistics.

We present a few images of primes distribution across the sunset/moonset transition (Fig. 1a–d). Only prime numbers were mapped and colored. In fact, in the pure atmosphere (without clouds) saturated with water vapor the primes follow the real number line in very strong order. This ring-like structure is not only straightforwardly visible (with very small "interference" of the consecutive primes, without any noisy counterparts of images. Notwithstanding, any original image transformation (change of size, gamma of colors, etc.) is following by the change of primes distribution across the number matrix. Thus, *primes are really everywhere and nowhere*.

The similar but more complex prime numbers patterns were extracted from images of Atmosphere–Water–Solid Surface Interfaces, where the Roughness was statistically encrypted by prime numbers.

3.2 Time Series Patterns Encrypted by Prime Numbers

During any geophysical exploration, the volume of information workflow is always the huge problem. We are successfully using the primes patterns in order to drastically decrease the data volume. In Fig. 2a–c, the example of oil pressure representation, the primes decrease the length of the original time series (more than 19,000 values) to only 6% without any statistically significant change in graphic shape and behavior. The zoom applied to the curve shows the strong similarity between the original curve and its primes copy.

4 Conclusions

There are two main questions we are trying to answer now. First of all, we are looking for the distribution of gaps between the primes extracted from the natural scenes under contrasting physical conditions, as well as from the corresponding time series (for instance, the radar and laser data matrices). The other question is: *from where the primes came to the real physical word and where these numbers disappeared*?

In order to answer the last question, we have analyzed a set of NASA images, taken from the cosmic space. On the prime numbers map (Fig. 3b) extracted from the NASA original image (Fig. 3a), the prime numbers disappeared on the bound of atmosphere. We conclude that *the prime number distribution is closely related with the nature of energy source*.

Acknowledgements This paper was financially supported by the project SENER-CONACYT-Hidrocarburos, Yacimiento Petrolero como un Reactor Fractal, N 168638 and by the Consejo Nacional de Ciencia y Tecnología (CONACYT), Mexico, under grant 312-2015, Fronteras de la Ciencia.

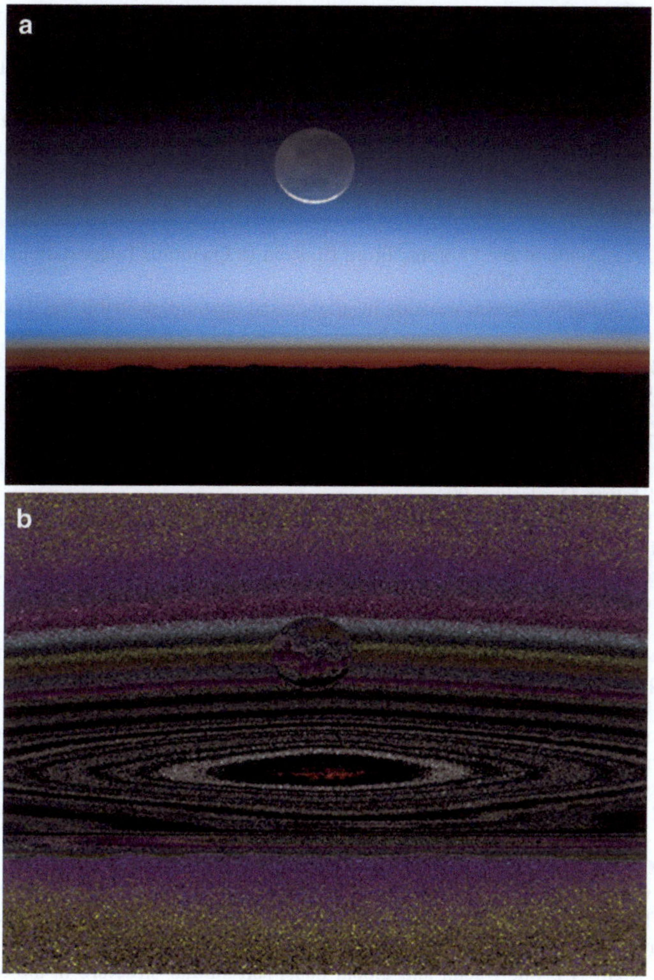

Fig. 3 (**a**) NASA image taken from the cosmic space. (**b**) Prime number skeleton corresponding to the image of (**a**)

References

1. Merkle, R., Hellman, M.: Hiding information and signatures in trapdoor knapsacks. IEEE Trans. Inf. Theory. **24**, 525–530 (1978)
2. Crandall, R., Pomerance, C.: Prime Numbers: A Computational Perspective. Springer, New York (2005)
3. Shankar, S., Udupi, V.R.: A dynamic security protocol for face recognition systems using seismic waves. Int. J. Image Graph. Signal Process. **7**(4), 28–34 (2015)
4. Lamb, E.: Peculiar pattern found in 'random' prime numbers: last digits of nearby primes have 'anti-sameness' bias. Nature News. http://www.nature.com/news/peculiar-pattern-found-in-random-prime-numbers-1.19550 (2016)

5. Nicomachus of Gerasa: Introduction to Arithmetic (Arithmetike eisagoge). Macmillan, London (1926)
6. du Sautoy, M.: The Music of the Primes: Searching to Solve the Greatest Mystery in Mathematics. Harper Collins, New York (2004)
7. Wheeler, J.A.: Information, physics, quantum: the search for links. In: Zurek, W.H. (ed.) Complexity, Entropy, and the Physics of Information. Addison-Wesley, Redwood City, California (1990)
8. Zeilinger, A.: A foundational principle for quantum mechanics. Found. Phys. **29**, 631–643 (1999)
9. Zeilinger, A.: Dance of the Photons: From Einstein to Quantum Teleportation. Farrar, Straus and Giroux, New York (2010)
10. Brukner, C., Zeilinger, A.: Information invariance and quantum probabilities. Found. Phys. **39**, 677–689 (2009)
11. Plotnitsky, A., Khrennikov, A.: Reality without realism: on the ontological and epistemological architecture of quantum mechanics. Found. Phys. **45**, 1269–1300 (2015)
12. Volovich, I.: Number theory as the ultimate physical theory. P-Adic Numbers Ultrametric Anal. Appl. **2**, 77–87 (2010). preprint 1987
13. Khrennikov, A.: P-adic Valued Distributions and their Applications to the Mathematical Physics. Kluwer, Dordreht (1994)
14. Dragovich, B., Khrennikov, A., Kozyrev, S.V., Volovich, I.V.: On p-adic mathematical physics. P-Adic Numbers Ultrametric Anal. Appl. **1**(1), 1–17 (2009)
15. Khrennikov, A., Kozyrev, S.V., Oleschko, K., Jaramillo, A.G., de Jesus Correa Lopez, M.: Application of p-adic analysis to time series. Inf. Diml. Anal. Quant. Prob. Related Topics. **16**, 1350030 (2013)
16. Khrennikov, A., Oleschko, K., de Jesús Correa Lopez, M.: Modeling fluid's dynamics with master equations in ultrametric spaces representing the treelike structure of capillary networks. Entropy. **18**, 249 (2016)
17. Khrennikov, A.Y., Oleschko, K., de Jesús Correa López, M.: Applications of p-adic numbers: from physics to geology. Contemp. Math. **665**, 121–131 (2016)
18. Khrennikov, A., Oleschko, K., de Jesus Correa Lopez, M.: Application of p-adic wavelets to model reaction-diffusion dynamics in random porous media. J. Fourier Anal. Appl. **22**, 809–822 (2016)
19. Cattani, C.: Fractal patterns in prime numbers distribution. In: Lecture Notes in Computer Science, vol. 6017. Springer, Berlin, Heidelberg (2010)
20. Cloitre, B.: On the fractal behavior of primes. http://bcmathematics.monsite-orange.fr/FractalOrderOfPrimes.pdf (2011)
21. Batchko, R.G.: A prime fractal and global quasi-self-similar structure in the distribution of prime-indexed primes. arXiv:1405.2900 [math.GM] (2014)
22. Wolf, M.: Some conjectures on the gaps between consecutive primes. In: The Distribution of Prime Numbers, Proceedings of the 8th Joint EPS-APS International Conference Physics (1997)
23. M. Wolf, Random walk on the prime numbers. P.R.G. Batchko, A Prime Fractal and Global Quasi-self-similar Structure in the Distribution of Prime-indexed Primes. arXiv:1405.2900 [math.GM] (2014)
24. Scafetta, N., Imholt, T., Roberts, J.A., West, B.J.: An intensity-expansion method to treat non-stationary time series: an application to the distance between prime numbers. Chaos Solitons Fractals. **20**, 119–125 (2004)
25. Julia, B.L.: Statistical theory of numbers. Physica A. **203**, 425 (1994)
26. Wolf, M.: Applications of statistical mechanics in number theory. Physica A. **274**, 149–157 (1999)
27. Oleschko, K., Korvin, G., Figueroa, B., Vuelvas, M.A., Balankin, A.S., Flores, L., Carreon, D.: Fractal radar scattering from soil. Phys. Rev. E. **67**, 041403 (2003)
28. Oleschko, K., Parrot, J.-F., Ronquillo, G., Shoba, S., Stoops, G., Marcelino, V.: Weathering: toward a fractal quantifying. Math. Geol. **36**, 607–627 (2004)

29. Kozyrev, S.V.: Dynamics on rugged landscapes of energy and ultrametric diffusion. P-Adic Numbers Ultrametric Anal. Appl. **2**, 122–132 (2010)
30. Murtagh, F.: On ultrametric algorithmic information. Comput. J. **53**, 405–416 (2010)
31. Murtagh, F.: The new science of complex systems through ultrametric analysis: application to search and discovery, to narrative and to thinking. P-Adic Numbers Ultrametric Anal. Appl. **5**, 326–337 (2013)
32. Kornei, K.: Synopsis: quantum cryptography goes a long way. Physics. (2016)
33. Malachovskiy, V.C.: The Known and Unknown Numbers, p. 182. FGUIPP, Yantarniy Skaz, Kaliningrad (2004)
34. Cowley, L., Laven, P., Vollmer, M.: Rings around the sun and moon: coronae and diffraction. Phys. Educ. **40**, 51–59 (2005)

The Logical Combinatorial Approach Applied to Pattern Recognition in Medicine

Martha R. Ortiz-Posadas

Abstract The logical combinatorial approach of the pattern recognition theory works with the description of objects in terms of a combination of quantitative and qualitative variables, giving the possibility to consider absent information for the values of some variables in the object description. This approach uses supervised classification algorithms, which are based on the concept of partial precedence, that is, partial analogies (an object can be alike to another object, not in its totality). These characteristics are suitable to model classification problems in Medicine. The objective of this work is to show the usefulness of the logical combinatorial approach to solve problems of pattern recognition in Medicine by illustrating three case studies: the differential diagnosis of Glaucoma, a method for comparing somatotypes (human body types in terms of physical structure), and the prognosis of rehabilitation of patients with cleft lip and palate.

Keywords Medical pattern recognition • Logical combinatorial approach • Supervised classification

1 Introduction

The logical combinatorial approach of pattern recognition theory [1] works with the descriptions of the objects in terms of a combination of quantitative and qualitative variables. These variables can be processed by numeric functions in a differential manner, depending on their nature. Furthermore, this approach gives the possibility to consider absent information for the values of some variables in the object descriptions. For classifying objects, there are algorithms based on the concept of partial precedence, that is, partial analogies. An object can be similar to another object but not in its totality; those parts which do look, although not of the same

M.R. Ortiz-Posadas (✉)
Electrical Engineering Department, Universidad Autónoma Metropolitana-Iztapalapa, Av. San Rafael Atlixco No. 186, Col. Vicentina, Delegación Iztapalapa, C. P. 09340, Ciudad de México, Mexico
e-mail: posa@xanum.uam.mx

B. Toni (ed.), *New Trends and Advanced Methods in Interdisciplinary Mathematical Sciences*, STEAM-H: Science, Technology, Engineering, Agriculture, Mathematics & Health, DOI 10.1007/978-3-319-55612-3_8

magnitude, can give information about possible regularities. All these characteristics are suitable to model classification problems in soft sciences such as Medicine.

The logical combinatorial approach has been applied to several medical problems. For example, a new approach for the differential medical diagnosis was proposed; based on the change in the characterization of the disease period, it offered k-admissible values for patient signs and symptoms in order to represent different degrees in symptom manifestation, and this provides more flexibility in making differential diagnosis for physicians [2]. Later, a new method for a complete description of the primary and secondary cleft palates was proposed, incorporating elements related to the palate, lip, and nose, that were also reflecting the surgical complexity of this medical problem. Using this method, it is possible to incorporate elements that are not considered in other approaches and to describe all possible clefts that may exist [3]. Other work shows a similarity function modeled by this approach for comparing the condition of patients with this kind of malformations. The function includes the importance of every variable as well as a weight which reflects the surgical complexity of the cleft [4], and a computational tool for the prognosis of the rehabilitation of these patients was developed [5]. It is important to say that these kinds of congenital malformations are treated by a Cleft Palate Team encompassing four medical specialties: reconstructive surgery, orthodontics, speech therapy, and psychology. In this sense, there are some works applying the logical combinatorial approach for modeling each aspect. There is a work [6] that shows how a similarity function modeled by this approach can be used to compare the orthodontic condition of patients with cleft lip and/or palate; and how this function allows to assess the effect of treatment in a single patient, hence providing valuable auxiliary criteria for medical decision-making as to the patient's rehabilitation. There is also a method to evaluate the speaking quality of patients with cleft palate [7] by a similarity function modeled by this approach. This method allows the specialists to know the speaking quality of their patients performed during rehabilitation and they can modify or continue with their planned therapeutic strategy.

There are other biological fields where this approach has been applied. Such is the case for the cytogenetic analysis. Cytogenetics is a branch of genetics that is concerned with the study of the structure and function of the cell, especially the chromosomes. The identification of chromosomes must be performed in the process of cellular division (metaphase). Related with this topic, there is a mathematical model that considers three characteristics related to the chromosome form: size, centromere position, and secondary constrictions; and a computational tool was developed based in such model. This tool helps in the development of new techniques applied in automated systems for the identification of chromosomes allowing better precision in this task [8].

Health technology management is another field where the logical combinatorial approach has been applied. For example, due to the wide variety of medical equipments existing in clinical areas, there is an important question: which and how often electrical safety tests must be applied to the medical equipment. There are important differences about the electrical safety, such as the electrical insulation or the hazard considering the connection between patient and electrical instrumentation. In this sense, an electrical safety priority index for medical equipment was established [9], involving different classifications related with electrical safety, in order to provide a numeric code indicating the priority and frequency for applying the electrical safety tests to medical equipment. Also a strategy for the distribution of the imaging services of the Costa Rican Public Health System was developed, for a more equitable access to the entire population of the country [10]. That for, different aspects were considered such as the type and number of the imaging equipment available in the health care centers, as well as the distance and time that take to travel to get the service in the Capital, among others. These aspects were weighted and included into a mathematical function, whose result is an indicator that considers the conditions of the current imaging service and the community access to it. The function was applied to eight different regions, and the results showed in which ones it was necessary to strengthen the infrastructure and resources of its imaging services, in order to provide a more equitable service across the country.

The objective of this work is to show the usefulness of the logical combinatorial approach to deal with problems of pattern recognition in Medicine by illustrating three case studies: the differential diagnosis of glaucoma, a method for comparison of somatotypes, and the prognosis of the rehabilitation of patients with cleft lip and palate. I present not just the mathematical model of each clinical problem but also the solution of the problem classification using an algorithm based on partial precedence, tested with a patient sample.

2 Mathematical Model

Let U be a universe of objects, and let us consider a given finite sample $\boldsymbol{O} = \{O_1, \ldots, O_m\}$ of such (descriptions of the) objects. We shall denote by $X= \{x_1, \ldots, x_n\}$ the set of features or variables used to study these objects. Each of these features has associated a set of admissible values (its domain of definition) *Mi*, i= 1, ..., n. These sets of values, in contrast to other approaches, can be of any nature: variables can be quantitative and qualitative simultaneously. Each of these sets contains a special symbol denoting the absence of information (missing data). Thus, some variables are numeric; others, symbolic; incomplete information about some objects is allowed. This will turn out to be a fundamental feature of this pattern recognition paradigm. By a description of an object O, we understand an n-tuple $I(O)=(x_1(O), \ldots, x_n(O))$, where x_i: $M \rightarrow M_i$, for i=1, ..., n, are the variables of features used to describe it. Over M_i, no algebraic or topologic structure is assumed. Any of the pattern recognition problems mentioned above is formulated from a set of descriptions of *m* such objects [11].

Definition 1 Let the initial space representation (ISR) be the object space representation defined by the Cartesian product of M_i sets:

$$I(O) = (x_1(O), \ldots, x_n(O)) \in M_1 \times \ldots \times M_n$$

Remark There is no norm or algebraic operation over M_i defined a priori. But this does not mean that they cannot be present in ISR. Sometimes one can consider a function which does not satisfy the norm properties over M_i (over ISR).

Definition 2 Let $\boldsymbol{C} = \{C_1, \ldots, C_n\}$ be a set of functions called comparison criteria for each variable $x_i \in X$ such as: C_i: $M_i \times M_i \rightarrow \Delta_i$; $i=1, \ldots, n$ where Δ_i can be of any nature; it is an ordered set and can be finite or infinite.

Remark Comparison criteria can denote similarity or difference. For the sake of illustration, we present the following examples of an ordered set Δ_i:

- If $\Delta_i = [0, 1]$, the value 1 represents maximum similarity/difference and the value zero the minimum. Any intermediate value represents a grade of similarity/difference between compared values.
- If the comparison criterion is finite valued, that is: $\Delta_i = \{0, 1, 2, \ldots, m\}$, it is not difficult to transform Δ_i into: $\Delta_i = \{0, 1/m, 2/m, \ldots, (m-1)/m, 1\}$. Obviously, this is a subset of the interval [0,1]. Then, in this case, the intermediate values Δ_i also represent a grade of similarity/difference between compared values.

The characteristics of each comparison criterion (C_i) depend on the problem being modeled. However, it is important to remark that every C_i is designed individually to reflect the nature and interpretation of each feature x_i. In this sense, the set $\boldsymbol{C}$ allows the differentiation and nonuniform treatment of the features that describe the objects. Furthermore, it also gives the possibility to consider "absent information" in some feature values in the objects descriptions. It is important to mention that all comparison criteria must be defined jointly with the experts, in order to incorporate his/her expertise. In the context of medical problems, the experts will be the physicians, surgeons, etc. with knowledge, expertise, and the ability to provide criteria about medical problem being modeled.

3 Supervised Classification

Many problems in Medicine, such as differential diagnosis of diseases, to determine the patient's treatment, or to prognosticate the patient's rehabilitation, are problems of supervised classification. In general, physicians reach their conclusions for diagnosis or prognosis on the basis of the analogies found between patients by accumulating knowledge through their experiences and observations. Thus, the analogy concept is present in almost all the reasoning of the medicine specialists. The availability of additional methods for the evaluation of such analogies or similarities, considering their description, is important to provide auxiliary criteria for medical decision-making [4, 6]. In this sense, it is convenient to introduce the concept of analogy, which is a fundamental methodological tool in Medicine.

In supervised classification problems, we assume that the universe U is structured in a finite number $K_1, \ldots, K_r$ of classes, and from each of them we have a sample of descriptions of objects, the so-called learning matrix $LM = K_1 \cup \ldots \cup K_r$.

The problem is to find the membership relations of a new object from U (outside the given samples) with r classes. This relationship does not have to be all or nothing. The logical combinatorial approach deals with spaces without algebraic (or any other kind of) structure. The representation space is simply a Cartesian product, which also has the peculiarity of being heterogeneous, that is, each of the sets forming it can be of different nature: a set of real numbers, a set of labels, a set of truth values, etc. [11]. An example of this appears in medical diagnosis problems, where patients descriptions take the form $I(O)$ = (female, 45, 38.6, *, slight, 2). In soft sciences such as Medicine, objects are described in terms of qualitative and quantitative variables. In the logical combinatorial approach, most algorithms of supervised classification are based on partial precedence as the Voting algorithm, described below.

3.1 Voting Algorithm

This algorithm comprises six steps [1]: (1) defining the system of support sets; (2) defining the similarity function; (3) row evaluation, given a fixed support set; (4) class evaluation for a fixed support set; (5) class evaluation for all the system of support sets, and (6) resolution rule. Thus, to define a voting algorithm is to define a set of parameters for each of the above six steps.

(1) Defining the system of support sets. A support set is a nonempty subset of features which shall be used to analyze the objects, defined as:

Definition 3 Let $\omega \subseteq X$ be a support set, where $\omega \neq \varnothing$. A system of support sets is defined as $\Omega = \omega_1, \ldots, \omega_s$. By ωO, we denote the ω-part of O formed by the variables $x_j \in \omega_m$, m=1, ..., s.

The system of support sets Ω will allow analysis of the objects to be classified, by paying attention to different parts or sub-descriptions of the objects, and not analyzing the complete descriptions. Examples of systems of support sets are combinations with a fixed cardinality, combinations with variable cardinality, the power set of features, etc.

(2) Defining the similarity function. The analogy between two objects is formalized by the concept of similarity function. This function is based on the comparison criterion C_i generated for each variable x_i. It is important to mention that the similarity function can evaluate the similarity or difference between two objects, i.e., between their descriptions.

Definition 4 Let $\beta:(M_i \times M_i)^2 \rightarrow \Delta$ be the similarity function, where Δ (as in the comparison criterion function) can be of any nature; it is an ordered set and can be finite or infinite. For $I(O_i)$ and $I(O_j)$ being two object descriptions in the domain $(M_1 \times \ldots \times M_n)$, $\beta(I(Oi), I(O_j))$ is defined by:

- $\beta((C_1(x_1(O_i), x_1(O_j)), \ldots, C_n(x_n(O_i), x_n(O_j))))$, if C_i denotes similarity
- $1 - \beta((C_1(x_1(O_i), x_1(O_j)), \ldots, C_n(x_n(O_i), x_n(O_j))))$, if C_i denotes difference

Remark Both, the comparison criterion and the similarity function, are not necessarily symmetric and, in general, these functions are not defined as positive, they do not have to satisfy the triangle inequality, and there is no a priori metric considered.

Definition 5 Let β_ω be a partial similarity function defined by:

$$\beta_\omega \left(I\left(O_i\right), I(O)\right) = 1 - \sum_{x_i \in \omega} \rho_i C_i \left(x_i\left(O_i\right), x_i(O)\right)$$

where ω represents a support set and ρ_i is the relevance parameter associated with each variable x_i defined by the expert.

(3) Row evaluation, given a fixed support set. When the system of support sets and the similarity function have been defined, the voting process starts in the stage of row evaluation. The similarity between the different parts of the objects already classified and those to be classified is analyzed. Each row of *LM* (each object $Oi \in LM$) is compared with the object *O* to be classified using the similarity function β_ω. This evaluation is a function of the similarity values among the different parts being compared. An example of this evaluation is: $\Gamma_\omega(O_i,O) = \rho(\omega)\beta(\omega I(Oi), \omega I(O))$, where $\rho(\omega)$ is the weight of the support set ω and $\beta(\omega I(O_p),\omega I(O))$ is the similarity value of the compared objects.

(4) Class evaluation for a fixed support set ω. This consists in totaling the evaluations obtained for each of the objects in *LM* with respect to the object *O* to be classified. This total evaluation is a function of the row evaluations already obtained. An example of this evaluation is: $\Gamma_\omega^j(\mathrm{O}) = 1/\left|K_j\right| \left[\sum_{t=1,\ldots,\left|K_j\right|} \Gamma_\omega\left(\mathrm{O_i}, \mathrm{O}\right)\right]$. The upper index refers to the class K_j.

(5) Class evaluation for all the system of support sets Ω. Evaluations are totaled for all the system of support sets. Following our example, this step could be expressed as follows:

$$\Gamma^{\mathrm{j}}(\mathrm{O}) = \frac{1}{|\Omega|} \sum_{\omega \in \Omega} \Gamma_\omega^j(O)$$

(6) Resolution rule. It is a function that establishes a criterion which considers each voting thus obtained and reaches a decision concerning the relations of the object to be classified with every class of the posed problem.

4 Mathematical Modeling Applied to the Clinical Problems

4.1 Differential Diagnosis of Glaucoma

Glaucoma is a set of ocular disorders that leads to optic's nerve damage, characterized by abnormally high intraocular fluid pressure, damaged optic disk, hardening of the eyeball, and partial to complete loss of vision. It is the main cause for irreversible blindness around the world. The World Health Organization estimates 37 million affected people, prognosticating 80 million by 2020 [12]. Accurate diagnosis of glaucoma allows initiating drug and surgical treatment to avoid vision loss. In the case of congenital glaucoma (incorrect or incomplete development of the eye's drainage canals during the prenatal period), its differential diagnosis is very important because prescription of corticoids in young people is contraindicated, because it interferes with normal child development. The objective of this work was to develop a method for the differential diagnosis of glaucoma using the logical combinatorial approach.

4.1.1 Mathematical Model

Six variables were defined: intraocular pressure (IOP), optic nerve/disc ratio (DR), iridocorneal angle (ICA), visual acuity (VA), presence of scotomas (a partially diminished or entirely degenerated visual acuity which is surrounded by a field of normal—or relatively well-preserved—vision) (S), and age (A) [13]. Likewise, six comparison criteria of difference were modeled considering Definition 2 (Section 2). The zero value means that compared values are equal (there is no difference), and the 1 value means that compared values have the greatest difference. The variables, their domain, and their comparison criterion are shown in Table 1. Observe that the comparison criterion function considers the absence of information denoted as (*).

The comparison between the descriptions of two eyes was determined considering Definition 4 (Section 3.1) by the Glaucoma similarity function β_{Glaucoma} as:

$$\beta_{Glaucoma}\left(I\left(P_1\right),I\left(P_2\right)\right)=1-\left(\sum_{i=1}^{6}\rho_i C_i\left(x_i\left(P_1\right),x_i\left(P_2\right)\right)/\sum_{i=1}^{6}\rho_i\right) \quad (1)$$

where ρ_i is the relevance parameter associated with each variable x_i defined by the expert. In this case: $\rho_{IOP}=0.375$, $\rho_{DR}=0.25$, $\rho_{ICA}=\rho_{VA}=0.125$, and $\rho_S=\rho_A=0.0625$.

4.1.2 Classification

The method was tested with a sample of 58 different clinical cases of glaucoma diagnosis, reported in the medical literature [13]. Some cases were about bilateral glaucoma, so the information was considered for both eyes. The learning matrix (LM) was constructed with 48 eyes, and the control matrix (CM) with 10 eyes. The following five classes were defined:

K_1: Open-angle Glaucoma (OAG); 12 eyes in LM and 2 in CM.
K_2: Close-angle Glaucoma (CAG); 14 eyes in LM and 2 in CM.
K_3: Congenital Glaucoma (CG); 7 eyes in LM and 2 in CM.

Table 1 Domain and comparison criteria of six glaucoma variables

x_i	M_i	C_i					
1. IOP			N	LH	HI	HII	Vh
	Normal	N	0	0.7	0.8	0.9	1
	Lightly high	LH		0	0.7	0.8	0.9
	High I	HI			0	0.7	0.8
	High II	HII				0	0.7
	Very high	VH					0
2. DR			N	S	*	D	
	Normal	N	0	0.4	0.5	1	
	Suspect	S		0	0.4	0.5	
	*	*			0	0.4	
	Damage	D				0	
3. ICA			G0	GI	GII	GIII	GIV
	Grade 0	G0	0	0.1	0.3	0.8	1
	Grade I	GI		0	0.1	0.3	0.8
	Grade II	GII			0	0.1	0.3
	Grade III	GIII				0	0.1
	Grade IV	GIV					0
4. VA	{[0, 1], *]}	|x−y| (* = 1)					
5. S	{Yes, No, *}		Yes	No	*		
		Yes	0	1	1		
		No		0	1		
		*			0		
6. A			a	b	c	d	e
	a = [0, 30] years	a	0	0.25	0.5	0.75	1
	b = (31, 45] years	b		0	0.25	0.5	0.75
	c = (45, 60) years	c			0	0.25	0.5
	d = [60, 70) years	d				0	0.25
	e = [70, 100] years	e					0

K_4: Normal pressure Glaucoma (NPG); 9 eyes in LM and 2 in CM.

K_5: Eye hypertension (EHT); 6 eyes in LM and 2 in CM.

For the classification of 10 eyes in the CM, we used Eq. (1) for calculating the similarity of each of these eyes with the 48 eyes in LM, and for the classification, the voting algorithm described above. The results are shown in Table 2. Note that the values highlighted in bold correspond to the higher values obtained by the Voting algorithm for the classification.

The algorithm efficiency was 90%, because eye number 4 was incorrectly classified. This eye obtained the great majority of votes for K_5, classifying it as EHT, leaving K_1 (OAA) in the second place, which is the class where the eye really belongs to. The algorithm determined its similarity to this class probably because this eye is an atypical case of GAA, being more alike to an HTO case.

Table 2 Classification of 10 eyes in control matrix

	Eye	K_1	K_2	K_3	K_4	K_5
K_1	4	0.344	0.263	0.317	0.309	**0.372**
	11	**0.336**	0.245	0.334	0.305	0.332
K_2	19	0.223	**0.313**	0.301	0.213	0.250
	24	0.223	**0.373**	0.346	0.218	0.219
K_3	33	0.272	0.266	**0.372**	0.254	0.280
	39	0.215	0.274	**0.363**	0.215	0.202
K_4	42	0.222	0.136	0.225	**0.409**	0.280
	48	0.229	0.146	0.247	**0.436**	0.257
K_5	52	0.242	0.168	0.229	0.264	**0.448**
	56	0.284	0.193	0.224	0.248	**0.431**

4.1.3 Conclusion

This method provides a new approach for the diagnosis of glaucoma, because it allows the absence of information in the values of some variables, and this is a common situation in patient's description. Sometimes, the patient does not take all the required studies and all the information needed to do the diagnosis is not available.

The method proposed here is useful in making the diagnosis of glaucoma at an earlier stage and is also useful to determine the progression of damage at an earlier stage than the traditional tests, because the similarity function allows evaluating the likelihood not only of different eyes, but of the same eye in different points in time.

4.2 A Method for Comparing Somatotypes

The term somatotype is defined as the quantification of the present shape or physical structure and composition of a human body and is one of the most frequent tasks of Kineanthropometry, the discipline that studies the human body through the measurements and assessments of their size, shape, proportionality, composition, biological maturation, and body functions. The technique of somatotyping is used to appraise body shape and composition. Somatotype concept is very useful in different areas of healthcare, such as diet monitoring, effect of ergogenic aids, eating disorders, and/or sport sciences, in order to compare an athlete somatotype with his/her team, or with a standard reference, or with a normal population, or itself at different stages of the training [14]. It is expressed in a three-number rating, always in the same order, representing the components endomorphy, mesomorphy, and ectomorphy, respectively. Endomorphy is the relative fatness, mesomorphy is the relative muscle-skeletal robustness, and ectomorphy is the relative linearity or slenderness of a physique [15] (Fig. 1).

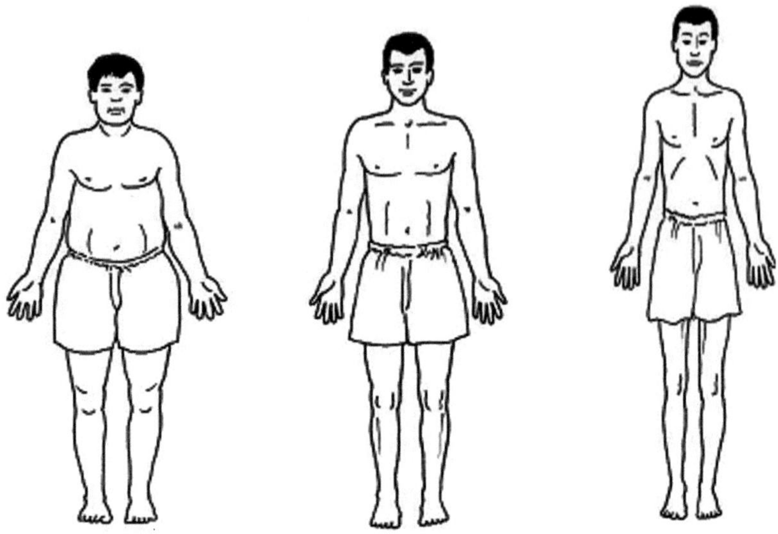

Fig. 1 The three main somatotypes [15]

The Heath–Carter Method of somatotyping is most commonly used today, and it is necessary to calculate the three components mentioned above. The equation for endomorphy [14] is a third-degree polynomial (Eq. 2).

$$Endomorphy = -0.7182 + 0.1451\ (\mathrm{X}) - 0.00068\ \left(\mathrm{X}^2\right) + 0.0000014\ \left(\mathrm{X}^3\right) \quad (2)$$

where X = (sum of triceps, subscapular, and supraspinale skinfolds) multiplied by 170.18/height in cm.

The Eq. (3) calculates mesomorphy [14]:

$$Mesomorphy = 0.858 \times \text{humerus breadth} + 0.601 \times \text{femur breadth}$$
$$+0.188 \times \text{corrected arm girth} + 0.161 \times \text{corrected calf girth} - \text{height } 0.131 + 4.5 \quad (3)$$

Three different equations are used to calculate ectomorphy [14] according to the height–weight ratio (Eqs. 4, 5, and 6):

If HWR is greater than or equal to 40.75, then

$$Ectomorphy = 0.732\ \mathrm{HWR} - 28.58 \quad (4)$$

If HWR is less than 40.75 but greater than 38.25, then

$$Ectomorphy = 0.463\ \mathrm{HWR} - 17.63 \quad (5)$$

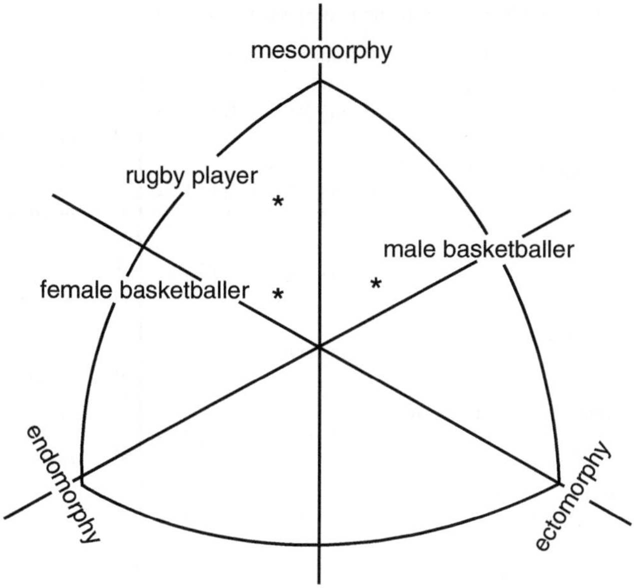

Fig. 2 Scheme of the somatochart used in Heath–Carter Method (Adapted from [17])

If HWR is equal to or less than 38.25, then

$$Ectomorphy = 0.1 \tag{6}$$

Traditionally, the three-number somatotype rating is plotted on a two-dimensional somatochart using X,Y coordinates derived from the rating (Fig. 2).

The objective of this work [16] was to develop a direct method for somatotype classification, without using the equations and the somatochart by the Heath–Carter Method, considering the three main somatotype classes: endomorph, mesomorph, and ectomorph.

4.2.1 Mathematical Model

We used the 10 anthropometric dimensions (variables) proposed in the Heath–Carter Method [14] and we defined their domain and their difference comparison criterion considering Definition 2 in Section 2 (Table 3). In this criterion, 0 means that there is no difference, and 1 represents the greatest difference between the two values compared.

We defined three support sets (Definition 3, Section 3.1), one for each somatotype component: $\omega_{endo} = \{x_1, x_2, x_3, x_{10}\}$; $\omega_{meso} = \{x_3, x_4, x_5, x_6, x_7, x_8\}$; and $\omega_{ecto} = \{x_9, x_{10}\}$. So we defined three partial similarity functions considering Definition 5 (Section 3.1) as follows:

Table 3 Somatotype variables, domain, and comparison criterion

Variable	Domain	Comparison criteria
x_1: Supraspinale skinfold	[6, 55]	$C_2(x_2(O_i), x_2(O_j)) = \begin{cases} 1 & if\ \frac{\lvert x_2(O_i)-x_2(O_j)\rvert}{49} \leq 0.1 \\ 0 & in\ other\ case \end{cases}$
x_2: Subscapular skinfold	[8, 41]	$C_2(x_2(O_i), x_2(O_j)) = \begin{cases} 1 & if\ \frac{\lvert x_2(O_i)-x_2(O_j)\rvert}{33} \leq 0.1 \\ 0 & in\ other\ case \end{cases}$
x_3: Triceps skinfold	[4, 25]	$C_2(x_2(O_i), x_2(O_j)) = \begin{cases} 1 & if\ \frac{\lvert x_2(O_i)-x_2(O_j)\rvert}{21} \leq 0.1 \\ 0 & in\ other\ case \end{cases}$
x_4: Medial calf skinfold	[0.5, 3.0]	$C_2(x_2(O_i), x_2(O_j)) = \begin{cases} 1 & if\ \frac{\lvert x_2(O_i)-x_2(O_j)\rvert}{2.5} \leq 0.1 \\ 0 & in\ other\ case \end{cases}$
x_5: Calf girth, right	[30.0, 43.0]	$C_2(x_2(O_i), x_2(O_j)) = \begin{cases} 1 & if\ \frac{\lvert x_2(O_i)-x_2(O_j)\rvert}{13} \leq 0.5 \\ 0 & in\ other\ case \end{cases}$
x_6: Upper arm girth, elbow flexed and tensed	[29.0, 41.0]	$C_2(x_2(O_i), x_2(O_j)) = \begin{cases} 1 & if\ \frac{\lvert x_2(O_i)-x_2(O_j)\rvert}{12} \leq 0.5 \\ 0 & in\ other\ case \end{cases}$
x_7: Biepicondylar breadth of the femur	[8.0, 12.0]	$C_2(x_2(O_i), x_2(O_j)) = \begin{cases} 1 & if\ \frac{\lvert x_2(O_i)-x_2(O_j)\rvert}{4} \leq 0.5 \\ 0 & in\ other\ case \end{cases}$
x_8: Biepicondylar breadth of the humerus	[5.0, 8.0]	$C_2(x_2(O_i), x_2(O_j)) = \begin{cases} 1 & if\ \frac{\lvert x_2(O_i)-x_2(O_j)\rvert}{3} \leq 0.5 \\ 0 & in\ other\ case \end{cases}$
x_9: Body mass (weight)	[51.0, 120.0]	$C_2(x_2(O_i), x_2(O_j)) = \begin{cases} 1 & if\ \frac{\lvert x_2(O_i)-x_2(O_j)\rvert}{69} \leq 0.1 \\ 0 & in\ other\ case \end{cases}$
x_{10}: Stature (height)	[157.0, 190.0]	$C_2(x_2(O_i), x_2(O_j)) = \begin{cases} 1 & if\ \frac{\lvert x_2(O_i)-x_2(O_j)\rvert}{33} \leq 0.1 \\ 0 & in\ other\ case \end{cases}$

$$\beta_{endo}\left(\Omega I(O_i), \Omega I(O_j)\right) = 1 - \sum_{t=1,2,3,10} \frac{C_t\left(x_t(O_i), x_t(O_j)\right)}{4} \tag{7}$$

$$\beta_{meso}\left(\Omega I(O_i), \Omega I(O_j)\right) = 1 - \sum_{t=3}^{8} \frac{C_t\left(x_t(O_i), x_t(O_j)\right)}{6} \tag{8}$$

$$\beta_{ecto}\left(\Omega I(O_i), \Omega I(O_j)\right) = 1 - \sum_{t=9}^{10} \frac{C_t\left(x_t(O_i), x_t(O_j)\right)}{2} \tag{9}$$

Finally, the total similarity function was composed by the three partial similarities as follows:

$$\beta_{total}\left(I(O_i), I(O_j)\right) = \frac{\beta_{endo} + \beta_{meso} + \beta_{ecto}}{3} \tag{10}$$

Table 4 Classification results of 9 somatotypes

Subject	Γ_{endo}	Γ_{meso}	Γ_{ecto}
O_1	**0.12**	0.02	0
O_2	**0.32**	0.03	0.03
O_3	**0.04**	0.02	0
O_4	0.1	**0.12**	0.09
O_5	0.1	**0.35**	0.18
O_6	0	**0.28**	0
O_7	0.06	0.15	**0.31**
O_8	0	0.08	**0.63**
O_9	0	0	**0.38**

All similarity functions were bounded in the interval [0, 1], where 0 means that there is no similarity (greatest difference) and 1 corresponds to identical somatotypes.

The procedure to calculate the similarity between two somatotypes using the similarity function is described as follows: First, the partial similarity between the three components (β_{endo}, β_{meso}, and β_{ecto}) of each pair of somatotypes is calculated using Eqs. (7), (8), and (9) respectively; and second, the overall similarity between both somatotypes is calculated using Eq. (10).

4.2.2 Classification

For testing the classification method, there was a sample of 38 subjects previously classified by the Heath–Carter Method into the three classes mentioned [16]. The LM was performed with 29 subjects: 10 in the endomorph class, 11 in the mesomorph class, and 8 in the ectomorph class. The CM was constructed with the 9 remaining subjects, 3 in each class mentioned. The classification results are shown in Table 4. Note that the values highlighted in bold correspond to the higher values obtained by the Voting algorithm for the classification. Subjects O_1, O_2, and O_3 obtained the major voting (similarity) for the endomorph class, and according to the general solution rule these were classified into this class. In the same way, subjects O_4, O_5, O_6 were included in the mesomorph class and finally, subjects O_7, O_8, O_9 were classified in the ectomorph class. In this way, 100% of the subjects were correctly classified.

4.2.3 Conclusion

The method proposed offers a new perspective for somatotype comparison; it does not need the three components rating neither the somatochart. It just uses the 10 body measurements from the individual somatotype description and the similarity function, in order to compare somatotypes. Furthermore, sometimes in biotypological research or sport sciences, it is necessary to analyze one of the three components in a separate way. Our method allows comparing (analyzing) each component in separate way by defining the support sets and the partial similarity function. Finally we have shown that our method is effective for comparing somatotypes and estimates the similarity between them, and all these characteristics make it simpler than the traditional methods.

4.3 Prognosis of the Rehabilitation of Patients with Cleft Lip and Palate

Cleft lip and palate consist of congenital malformations in the lip and/or palate, which are called cleft-primary palate and/or cleft-secondary palate, respectively (Fig. 3). Primary palate is formed by the prolabium, the premaxilla, and columella [18]. This is the "visible" part of these kinds of malformations. The secondary palate begins at the incisive foramen and extends posteriorly. It includes the horizontal portion of the premaxilla, horizontal portion of the palatine bones, and

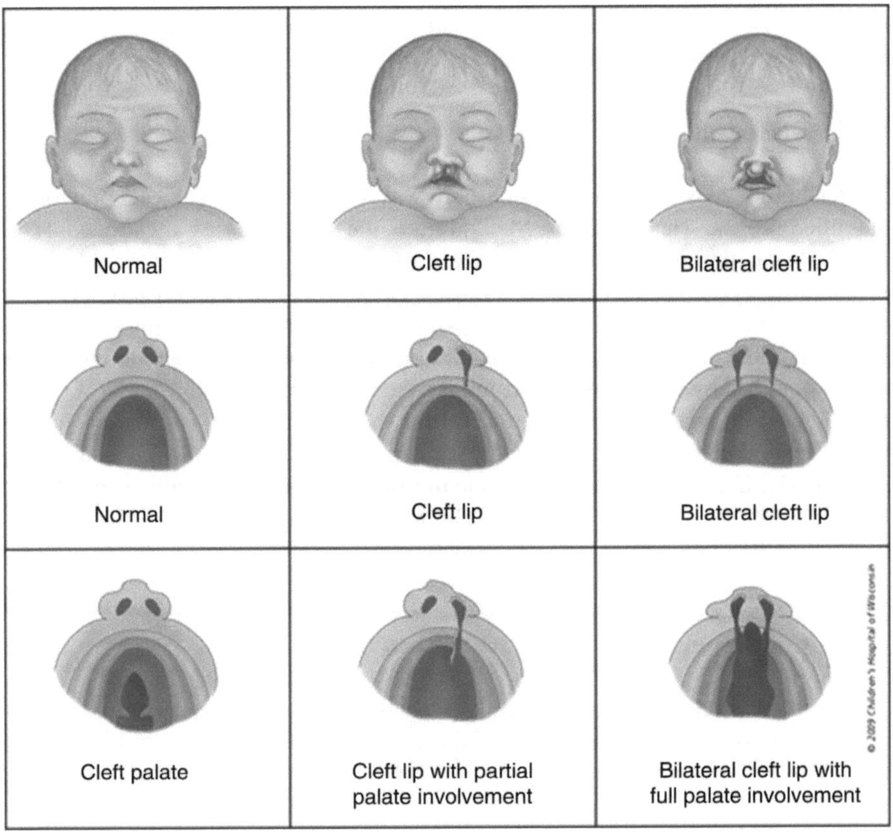

Fig. 3 Cleft lip and palate (unilateral and bilateral) [21]

soft palate [18]. The worldwide incidence of these congenital abnormalities is 1.7 per 1000 live born babies [19]. In Mexico, the incidence is 0.81 per 1000 live born babies [20], meaning more than 90,000 actual cases.

Surgical complexity for cleft reconstruction will depend on fissure complexity involving lip, nose, and/or palate. Cleft correction translates into a very slow and complex process; because it is related to the growth and development of the patient and it requires at least one surgical procedure. The importance of prognosis of the patient's rehabilitation, and subsequent evaluation of the surgical result, is the physician's self-feedback during all the rehabilitation process. The physician will learn if the work for patient rehabilitation is adequate, or if it can be improved. This has a direct consequence in the future quality of patient's life.

The objective of this work was to develop a method to help physicians to assess and prognosticate the rehabilitation of patients with this kind of congenital malformations.

4.3.1 Mathematical Model

It was necessary to define, in conjunction with surgeons, 18 variables related to the three anatomical structures affected [3]: 2 variables for cleft (primary palate and secondary palate); 9 for lip (symmetry of nasal floor, symmetry of nostril archs, symmetry of nostrils (vertical and anteroposterior plane), nasal septum deviation, length of columella, and width of nasal base), and 7 for nose (malocclusion by maxillary retrusion, malocclusion by protrusive mandible, contact of the segments bones, collapse of the maxillary (left, right, and anteroposterior), dental occlusion, overbite (vertical and horizontal), and premaxilla (horizontal and vertical plane, and centric)). These variables describe the cleft of the patient [3] and the model includes, further, the importance of every variable. Likewise, six fuzzy comparison criteria were modeled considering Definition 2 (Section 2). All of them were of absolute difference type. That is, the minimum value of its domain means that the compared values are equal (there is no difference), and the maximum value means that the compared values are different [4].

The similarity function was defined considering the partial similarity (Definition 5, Section 3.1) related with the different structures considered in cleft. Three support sets (Definition 3, Section 3.1) were defined: $\Omega = \{\omega_{\text{cleft}}, \omega_{\text{lip}}, \omega_{\text{nose}}\}$, where $\omega_{\text{cleft}} = \{x_1, x_2\}$, $\omega_{\text{lip}} = \{x_3, \ldots, x_{11}\}$, and $\omega_{\text{nose}} = \{x_{12}, \ldots, x_{18}\}$. The three partial similarity functions were defined [4] by:

$$\beta_{cleft}\left(I\left(P_1\right), I\left(P_2\right)\right) = 1 - \sum_{t=1}^{2} \rho_t C_t\left(x_t\left(P_1\right), x_t\left(P_2\right)\right) \tag{11}$$

where: $\rho_{x1} = 0.65$, $\rho_{x2} = 0.35$

$$\beta_{lip}\left(I\left(P_1\right), I\left(P_2\right)\right) = 1 - \sum_{t=3}^{11} \rho_t C_t\left(x_t\left(P_1\right), x_t\left(P_2\right)\right) \tag{12}$$

where: $\rho_t = \{0.16, 0.15, 0.14, 0.15, 0.08, 0.12, 0.10, 0.05, 0.05\}$

$$\beta_{nose}\left(I\left(P_1\right), I\left(P_2\right)\right) = 1 - \sum_{t=12}^{18} \rho_t C_t\left(x_t\left(P_1\right), x_t\left(P_2\right)\right) \quad (13)$$

where: $\rho_t = \{0.17, 0.25, 0.10, 0.10, 0.11, 0.15, 0.12\}$.
The total similarity function for cleft lip and palate β_{TC} was defined by:

$$\beta_{TC}\left(I\left(P_1\right), I\left(P_2\right)\right) = 0.60\left[\beta_{cleft}\left(I\left(P_1\right), I\left(P_2\right)\right)\right] + 0.20\left[\beta_{lip}\left(I\left(P_1\right), I\left(P_2\right)\right)\right] + 0.20\left[\beta_{nose}\left(I\left(P_1\right), I\left(P_2\right)\right)\right] \quad (14)$$

where β_{cleft}, β_{lip}, and β_{nose} are the similarity functions corresponding to the affected structures cleft, lip, and nose, with a relevance parameter of 0.60, 0.20, and 0.20, respectively.

4.3.2 Classification

The rehabilitation prognosis of patients with cleft palate is carried out by considering the original condition of the patient and taking into account the degree of rehabilitation attained by previous patients cared for in the hospital. Prognosis is conceived as a result from a supervised classification problem and uses a learning matrix (LM) made from cleft descriptions from patients who already finished their rehabilitation, and the voting classification algorithm. The LM was divided into three postsurgical classes: excellent (E), very good (VG), and good (G), determined from the evaluation of the surgical result of each patient. These classes provide the expert with a criterion for the evaluation (classification) of the degree of rehabilitation accomplished by the patient. Each patient is prognosticated (classified) by comparing his/her initial description with the initial descriptions of patients already included in the LM. The most relevant patients for the prognosis will be those who are the most similar to the patient one is about to classify. This means that the prognosis corresponds to the class that includes the patients most similar to the subject that will be classified. In this way, a patient will be predicted as VG if his/her description is most similar to patients from the LM included in the "very good" class. In the same way, evaluation of rehabilitation advance is made using the patient's postsurgical description, and applying the expert criteria which defined the postsurgical classes mentioned above. The classification will correspond to the patient's rehabilitation advance.

The method was tested with a sample of 95 patients attended by the cleft palate team at the reconstructive surgery service of the Pediatric Hospital of Tacubaya, which belongs to the Health Institute of the Federal District in Mexico City. For acquiring patient data, we designed a patient's registration form given to the surgeons, in order to fill it out with the cleft description by the variables defined for. The LM was constructed with 32 patients, distributed as follows: 10 in E class, 14

Table 5 Classification results for 63 cleft palate patients

Class (algorithm)	E	VG	G	Total
Class (inference)				
Excellent (E)	**17**	2	0	19
Very good (VG)	0	**26**	3	29
Good (G)	0	1	**14**	15
Total	17	29	17	63

in VG class, and 8 in G class, and the CM (a patient's sample already classified by the physician that is going to be classified by the voting algorithm) was built with 63 patients: 19 in E, 29 in VG, and 15 in G.

The patients correctly classified are shown in the matrix diagonal of Table 5 (values highlighted in bold). Observe that from the 19 patients in the E class, 17 were correctly classified by the algorithm and 2 were incorrectly placed in the VG class. From the 29 patients grouping in VG class, 26 were correctly classified by the algorithm and 3 were incorrectly classified in G class. For those patients in G class, 14 were correctly classified and just one was incorrectly placed into VG class. In total, there were 57 patients correctly classified, which represents 90.5%.

4.3.3 Conclusion

We developed a method to prognosticate the rehabilitation of patients with cleft lip and/or palate, considering the original condition of the patient described in terms of the variables defined for, and the degree of rehabilitation reached by previous patients cared for in the hospital. Such prognosis is the result of a supervised classification of the patients, taking into account the expertise (knowledge) of the surgeons. It is important to say that, although the method was developed for a particular medical problem, it can be useful in other fields of knowledge. This is because we developed a method to assess the evolution of a process from its initial condition to its conclusion, considering information about the process performance itself. Furthermore, if we defined expert criteria about the expected results of the process, it is feasible to evaluate the process advance at any time.

5 Conclusion

In this work, it is shown how the logical combinatorial approach is very useful in the mathematical modeling and supervised classification of medical problems. This is because this approach allows objects description including both qualitative and quantitative variables simultaneously, and it can process them depending on the variable type in an individual differential way by the comparison criterion function. Furthermore, it is also possible to compare partial (or total) object description to include both support set concept and the partial (or total) similarity function. As well as the use of the Voting classification algorithm based on the concept of partial precedence.

This paper serves as an illustration of the application of the logical combinatorial approach to pattern recognition in three medical problems conceived as a supervised classification one, in order to show that this approach is adequate for problem modeling in soft sciences.

6 Future Work

It is important to say that even though there are more and better available health technologies, physicians sometimes require specific information about a particular diagnosis or treatment that commercial equipment does not provide. Hence, the importance of developing knowledge-based computational tools to aid in medical decision-making. In this sense, the logical combinatorial approach of pattern recognition theory is currently being applied in two medical diagnostic problems: liver fibrosis and cervical cancer.

Cervical Cancer (CC) worldwide is the second most prevalent neoplasm and the fourth leading cause of cancer death in women [22]. Annually about 530,000 new cases are detected, approximately half of which will end with the death of the patient [23]. Most CC cases occur in underdeveloped or developing countries such as Latin American and Caribbean countries. Particularly in Mexico, an incidence of 23.3 cases per 100,000 women and a mortality rate of 8.1 cases per 100,000 women are estimated [24].

At present, there are technologies that can make a high resolution scan of the slides that have the biopsies, to obtain a digital image. These technologies allow to share the digital images with other pathologists around the world, in order to be trained by an expert pathologist for training how to evaluate positive patterns of immunohistochemistry with respect to the pixels of the image. However, there is no mathematical tool that allows the analysis of biopsy images which can differentiate types of epithelia and assign a diagnosis itself. Therefore, the development of a computational tool for the analysis of cellular patterns in preneoplastic and neoplastic cervical lesions, which provides an accurate and opportunistic diagnosis itself, would be very useful to decongest the reading system of biopsies, to do objective diagnosis, and deliver a correct result to the patient. This is the purpose of this research.

On the other hand, liver fibrosis is the essential component of chronic liver diseases and its final stage leads to cirrhosis. This is a condition with a high rate of morbidity and mortality and it constitutes one of the main health problems in the world. In some Latin American countries, cirrhosis of the liver occupies between 5° and 6° place as a general cause of death [25]. In Mexico, liver cirrhosis morbidity in 2005 was 11.7 cases per 100,000 inhabitants and 12.4% in 2006, which represents an increase of 6% [26].

In order to evaluate the progression of fibrosis, the most common test is the invasive one, percutaneous liver biopsy; however, it will be better for patients to have a noninvasive technique with high sensitivity and repeatability, which allows the characterization of biological tissue and makes a truthful and timely fibrosis diagnostic. That is why an initial study of the thermal characteristics of hepatic tissue with different degrees of disease progression was made [27], and now the combinatorial logic approach is currently being applied to classify tissue in the different stages of progression of liver fibrosis to cirrhosis. In this way, it will be possible to deepen the understanding of liver fibrosis and its development.

References

1. Martínez-Trinidad, J.F., Guzman-Arenas, A.: The logical combinatorial approach to pattern recognition, an overview through selected works. Pattern Recogn. **34**, 741–751 (2001)
2. Ortiz-Posadas, M.R., et al.: A new approach to differential diagnosis of diseases. Int. J. Biomed. Comput. **40**, 179–185 (1996)
3. Ortiz-Posadas, M.R., et al.: A new approach to classify cleft lip and palate. Cleft Palate Craniofac. J. **38**(6), 545–550 (2001)
4. Ortiz-Posadas, M.R., et al.: A mathematical function to evaluate surgical complexity of cleft lip and palate. Comput. Methods Prog. Biomed. **94**, 232–238 (2009)
5. Ortiz-Posadas, M.R., et al.: A computational tool for the prognosis of the rehabilitation of patients with cleft palate. In: Proceedings of the 5th Iberoamerican Symposium on Pattern Recognition, pp. 599–608 (2000)
6. Ortiz-Posadas, M.R., et al.: A similarity function to evaluate the orthodontic condition in patients with cleft lip and palate. Med. Hypotheses. **63**(1), 35–41 (2004)
7. Ortiz-Posadas, M.R., Lazo-Cortés, M.S.: A mathematical model to evaluate the speaking of patients with cleft palate. In: Proceedings of the 2nd European Medical & Biological Engineering Conference, p. 500 (2002)
8. Ortiz-Posadas, M.R., et al.: A mathematical model for classical chromosome identification using the logical combinatory approach. In: Proceedings of the 25th Annual International Conference of the IEEE Engineering in Medicine and Biology Society, pp. 1342–1345 (2003)
9. Ortiz-Posadas, M.R., Vernet-Saavedra, E.: Electrical safety priority index for medical equipment. In: Proceedings of the 28th Annual International Conference of the IEEE Engineering in Medicine and Biology Society, pp. 6614–6617 (2006)
10. Rosales-López, A., Ortiz-Posadas, M.R.: A distribution strategy for imaging centers in the costa rican public health system. In: Proceedings IFMBE World Congress on Medical Physics and Biomedical Engineering, pp. 12–15 (2009)
11. Ruiz-Shulcloper, J., Abidi, M.: Logical combinatorial pattern recognition: a review. In: Pandalai, S.G. (ed.) Recent Research Developments in Pattern Recognition, pp. 133–176. Transworld Research Networks, Kerala (2002)
12. World Health Organization: Prevention of blindness and visual impairment. Priority eye diseases. [Cited March 2011]. Available at http://www.who.int/blindness/causes/priority/en/index7.html (2011)
13. Olvera-Rocha, X.E., Ortiz-Posadas, M.R.: Differential diagnosis of glaucoma by the logical combinatorial approach of pattern recognition. In: Proceedings IFMBE V Latin-American Congress on Biomedical Engineering. (In Spanish) (2011)
14. Carter, J.E.L.: Part 1: The Heath-Carter Anthropometric Somatotype. Instruction Manual. Department of Exercise and Nutritional Sciences. San Diego State University. [cited January 2011]. Available at: http://www.somatotype.org/methodology.php (2002)

15. Shepherd, M.A.J.: Body type training – are we slaves to our 'body type' genes? [cited July 2011]. Available at http://www.pponline.co.uk/encyc/body-type-training-are-we-slaves-to-our-body-type-genes-39798 (2011)
16. Acosta-Pineda, I., Ortiz-Posadas, M.R.: Somatotype classification using the logical combinatorial approach of pattern recognition. In: Proceedings IFMBE V Latin-American Congress on Biomedical Engineering. (In Spanish) (2011)
17. Wood, R.: Somatotype and sport. [cited July 2011]. Available at http://www.topendsports.com/testing/somatotype.htm (2001)
18. Kernahan, D.A., Stark, R.B.: A new classification for cleft lip and cleft palate. Plast. Reconstr. Surg. **22**, 435 (1958)
19. Mossey, P.A.: Cleft lip and palate. Lancet. **374**(9703), 1773–1785 (2009)
20. Health Ministry of Mexico: Cleft lip and palate incidence by age. Unique Information System for Epidemiology Vigilance. Dirección General de Epidemiología. [cited may 2011]. Available at: http://www.dgepi.salud.gob.mx/anuario/html/anuarios.html
21. Children's Hospital of Wisconsin: Cleft lip and/or palate. [cited February 2009]. Available at http://www.chw.org/display/PPF/DocID/35472/Nav/1/router.asp
22. Ault, K.A.: Epidemiology and natural history of human papillomavirus infections in the female genital tract. Infect. Dis. Obstet. Gynecol. **2006**(40470), 1–5 (2006)
23. World Health Organization: Comprehensive cervical cancer control. In: A guide to essential practice. 2n edn. Geneve. WHO Library Cataloguing-in-Publication Data. [Cited may 2016]. Available at: http://apps.who.int/iris/bitstream/10665/144785/1/9789241548953_eng.pdf (2014)
24. Ferlay, J., et al.: Cancer incidence and mortality worldwide: sources, methods and major patterns in GLOBOCAN 2012. Int. J. Cancer. **136**, E359–E386 (2015)
25. Narro, J., et al.: Mortality by liver cirrhosis in Mexico: relevant epidemiologic characteristics. Salud Publica Mex. **34**(4), 378–387 (1992). (In Spanish)
26. Health Ministry: Health: Mexico 2006. Technical report, pp. 46–48. [cited April 2015]. Available at: http://www.salud.gob.mx/unidades/evaluacion/saludmex2006/SM06.pdf. (In Spanish)
27. Alemán-García, N., et al.: Fibrosis evaluation of animal liver tissue by thermal conduction. In: Proceedings IFMBE VII Latin American Congress on Biomedical Engineering, **60**, 674–677 (2016)

On the Uniqueness of Invariant Measures for the Stochastic Infinite Darcy–Prandtl Number Model

Rana D. Parshad and Brian Ewald

Abstract The infinite Darcy–Prandtl number model is an effective reduced model for describing convection in a fluid-saturated porous medium. It is well known that the deterministic model does not possess a unique invariant measure. In this work, we study the dynamics of the infinite Darcy–Prandtl number model, under an additive stochastic forcing of its low modes. This is the so-called stochastic infinite Darcy–Prandtl number model. We prove that the stochastically forced system, does indeed possess a unique invariant measure.

Keywords Stochastic infinite Darcy–Prandtl number model • Convection in porous media • Invariant measures • Stochastic partial differential equations

1991 Mathematics Subject Classification. Primary: 60H15; 60H30; 37L40 Secondary: 76S05

1 Introduction

Applied analysts are often interested in the analysis and modeling of physical phenomenon. Many of the models they investigate consist of a system of partial differential equations, with appropriate initial and boundary conditions. The equations depend on initial data intrinsic to the physical process, as well as parameters, which change according to physical situations. Varying these parameters, and taking appropriate limits with respect to them, sometimes leads to a simplification of the original model. This "reduced" model so to speak may be more tractable from an analysis and/or computational point of view, than the original model. It might

R.D. Parshad (✉)
Department of Mathematics, Clarkson University, Potsdam, NY 13699, USA
e-mail: rparshad@clarkson.edu

B. Ewald
Department of Mathematics, Florida State University, Tallahassee, FL 32306, USA

B. Toni (ed.), *New Trends and Advanced Methods in Interdisciplinary Mathematical Sciences*, STEAM-H: Science, Technology, Engineering, Agriculture, Mathematics & Health, DOI 10.1007/978-3-319-55612-3_9

also lend valuable insight into the properties of the physical process. One might then ask, what information is lost concerning the original model in the limiting process, or alternatively what is preserved? Also, in what sense are the original and reduced models close, if indeed they are so? A program to investigate such questions, in the context of convective fluid flow, modeled by the Boussinesq system, was initiated by Wang in [22–24]. We then continued this program in [15, 16], for convective fluid flow in porous media, where we considered the Darcy–Boussinesq system, and its "reduced" variant, the infinite Darcy–Prandtl number model. Both of these models have generated considerable interest lately. They find diverse applications in the fields of geothermal engineering, construction of thermal insulators, nuclear waste management, thermal enhanced oil recovery, groundwater flow, and hydraulic fracturing [2, 18, 19, 21]. Thus there is good reason to further investigate the aforementioned systems. However, for chaotic systems such as these, it is meaningless to talk of one single trajectory, or any information therein. This is due to sensitive dependence on initial data, as well as on parameters. Alternatively, one might focus on statistical properties, or "averages" of the system, such as invariant measures [24]. For the deterministic Darcy–Boussinesq system, there is no unique invariant measure. This is seen by constructing measures consisting of point masses centered at various steady states [25]. These multiple steady states then will give rise to multiple invariant measures. Our goal in the current manuscript is to show the existence of a unique invariant measure, for the deterministic Infinite Darcy–Prandtl number model, forced by a white noise. Although much of the forthcoming analysis involves stochastics, for the benefit of the reader, we will recap certain physical details of fluid flow in porous media, as well as details of the Darcy–Boussinesq system and infinite Darcy–Prandtl number model, before we proceed with the stochastic preliminaries.

Recall, a porous medium is a solid structure often called a matrix which consists of interconnected void spaces called pores. In many physical situations, there is fluid flow through such mediums [14], making it an important class of flows for physical applications. It is impossible to know a priori the exact physical structure of porous media at a local level. Hence, we consider averaging the flow over a representative pore scale. The point of this is to build a continuum on which we can then apply the regular principles of fluid mechanics. A porous medium has two important defining features: the porosity ϕ and the permeability K. The porosity is defined as the ratio of the void space to the whole space. The permeability K describes how easy or difficult it is for the fluid to flow through a medium. The literature on modeling convective flow in porous media is very extensive and dates back to the work of Horton and Rogers [7] and Lapwood [8].

An effective model for convection in a porous medium is the Darcy–Boussinesq system [14]. The physical space for the problem consists of a fluid-saturated porous medium, confined between two plates, a distance of h units apart. The bottom plate is heated to a temperature say T_2 and the top plate is cooled to a temperature say T_1 where $T_2 > T_1$. In order to non-dimensionalize the problem, we measure length in units of the layer thickness h, and time in units of the thermal diffusion timescale $\frac{h^2}{\kappa}$. Thus the fluid occupies the non-dimensional region

$$\Omega = [0, L_x] \times [0, L_y] \times [0, 1]. \tag{1}$$

The differential heating induces convection in the fluid, which is modeled by a system of coupled partial differential equations, see [14]. In non-dimensional form, the equations are

$$\frac{1}{Pr_D}\frac{\partial \mathbf{u}}{\partial t} + \mathbf{u} + \nabla p = Ra_D\,\mathbf{k}T, \nabla \cdot \mathbf{u} = 0, u_3|_{z=0,1} = 0, \mathbf{u}|_{t=0} = \mathbf{u}_0, \tag{2}$$

$$\frac{\partial T}{\partial t} + \mathbf{u} \cdot \nabla T = \Delta T, T|_{z=0} = 1,\ T|_{z=1} = 0, T|_{t=0} = T_0. \tag{3}$$

On the sidewalls, periodic boundary conditions are imposed for convenience. Here $\mathbf{u}$ is the seepage velocity, T is the temperature field, and $\mathbf{k}$ is the upward pointing unit vector. The first parameter in the system is the Darcy–Rayleigh number defined as

$$Ra_D = \frac{g\alpha(T_2 - T_1)Kh}{\nu\kappa}. \tag{4}$$

where ν is the kinematic fluid viscosity, κ is the coefficient of thermal diffusion, α is the volume expansion coefficient of the fluid, g is the gravitational constant, and K is the Darcy permeability coefficient. The second parameter in the system is the Darcy–Prandtl number defined as

$$Pr_D = \frac{\nu h^2}{\kappa K}. \tag{5}$$

As mentioned earlier, K measures the ability of the porous medium to transport fluid and can vary quite drastically within different materials. We illustrate with a table providing ranges in permeability of some common rocks [9],

Material	Range of permeability (m/day)
Sand	10^{-1} to 5×10^{-3}
Clay	5×10^{-7} to 10^{-3}
Limestone	5×10^{-6} to 1
Sandstone	5×10^{-5} to 20
Dense crystalline rock	5×10^{-8} to 10^{-5}

Thus we see that when the porous medium is say certain limestone or sandstone, the permeability K is extremely small, in comparison to the thickness h of the medium, which is of order 1. This situation leads to a very large Pr_D or Darcy–Prandtl number. Under a large Darcy–Prandtl number assumption, this system can be reduced to the infinite Darcy–Prandtl number model [5], via formally taking the limit as $Pr_D \to \infty$ in the Darcy–Boussinesq system [5]. Thus we can interpret the physical meaning of an infinite Darcy–Prandtl number, as a situation where the ratio of the layer thickness to the permeability of the medium is very large, or $\frac{h}{K} >> 1$, which results in $Pr_D >> 1$, so $Pr_D \to \infty$ is a reasonable approximation. The resulting reduced model is described by the following set of coupled partial differential equations along with a free-slip set of boundary conditions:

$$\mathbf{u} + \nabla p = Ra_D\, \mathbf{k}T, \tag{6}$$

$$\frac{\partial T}{\partial t} + \mathbf{u} \cdot \nabla T = \Delta T, \tag{7}$$

$$u_3|_{z=0,1} = 0, \ \left.\frac{\partial u_1}{\partial z}\right|_{z=0,1} = \left.\frac{\partial u_2}{\partial z}\right|_{z=0,1} = 0, \tag{8}$$

$$\nabla \cdot \mathbf{u} = 0, \tag{9}$$

$$T|_{z=0} = 1, T|_{z=1} = 0, \mathbf{u}|_{t=0} = \mathbf{u}_0, T|_{t=0} = T_0. \tag{10}$$

On the sidewalls, periodic boundary conditions are imposed for convenience. There are a number of works in the literature, on the infinite Darcy–Prandtl number model [11, 17]. The model possesses strong solutions which are C^∞ in space. In fact, the solutions are actually Gevrey regular in space, and analytic in time. For a detailed proof of global existence of strong solution, as well as Gevrey regularity of solution, the reader is referred to [11].

Note that the infinite Darcy–Prandtl number model is not equipped with homogeneous boundary conditions. This is circumvented by introducing a change of variable

$$T = \theta + \gamma(z). \tag{11}$$

Here $\gamma(z)$ is a background temperature profile: see [4]. We require that

$$\gamma(1) = 0, \ \gamma(0) = 1. \tag{12}$$

Thus θ will satisfy homogenous boundary conditions,

$$\theta|_{z=0,1} = 0. \tag{13}$$

Inserting the above into (6) and (7) yields

$$\nabla p + \mathbf{u} = Ra_D\, \mathbf{k}\theta, \tag{14}$$

$$\nabla \cdot \mathbf{u} = 0, \tag{15}$$

$$\frac{\partial \theta}{\partial t} + \mathbf{u} \cdot \nabla\theta + u_3\gamma'(z) = \Delta\theta + \gamma''(z), \tag{16}$$

$$\theta|_{z=0,1} = 0. \tag{17}$$

We will suppose that the temperature profile is linear so $\gamma(z) = 1 - z$. Thus

$$\gamma'(z) = -1, \tag{18}$$

also

$$\gamma''(z) = 0. \tag{19}$$

Notice from (14), the velocity is completely determined by the temperature field modulo the pressure. We will use this to our advantage and convert (14)–(16) to a single equation. We apply the Leray–Hopf projector, denoted P, [20], to (14) and insert the result into (16) to yield the following equation

$$\frac{\partial \theta}{\partial t} = \Delta\theta - Ra_D P(\mathbf{k}\theta)\cdot\nabla\theta - Ra_D \gamma'(z)\,(P(\mathbf{k}\theta))_3, \tag{20}$$

$$\theta|_{z=0,1} = 0. \tag{21}$$

Note that using the 3D Helmholtz decomposition [6], one can show that $(P(\mathbf{k}\theta))_3 = \theta$, as long as θ is sufficiently regular. This is not an issue here, as the solutions to the infinite Darcy–Prandtl number model are very regular [11]. Thus using the abovementioned decomposition, and adding an appropriate white noise to (20), we obtain the stochastic infinite Darcy–Prandtl number model

$$\mathrm{d}\theta = (\Delta\theta - Ra_D P(\mathbf{k}\theta)\cdot\nabla\theta - Ra_D \gamma'(z)\,\theta)\,\mathrm{d}t + \mathrm{d}\mathbb{W}, \tag{22}$$

$$\theta|_{z=0,1} = 0. \tag{23}$$

Heuristically, the noise will connect the various branches of the attractor, which would not be able to interact without the presence of the noise. Essentially, the stochastic forcing introduces a probability of being "kicked" from one steady state to another, an impossibility without the noise. The noise will thus connect all disconnected branches of the attractor to yield an invariant measure. We pay the price however by having to work in a probabilistic setting on a Hilbert space.

Techniques of adding noise to deterministic partial differential equations have become quite popular. There is a vast literature on these so-called stochastic partial differential equations. See [3] for a detailed treatment of such techniques. The breadth of techniques covered in [3] is of a very technical nature. The authors treat the case of infinite-dimensional noise. For our case however, a finite-dimensional noise will suffice. Although there is no general theory to treat equations forced by degenerate or finite-dimensional noise, there are various results that provide us with valuable insight into the existence and uniqueness of invariant measures for dissipative stochastic systems. We recall certain prominent results relevant in our setting.

Mattingly, in his PhD dissertation, considered the 2D stochastic Navier–Stokes equations forced by degenerate white noise [12]. He proved uniqueness of the invariant measure for this equation. Later, along with E and Sinai he studied this problem further, deriving various results that were reported in [13]. The methodology developed by them was extended to a host of other stochastic partial differential equations such as the stochastic Ginzburg–Landau and stochastic Cahn–Hilliard.

These equations were considered by E and Liu [10]. Recently these techniques were also applied to the 3D truncated stochastic Boussinesq system by Wu and Lee [26].

Inspired by the above results, we rigorously investigate the stochastic infinite Darcy–Prandtl number model (22). We primarily ask the following question. Does there exist a unique invariant measure for (22)? Our contribution in the current manuscript is to apply results directly from [10] to the stochastic infinite Darcy–Prandtl number model, in order to answer the above question in the affirmative. E and Liu in [10] prove the existence of a unique invariant measure for a general dissipative stochastic PDE, under certain suitable conditions. We will systematically check that the stochastic infinite Darcy–Prandtl number model satisfies each one of these. We have organized our manuscript as follows. In Section 2, we describe the mathematical formulation of the problem. Section 3 is aimed at validating the first two conditions as posed in [10]: see the appendix. In Section 4, we make various probabilistic estimates aimed at validating the remaining conditions. The results of these sections are brought together in Section 5, where we state our main result via Theorem 2. We then offer some concluding remarks in Section 6. In all estimates made henceforth, C is a generic constant that can change in its value from line to line, and sometimes within the same line, if so required.

2 The Mathematical Formulation

2.1 Stochastic Preliminaries

In this section, we discuss the stochastic framework on which much of the subsequent analysis relies. The key idea is to perform stochastic analysis on a Hilbert space. To this end, we would like to discuss the structure of the noise and spaces used. For details, the reader is referred to [3]. We will formulate the preliminaries in terms of a general equation of the form

$$d\theta = (\Delta\theta + R(\theta))\,dt + d\mathbb{W},\ t \geq 0,\ \theta(0) = \theta_0. \tag{24}$$

which is defined on a separable Hilbert space H. For us is $H = L^2(X)$. This is equipped with the standard inner product $\langle \cdot, \cdot \rangle_H$. $-\Delta$ is a self-adjoint linear operator on the domain $D(-\Delta) \subset H$, with eigenvalues $0 \leq \lambda_1 \leq \lambda_2 \leq \ldots \leq \lambda_N \ldots$, $\lim_{k\to\infty} \lambda_k = \infty$ and a complete orthonormal system of eigenvectors $e_1, \ldots e_N \ldots$ such that $-\Delta e_i = \lambda_i e_i$. Furthermore, $-\Delta : D(-\Delta) \mapsto H$ is the generator of a C_0 semigroup $S(t)$, $t \geq 0$, of linear operators on H.

For our specific application, the nonlinear term is

$$R(\theta) = R(\theta, \nabla\theta, z) = -Ra_D\, P(\mathbf{k}\theta) \cdot \nabla\theta - Ra_D\, \gamma'(z)\theta. \tag{25}$$

such that $R : D(R) \subset H \mapsto H$. However, we will formulate things for a general R.

The noise $\mathbb{W}(\cdot, t)$ is of the form

$$\mathbb{W}(x, \omega, t) = \sum \sigma_k \omega_k(\omega, t) e_k(x), \tag{26}$$

where the ω_k's are independent standard Brownian motions which generate the filtered probability space $(\Omega, \mathcal{F}, \mathcal{F}_t, \mathbb{P})$. The $\sigma_k \in \mathbb{R}$ are coordinate-wise standard deviations, and the $e_k(x)$ are as described earlier, and form a basis for $L^2(X)$. Expectations are taken with respect to $\mathbb{P}$. Defining noise with this structure facilitates taking inner products. This becomes important in the energy estimates performed henceforth. We will assume that (24) possesses a unique strong solution for almost all $\omega \in \Omega$. Here we pause and define the notions of strong and weak solution.

Definition 2.1. A $D(-\Delta)$ valued predictable process $\theta(t)$, $t \in [0, T]$, is called an analytically strong solution of (24) if

$$\theta(t) = \theta_0 + \int_0^t (\Delta\theta(s) + R(\theta))ds + W(t), \ \mathbb{P} - a.s \tag{27}$$

Definition 2.2. An H valued predictable process $\theta(t)$, $t \in [0, T]$, is called a mild solution of (24) if

$$\theta(t) = S(t)\theta_0 + \int_0^t S(t-s)R(\theta))ds + \int_0^t S(t-s)dW(s), \ \ \mathbb{P} - a.s \tag{28}$$

Remark 1. Note that we already have a probability space Ω, on which a Brownian motion process is defined. Thus solutions are assumed to be defined pathwise; that is, for each $\omega \in \Omega$, we would have a solution that $\theta(t, \omega)$ only really depends on ω via $W(\omega)$, the Brownian path. Since we apply Ito's formula along the paths, these paths have to be well defined. This is not an issue, since strong solution is being used here.

Furthermore, we have that (24) defines a stochastic semiflow

$$\phi^{\omega}_{s,t} u_0 = u(s, t; \omega, u_0). \tag{29}$$

This semiflow generates a continuous Markovian semigroup. Next we define an invariant measure.

Definition 2.3. A probability measure μ on the phase space H equipped with the Borel σ-algebra is invariant if and only if

$$\int_H F(\theta)\mu(\mathrm{d}\theta) = \int_H \mathbb{E}[F(\phi^{\omega}_t \theta)]\mu(\mathrm{d}\theta), \tag{30}$$

for all continuous bounded functions F on H and all $t \geq 0$.

An invariant measure μ can be extended to a measure μ_p on the path space $C((-\infty,0],H)$. First, we define a cylinder set

$$A = \{\theta(s) \in C((-\infty,0],H), \theta(t_i) \in A_i, i = 0, \ldots n\}, \tag{31}$$

where $t_0 < t_1 < t_2 < \ldots < t_n < 0$, and the A_i are Borel subsets of H. Define $B \subset H \times \Omega$ by

$$B = \left\{(\theta, w), \theta \in A_0, \phi^{\omega}_{t_0,t_i} \in A_i, i = 1, 2, ..n\right\}, \tag{32}$$

and define $\mu_p(A) = (\mu \times \mathbb{P})(A)$. Then μ_p is consistent on cylinder sets and can be extended to the natural σ-algebra by Kolmogorov's extension theorem.

We also briefly recount the concept of Gibbsian dynamics of the low modes. We partition the phase space $H = L^2(X)$ into a space of high modes and low modes:

$$H_l = \operatorname{span}\{e_k(x), k \le N\}, \; H_h = \operatorname{span}\{e_k(x), k > N\}, \tag{33}$$

where $H = H_l \oplus H_h$. We denote by P_l the projection operator from H to H_l and by P_h the projection operator from H to H_h. Thus a solution to (24) is written as $\theta(t) = (l(t), h(t))$ where $P_l(\theta(t)) = l(t)$ and $P_h(\theta(t)) = h(t)$. We can thus rewrite (24) in terms of its high-mode and low-mode components as

$$\mathrm{d}l(t) = \Delta l(t)\,\mathrm{d}t + P_l(R(\theta))\,\mathrm{d}t + \mathrm{d}\mathbb{W}(t), \tag{34}$$

$$\frac{\mathrm{d}h(t)}{\mathrm{d}t} = \Delta h(t) + P_h(R(\theta)). \tag{35}$$

Next a number of conditions are imposed on (24):

Condition 1: There exist constants $\eta \ge 0$ and $K_0 \ge 0$ such that

$$-\langle Ax, x\rangle_H + \langle Rx, x\rangle_H \le -\eta|x|^2_H + K_0. \tag{36}$$

Condition 2: Let $\theta_1, \theta_2 \in H$ and let $\rho = \theta_1 - \theta_2$. There exists a constant $\alpha \in [0,1)$ and a nonnegative function $K(\theta)$ on H such that

$$\langle R(\theta_1) - R(\theta_2), \rho\rangle_H \le \alpha\,\langle A\rho, \rho\rangle_H + K(\theta_1)|\rho|^2_H. \tag{37}$$

Furthermore,

$$\int_H K(\theta)\,\mathrm{d}\mu(\theta) \le \beta, \tag{38}$$

for some constant β independent of the invariant measure μ.

Condition 3: For all $L^0 \in P$, $\bar{h}_0$, and for all $a \in (0,1)$ and $T \ge 0$, there exists $K \ge 0$ such that

$$\mathbb{P}\left\{\int_0^{\infty} |D(l(t), h(t), \bar{h}(t))|^2_H\,\mathrm{d}s < K\right\} \ge 1 - a > 0. \tag{39}$$

Condition 4[1]: For all $L^0 \in P$, $a \in (0,1)$, and $T \geq 0$, there exists $K \geq 0$ such that

$$\mathbb{P}\left\{\int_0^T |G(l(s), h(s))|_H^2 \, \mathrm{d}s < K\right\} \geq 1 - a > 0. \tag{40}$$

If the above are satisfied, one obtains the following,

Theorem 1 (E and Liu, Journal of Statistical Physics, 2002). *Suppose that the stochastic partial differential equation* (24) *satisfies conditions 1–4 and N is chosen large enough, then Eq.* (24) *has a unique invariant measure.*

Note that given an ergodic invariant measure μ, for μ_p-almost all $\theta(\cdot) \in C((-\infty,0], H)$ we have that

$$\lim_{t_0 \to -\infty} \frac{1}{t - t_0} \int_{t_0}^t K(\theta(s)) \, \mathrm{d}s \leq \beta, \tag{41}$$

Here K (see condition 2 in the appendix) is required to be a positive function and will depend explicitly on the form of the equation considered.

We next define the set $U \subseteq C((-\infty,0], H)$ to consist of all $v : (-\infty, 0] \to H$ such that v satisfies the estimates derived in Lemma 4.4 and the integral estimate derived above. Then due to conditions 1 and 2, which we prove henceforth, and the ergodicity assumption, we have that $\mu_p(U) = 1$. Also we will use $l(t)$ to denote the value of the low modes of the solution at time t, and L^t to denote the entire trajectory of the low modes from $-\infty$ to t. Therefore we have that $l(t) \in H_l$ and $L^t \in C((-\infty, t], H_l)$, and that $l(s) = L^t(s)$ for $0 \leq s \leq t$. We define a map $\Phi_s(L^t, h_0)$ which is a solution to (35) at time s with initial condition h_0 and low mode forcing L^t.

Remark 2. We consider measures on the function spaces with time set $(-\infty, t]$, as we will show that for statistically invariant solutions of (22) existing for time from $-\infty$ to $+\infty$, h is uniquely determined by the past history of l from $-\infty$ to 0 for almost all $u(t)$.

Note that $\Phi_s(L^t, h_0)$ only depends on the information of L^t between 0 and s. Therefore we can define $\Phi_{t_0,s}(L^t, h_0)$ for solutions starting from t_0 rather than time 0. We will suppose that N is large enough so that the requisite condition from Lemma 5.1 holds. Then we can solve for the future of l using the Gibbsian dynamics

$$\mathrm{d}l(t) = [\Delta l(t) + G(l(t), \Phi_t(L^t))] \, \mathrm{d}t + \mathrm{d}\mathbb{W}(t). \tag{42}$$

Here

$$G(l, h) = P_l(Ra_D P(\mathbf{k}(l+h)) \cdot \nabla(l+h) + Ra_D(l+h)). \tag{43}$$

[1]For the notation in conditions 3 and 4, please refer to Section 2.3 of [10].

Thus we have a closed form for the dynamics of the low modes given an initial past. The following difference operator also appears often:

$$D(f, g_1, g_2) = G(f, g_1) - G(f, g_2). \quad (44)$$

Remark 3. We only consider the case of finite-dimensional noise, that is when only the low modes are forced. This facilitates making the energy estimates. However, the arguments we make here apply when the high modes are also randomly forced. All we require is that the forcing amplitudes on the high modes decay fast enough.

3 Proof of the Monotonicity and Coercivity of the Drift Operator

We now proceed systematically to verify conditions 1–4, via (36), (37), (39), and (40), from [10]. Essentially, we want to follow the ideas in [10] and reduce the infinite-dimensional dynamics of the stochastic infinite Darcy–Prandtl number model to the finite-dimensional Gibbsian dynamics. This will facilitate the use of Girsanov's Theorem to yield a unique invariant measure. In this section, we state and prove two Lemmas, of which the first two conditions (36) and (37) will be a direct consequence.

Lemma 3.1. *For the infinite Darcy–Prandtl number model, there exist constants* $\eta > 0$ *and* $k_0 \geq 0$ *such that*

$$\langle \Delta\theta, \theta \rangle_2 + \left\langle -Ra_D\, P(\mathbf{k}\theta) \cdot \nabla\theta - Ra_D\, \gamma'(z)\theta, \theta \right\rangle_2 \leq -\eta |\theta(t)|_2^2 + k_0. \quad (45)$$

Proof. From the form of (20), and the choice of $\gamma(z)$, that yields $\gamma'(z) = -1$, we have

$$\begin{aligned}
&\langle \Delta\theta, \theta \rangle_2 + \langle R(\theta), \theta \rangle_2 \\
&= \langle \Delta\theta, \theta \rangle_2 + \left\langle -Ra_D\, P(\mathbf{k}\theta) \cdot \nabla\theta - Ra_D\, \gamma'(z)\theta, \theta \right\rangle_2 \\
&\leq -|\nabla\theta|_2^2 + Ra_D |\theta|_2^2 \\
&\leq -\frac{1}{C}|\theta|_2^2 + Ra_D C_1
\end{aligned}$$

This follows via integration by parts, Poincaré's inequality

$$|\theta|_2^2 \leq C|\nabla\theta|_2^2, \quad (46)$$

and the estimates on the L^2 norm of θ via (63), and Lemma 4.2. Here $C_1 = |\theta(0)|_2^2 + \frac{\epsilon_0}{C_2}$. Thus the lemma is proved for, say, $\eta = \frac{1}{C}$ and $k_0 = Ra_D C_1$. □

Thus condition 1 (36) [10] is satisfied.

We now look at the difference of the nonlinear terms, $R(\theta_1 - \theta_2)$, where θ_1 and θ_2 are two different solutions to the infinite Darcy–Prandtl number model such that $\theta_1 = l + h_1$ and $\theta_2 = l + h_2$. Our motivation here is to put into effect the Gibbsian dynamics formalism via (33)–(35), and then (42)–(44). Essentially, the phase space has been split into a space of high modes and low modes, and we want to derive a decay estimate on the difference of high modes for two solutions, so that the Gibbsian dynamics via (42) is relevant. The next Lemma aims at precisely estimating the difference of high modes via (44), in our setting. We set $\rho = \theta_1 - \theta_2 = h_1 - h_2$ and state the following Lemma:

Lemma 3.2. *For the infinite Darcy–Prandtl number model, there exists a constant* $\alpha \in [0, 1)$ *and a nonnegative function* $K(\theta_1)$ *such that*

$$\langle R(\rho), \rho \rangle \leq \alpha \langle -\Delta\rho, \rho \rangle + K(\theta_1)|\rho|_2^2. \tag{47}$$

Furthermore, there exists a uniform constant C such that

$$\int_H K(\theta_1)\, \mathrm{d}\mu(\theta_1) \leq C. \tag{48}$$

Proof. Recall from the form of (22) we have

$$R(\rho) = -Ra_D\, P(\mathbf{k}\theta_1) \cdot \nabla\theta_1 + Ra_D\, P(\mathbf{k}\theta_2) \cdot \nabla\theta_2 + Ra_D\, \gamma'(z)\, \rho. \tag{49}$$

We will rewrite $R(\rho)$ as

$$R(\rho) = -Ra_D\, P(\mathbf{k}\theta_2) \cdot \nabla\rho - Ra_D\, \nabla\theta_1 \cdot P(\mathbf{k}\rho) + Ra_D\, \gamma'(z)\rho. \tag{50}$$

Without loss of generality, we assume $\theta_1 \geq \theta_2$. We multiply (50) through by $\rho = \theta_1 - \theta_2$ and integrate by parts to yield

$$\begin{aligned}
\langle R(\rho), \rho \rangle &= \big\langle -Ra_D\, P(\mathbf{k}\theta_2) \cdot \nabla\rho - Ra_D\, \nabla\theta_1 \cdot P(\mathbf{k}\rho) + Ra_D\gamma'(z)P(\rho), \rho \big\rangle \\
&= -Ra_D \int_H (\theta_1)_z \rho^2\, \mathrm{d}\mathbf{x} + Ra_D \int_H \gamma'(z)P(\rho)\rho\, \mathrm{d}\mathbf{x} \\
&= 2Ra_D \int_H \theta_1 \rho \rho_z + Ra_D \int_H \gamma'(z)P(\rho)\rho\, \mathrm{d}\mathbf{x} \\
&\leq 2Ra_D \int_H \theta_1 \rho \rho_z + Ra_D \int_H |\rho|^2\, \mathrm{d}\mathbf{x}
\end{aligned}$$

This follows via integration by parts and using the regularity of θ so that $P(\mathbf{k}\theta)_3 = \theta$. We next use Holders inequality and adopt the Ladyzhenskaya inequality, along with Young's inequality with ϵ to yield

$$
\begin{aligned}
&\frac{Ra_D}{2}\int_H \theta_1 \rho\rho_z + Ra_D\int_H |\rho|^2\,\mathrm{d}\mathbf{x}\\
&\le 2Ra_D|\theta_1|_4|\rho|_4|\nabla\rho|_2 + Ra_D|\rho|_2^2\\
&\le 2Ra_D|\theta_1|_4|\nabla\rho|_2^{\frac{3}{4}}|\rho|_2^{\frac{1}{4}}|\nabla\rho|_2 + Ra_D|\rho|_2^2\\
&\le \frac{(2Ra_D)^8}{8}|\theta_1|_4^8|\rho|_2^2 + \frac{7}{8}|\nabla\rho|_2^2 + Ra_D|\rho|_2^2\\
&= \frac{7}{8}\langle -\Delta\rho,\rho\rangle_2 + (Ra_D + \frac{(2Ra_D)^8}{8}|\theta_1|_4^8)|\rho|_2^2.
\end{aligned}
$$

Thus the lemma is proved for, say, $\alpha = \frac{7}{8}$ and $K(\theta_1) = (Ra_D + \frac{(2Ra_D)^8}{8}|\theta_1|_4^8)$. Note since we have estimates on $|\nabla\theta|_2^2$, via Sobolev embedding of $H_0^1 \hookrightarrow L^4$, we have the L^4 estimates also. This also follows via L^∞ boundedness of θ. In order to see how we obtain this estimate, that is,

$$
\int_H K(\theta_1)\,\mathrm{d}\mu(\theta_1) \le C, \tag{51}
$$

see the proof for Lemma 4.2. □

4 Estimates on the Growth Rate of Energy and Enstrophy

Our goal in this section is to derive estimates for the energy $\mathbb{E}[|\theta|_2^2]$ and the enstrophy $\mathbb{E}[|\nabla\theta|_2^2]$. This is necessary to derive certain lemmas which are crucial for proving the next set of conditions.

4.1 A Priori Estimates on the Energy

We begin with estimates on the energy. We apply Itô's lemma on the map

$$
\theta(t) \mapsto \frac{1}{2}|\theta|_2^2. \tag{52}
$$

This yields

$$
\begin{aligned}
\frac{1}{2}\mathrm{d}|\theta(t)|_2^2 = {} & [Ra_D\int_\Omega \gamma'(z)P(\mathbf{k}\theta(t))\theta(t)\,\mathrm{d}\mathbf{x} - |\nabla\theta(t)|_2^2 - \langle \mathbf{u}\cdot\nabla\theta,\theta\rangle_2]\,\mathrm{d}t\\
& - \langle\theta,\mathrm{d}\mathbb{W}\rangle_2 + \frac{1}{2}\epsilon_0\,\mathrm{d}t.
\end{aligned} \tag{53}
$$

Here $\epsilon_0 = \sum |\sigma_k|^2$. Via the appropriate choice for $\gamma(z)$, we obtain

$$\frac{1}{2}\mathrm{d}|\theta(t)|_2^2 \leq \left[\frac{1}{2C}|\theta(t)|_2^2 - |\nabla\theta(t)|_2^2\right]\mathrm{d}t + \epsilon_0\,\mathrm{d}t - \langle\theta, \mathrm{d}\mathbb{W}\rangle_2\,. \tag{54}$$

Define a stopping time T, for any given H, by

$$T = \inf\left\{t : |\theta(t)|_2^2 \geq H^2\right\}. \tag{55}$$

Integrate (54) from 0 to $t \wedge T$ and take expectations to yield

$$\frac{1}{2}\mathbb{E}[|\theta(t\wedge T)|_2^2] \leq \frac{1}{2}\mathbb{E}[|\theta(0)|_2^2] + \frac{1}{2C}\int_0^{t\wedge T}\mathbb{E}[|\theta(s)|_2^2]\,\mathrm{d}s - \int_0^{t\wedge T}\mathbb{E}[|\nabla\theta(s)|_2^2]\,\mathrm{d}s$$
$$-\mathbb{E}[\int_0^{t\wedge T}\langle\theta, \mathrm{d}\mathbb{W}\rangle_2] + \epsilon_0(t\wedge T). \tag{56}$$

Now define the quantity

$$M_t^T = \int_0^t \langle\theta(s\wedge T), \mathrm{d}\mathbb{W}\rangle_{L^2}\,. \tag{57}$$

We can show that the quadratic variation of M_t^T is finite, implying that M_t^T is a martingale and so

$$\mathbb{E}[M_t^T] = 0. \tag{58}$$

We can therefore use the optional sampling Theorem to conclude

$$E[M_{t\wedge T}^T] = 0. \tag{59}$$

Hence, we can take $H \to \infty$ so $T \to \infty$ and thus $t \wedge T \to t$. Incorporating these limits in (56) yields

$$\frac{1}{2}\mathbb{E}[|\theta(t)|_2^2] + \int_0^t \mathbb{E}[|\nabla\theta(s)|_2^2]\,\mathrm{d}s \leq \frac{1}{2}\mathbb{E}[|\theta(0)|_2^2] + \frac{1}{2C}\int_0^t \mathbb{E}[|\theta(s)|_2^2]\,\mathrm{d}s + \epsilon_0 t. \tag{60}$$

Thus application of Poincaré's inequality yields

$$\frac{1}{2}\mathbb{E}[|\theta(t)|_2^2] + \frac{1}{C}\int_0^t \mathbb{E}[|\theta(s)|_2^2]\,\mathrm{d}s \leq \frac{1}{2}\mathbb{E}[|\theta(0)|_2^2] + \frac{1}{2C}\int_0^t \mathbb{E}[|\theta(s)|_2^2]\,\mathrm{d}s + \epsilon_0 t. \tag{61}$$

Thus we have

$$\frac{1}{2}\mathbb{E}[|\theta(t)|_2^2] + \frac{1}{2C}\int_0^t \mathbb{E}[|\theta(s)|_2^2]\,\mathrm{d}s \leq \frac{1}{2}\mathbb{E}[|\theta(0)|_2^2] + \epsilon_0 t, \tag{62}$$

and an application of Gronwall's lemma yields

$$\mathbb{E}[|\theta(t)|_2^2] \leq e^{-2C_2 t}\mathbb{E}[|\theta(0)|_2^2] + \frac{\epsilon_0}{C_2}(1 - e^{-2C_2 t}). \tag{63}$$

Here $C_2 = \frac{1}{2C}$.

We now state a lemma that enables us to derive a uniform bound on the L^2-norm of θ.

Lemma 4.1. *Consider a θ that is a solution to the stochastic infinite Darcy–Prandtl number model. For an invariant measure μ on H, there exists a constant C such that the following estimate holds uniformly in the time parameter:*

$$\int_H |\theta|_2^2\,\mathrm{d}\mu(\theta) \leq C. \tag{64}$$

Proof. For any $\epsilon > 0$, there is a b_ϵ such that $\mu\left\{\theta : |\theta|_2^2 \leq b_\epsilon\right\} \geq 1 - \epsilon$. We define the set

$$B_\epsilon = \left\{\theta : |\theta|_2^2 \leq b_\epsilon\right\}. \tag{65}$$

Thus we have that for any H and $t > 0$,

$$\begin{aligned}
\int_H (|\theta|_2^2 \wedge H)\,\mathrm{d}\mu(\theta) &= \int_H \mathbb{E}(|\phi_{0,t}^{\omega}\theta|_2^2 \wedge H)\,\mathrm{d}\mu(\theta) \\
&\leq \int_{B_\epsilon} \mathbb{E}(|\phi_{0,t}^{\omega}\theta|_2^2)\,\mathrm{d}\mu(\theta) + H\epsilon \\
&\leq e^{-2C_2 t} b_\epsilon + \frac{\epsilon_0}{C_2}(1 - e^{-2C_2 t}).
\end{aligned}$$

This follows via the estimates derived in (63). We now let $t \to \infty$ and $H \to \infty$ and obtain the result, as ϵ was arbitrary. This proves the Lemma. □

4.2 A Priori Estimates on the Enstrophy

We will now derive estimates for $\mathbb{E}[|\nabla\theta(t)|_2^2]$. We apply Itô's lemma on the map

$$\theta(t) \mapsto \frac{1}{2}|\nabla\theta|_2^2. \tag{66}$$

This yields

$$\frac{1}{2}\mathrm{d}|\nabla\theta(t)|_2^2 = [-|\Delta\theta(t)|_2^2 + Ra_D \int_X \gamma'(z)\Delta\theta\theta\,\mathrm{d}\mathbf{x}]\,\mathrm{d}t$$
$$- \langle \Delta\theta, \mathrm{d}\mathbb{W}\rangle_2 + (\langle Ra_D P(\mathbf{k}\theta)\cdot\nabla\theta, \Delta\theta\rangle_2 + \epsilon_1)\,\mathrm{d}t. \tag{67}$$

Here $\epsilon_1 = \sum k^2|\sigma_k|^2$. We now define the stopping time T, for any given H, by

$$T = \inf\left\{t : |\nabla\theta(t)|_2^2 \geq H^2\right\}. \tag{68}$$

We use the Cauchy–Schwartz, Hölder's, Poincaré's, and Young's inequalities and choose

$$\gamma'(z) = -\frac{1}{8Ra_D C_3}, \tag{69}$$

where C_3 is the constant that arises in the embedding of $H^2(X) \hookrightarrow H^1(X)$, i.e.,

$$|\nabla\theta|_2^2 \leq C_3|\Delta\theta|_2^2, \tag{70}$$

This can easily be done via readjusting the temperature profile $\gamma(z)$, to still be linear but decay at a sharper rate. Thus we obtain

$$\frac{1}{2}\mathrm{d}|\nabla\theta(t)|_2^2 \leq [-|\Delta\theta(t)|_2^2 + Ra_D\,|\theta|_\infty|\nabla\theta|_2|\Delta\theta|_2 + \frac{1}{8C_3}\,|\nabla\theta|_2^2]\,\mathrm{d}t$$
$$- \langle \Delta\theta, \mathrm{d}\mathbb{W}\rangle_2 + \epsilon_1\,\mathrm{d}t. \tag{71}$$

Now recall the Gagliardo-Nirenberg interpolation inequality [1], for a function on a bounded domain $\Omega \subset \mathbb{R}^n$

$$|D^j\theta|_p \leq C_1|D^m\theta|_r^\alpha|\theta|^{1-\alpha}|\theta|_q^{1-\alpha} + C_2|\theta|_s, \tag{72}$$

such that $\frac{1}{p} = \frac{j}{n} + \left(\frac{1}{r} - \frac{m}{n}\right)\alpha + \frac{1-\alpha}{q}$, where $1 \leq q, r \leq \infty$. $s > 0$ is arbitrary. Using the above, we obtain

$$|\nabla\theta|_2 \leq C_1|\Delta\theta|_{\frac{4}{3}}^{\frac{1}{2}}|\theta|_4^{\frac{1}{2}} + C_2|\theta|_2^2, \tag{73}$$

we then use this to bound the nonlinear term:

$$\begin{aligned}
&-Ra_D \int_X (P(k\theta)\cdot\nabla\theta)\Delta\theta\,\mathrm{d}\mathbf{x} \\
&\leq C|\theta|_\infty|\nabla\theta|_2|\Delta\theta|_2 \\
&\leq C|\theta|_\infty|\left(C_1|\Delta\theta|_{\frac{4}{3}}^{\frac{1}{2}}|\theta|_4^{\frac{1}{2}} + C_2|\theta|_2^2\right)|\Delta\theta|_2 \\
&\leq C|\Delta\theta|_2^{\frac{3}{2}}|\theta|_4^{\frac{1}{2}} + C_2|\theta|_2^2|\Delta\theta|_2
\end{aligned}$$

$$\le C\left(\epsilon|\Delta\theta|_2^{\frac{3}{2}}\right)\left(\frac{1}{\epsilon}|\theta|_4^{\frac{1}{2}}\right)+\frac{1}{4}|\Delta\theta|_2^2+4(C_2)^2|\theta|_2^4$$

$$\le C\left(\epsilon|\Delta\theta|_2^{\frac{3}{2}}\right)^{\frac{4}{3}}+\left(\frac{1}{\epsilon}|\theta|_4^{\frac{1}{2}}\right)^4+\frac{1}{4}|\Delta\theta|_2^2+4(C_2)^2|\theta|_2^4$$

$$\le \frac{1}{4}|\Delta\theta|_2^2+C_5|\theta|_4^2+\frac{1}{4}|\Delta\theta|_2^2+4(C_2)^2|\theta|_4^2$$

This follows via the Sobolev embedding of $L^2 \hookrightarrow L^{\frac{4}{3}}$, Cauchy–Schwartz inequality, and Young's inequality with ϵ, where $\epsilon = \left(\frac{1}{4C}\right)^{\frac{4}{3}}$. Thus we obtain

$$\begin{aligned}
&\mathrm{d}|\nabla\theta|_2^2 \le \\
&2\left[-|\Delta\theta|_2^2+\frac{1}{2}|\Delta\theta|_2^2+\frac{1}{8C_3}\,|\nabla\theta|_2^2+C_5|\theta|_4^2+4C_2^2|\theta|_4^2\right]\mathrm{d}t \\
&-\langle\Delta\theta,\mathrm{d}\mathbb{W}\rangle_2+\epsilon_1\,\mathrm{d}t.
\end{aligned}\tag{74}$$

Now integrating from 0 to $t\wedge T$ and taking expectation on both sides yields

$$\begin{aligned}
&\mathbb{E}|\nabla\theta(t\wedge T)|_2^2 \\
&\le -\int_0^{t\wedge T}\mathbb{E}|\Delta\theta(s)|_2^2\,\mathrm{d}s+\frac{1}{4C_3}\int_0^{t\wedge T}\mathbb{E}[|\nabla\theta(s)|_2^2]\,\mathrm{d}s \\
&+\int_0^{t\wedge T}\left(C_5\mathbb{E}|\theta(s)|_4^2+4(C_2)^2\mathbb{E}|\theta(s)|_2^4\right)\,\mathrm{d}s \\
&-\int_0^{t\wedge T}\mathbb{E}\,\langle\Delta\theta,\mathrm{d}\mathbb{W}\rangle_2+\epsilon_1(t\wedge T).
\end{aligned}\tag{75}$$

We now define the quantity

$$M_t^T=\int_0^t\langle\Delta\theta(s\wedge T),\mathrm{d}\mathbb{W}(s)\rangle_2\,,\tag{76}$$

and we can show that the quadratic variation of M_t^T is finite, implying that M_t^T is a martingale and therefore $\mathbb{E}[M_t^T]=0$. Then the optional sampling Theorem yields $\mathbb{E}[M_{t\wedge T}^T]=0$. Hence, we can take $H\to\infty$, so $T\to\infty$ and thus $t\wedge T\to t$. Again incorporating these limits in (75) gives

$$\begin{aligned}
&\mathbb{E}[|\nabla\theta(t)|_2^2]-\frac{1}{4C_3}\int_0^t\mathbb{E}[|\nabla\theta(s)|_2^2]\,\mathrm{d}s+\int_0^t\mathbb{E}[|\Delta\theta(s)|_2^2]\,\mathrm{d}s \\
&\le\mathbb{E}[|\nabla\theta(0)|_2^2]+\int_0^{t\wedge T}\left(C_5\mathbb{E}|\theta(s)|_4^2+4(C_2)^2\mathbb{E}|\theta(s)|_2^4\right)]\,\mathrm{d}s+\epsilon_1 t.
\end{aligned}\tag{77}$$

Note the embedding of $H^2(X) \hookrightarrow H^1(X)$:

$$|\nabla\theta|_2^2 \leq C_3|\Delta\theta|_2^2. \tag{78}$$

Using the above in conjunction with the Poincaré's inequality yields

$$\begin{aligned} &\mathbb{E}|\nabla\theta(t)|_2^2 - \frac{1}{4C_3}\int_0^t \mathbb{E}[|\nabla\theta(s)|_2^2]\,\mathrm{d}s + \frac{1}{C_3}\int_0^t \mathbb{E}[|\nabla\theta(s)|_2^2]\,\mathrm{d}s \\ &\leq \mathbb{E}[|\nabla\theta(0)|_2^2] + \int_0^{t\wedge T} (C_5\mathbb{E}|\theta(s)|_4^2 + 4(C_2)^2\mathbb{E}|\theta(s)|_2^2)]\,\mathrm{d}s + \epsilon_1 t, \end{aligned} \tag{79}$$

and so

$$\mathbb{E}[|\nabla\theta(t)|_2^2] \leq C_4\int_0^t \mathbb{E}[|\nabla\theta(s)|_2^2]\,\mathrm{d}s + \mathbb{E}[|\nabla\theta(0)|_2^2] + (C_5 + \epsilon_1)t. \tag{80}$$

Here $C_4 = \frac{3}{4C_3}$ and C_5 is the uniform estimate on the sum of $\mathbb{E}[|\theta(s)|_2^2]$ via (63), and $\mathbb{E}[|\theta(s)|_4]$, via the L^∞ boundedness of θ. Hence, an application of Gronwall's lemma yields

$$\mathbb{E}[|\nabla\theta(t)|_2^2] \leq e^{-C_4 t}(\mathbb{E}[|\nabla\theta(0)|_2^2]) + \left(\frac{\epsilon_1 + C_5}{C_4}\right)(1 - e^{-C_4 t}). \tag{81}$$

We now state a lemma that enables us to derive a uniform bound on the L^2-norm of $\nabla\theta$.

Lemma 4.2. *Consider a θ which is a solution to the stochastic infinite Darcy–Prandtl number model. For an invariant measure μ on H, there exists a constant C such that if $|\nabla\theta(0)|_2^2 \leq C$, then the following estimate holds uniformly in the time parameter:*

$$\int_H |\nabla\theta|_2^2\,\mathrm{d}\mu(\theta) \leq C. \tag{82}$$

Proof. The proof could follow via mimicking the methods for the L^2-norm of θ. However, we provide an alternate proof. It follows from the a priori estimates on $\mathbb{E}[|\nabla\theta|_2^2]$ that

$$\frac{1}{2}\mathbb{E}[|\theta(t)|_2^2] + \int_{t_0}^t \mathbb{E}[|\nabla\theta(s)|_2^2]\,\mathrm{d}s \leq \frac{1}{2}\mathbb{E}[|\theta(t_0)|_2^2] + \frac{1}{2C}\int_{t_0}^t \mathbb{E}[|\theta(s)|_2^2]\,\mathrm{d}s + \epsilon_0(t-t_0). \tag{83}$$

This implies that

$$\lim_{t_0\to\infty}\frac{1}{t-t_0}\int_{t_0}^{t}\mathbb{E}[|\nabla\theta(s)|_2^2]\,\mathrm{d}s \le \lim_{t_0\to\infty}\frac{1}{t-t_0}\frac{1}{2}\mathbb{E}[|\theta(t_0)|_2^2] + \lim_{t_0\to\infty}\frac{1}{t-t_0}\frac{1}{2C}\int_{t_0}^{t}\mathbb{E}[|\theta(s)|_2^2]\,\mathrm{d}s + \epsilon_0 \le C,$$

using the uniform estimates on $\mathbb{E}[|\theta(s)|_2^2]$. Thus, from ergodicity, it follows that

$$\lim_{t_0\to\infty}\frac{1}{t-t_0}\int_{t_0}^{t}\mathbb{E}[|\nabla\theta(s)|_2^2]\,\mathrm{d}s = \int_H \mathbb{E}[|\nabla\theta(s)|_2^2]\,\mathrm{d}\mu(\theta) \le C. \tag{84}$$

However, an application of Fubini's Theorem for nonnegative integrands implies that

$$\int_H \mathbb{E}[|\nabla\theta(s)|_2^2]\,\mathrm{d}\mu(\theta) = \mathbb{E}\int_H [|\nabla\theta(s)|_2^2]\,\mathrm{d}\mu(\theta) \le C, \tag{85}$$

and from the definition of invariant measure we obtain

$$\mathbb{E}\int_H |\nabla\theta(s)|_2^2\,\mathrm{d}\mu(\theta) = \int_H |\nabla\theta(s)|_2^2\,\mathrm{d}\mu(\theta) \le C. \tag{86}$$

This proves the lemma. □

4.3 Probabilistic Estimate on Growth Rate of Energy and Enstrophy

The following lemma gives a probabilistic estimate of the growth rate of $|\theta|_2^2$ and $|\nabla\theta|_2^2$.

Lemma 4.3. *For all $a \in (0,1)$, $\delta > 1$, there exists a K_1 such that if $|\theta_0|_2^2 + |\nabla\theta_0|_2^2 \le C_0$, then, for any $t \ge 0$,*

$$\mathbb{P}\left\{\begin{array}{c}|\theta(t)|_2^2 + |\nabla\theta(t)|_2^2 + C(\int_0^t |\nabla\theta(s)|_2^2\,\mathrm{d}s + \int_0^t |\Delta\theta(s)|_2^2\,\mathrm{d}s)\\ \le C_1 + C_2 t + K_1(t+1)^\delta\end{array}\right\} \ge 1-a. \tag{87}$$

Proof. From (54) and (67), we have that

$$|\theta(t)|_2^2 + |\nabla\theta(t)|_2^2 + C\left(\int_0^t |\nabla\theta(s)|_2^2\,\mathrm{d}s + \int_0^t |\Delta\theta(s)|_2^2\,\mathrm{d}s\right)$$

$$\le C(|\theta(0)|_2^2 + |\nabla\theta(0)|_2^2) + (\epsilon_0 + C_5 + \epsilon_1)t + \int_0^t \langle\theta, \mathrm{d}\mathbb{W}\rangle_2 - \int_0^t \langle\Delta\theta, \mathrm{d}\mathbb{W}\rangle_2$$

$$\le C_1 + C_2 t + \int_0^t \langle\theta, \mathrm{d}\mathbb{W}\rangle_2 - \int_0^t \langle\Delta\theta, \mathrm{d}\mathbb{W}\rangle_2 .$$

Consider the processes

$$M_t = \int_0^t \langle\theta, \mathrm{d}\mathbb{W}\rangle_2 \tag{88}$$

and

$$M_t^1 = -\int_0^t \langle\Delta\theta, \mathrm{d}\mathbb{W}\rangle_2 . \tag{89}$$

For our purpose, it suffices to show that, for $t \ge 0$,

$$\mathbb{P}\left\{M_t + M_t^1 \le \frac{K_1}{2}(t+1)^\delta\right\} \ge 1 - a. \tag{90}$$

We consider the quadratic variations of the processes M_t and M_t^1 to obtain

$$[M, M]_t \le (\sigma^*_{\max})^2 \int_0^t |\theta(s)|_2^2 \,\mathrm{d}s \tag{91}$$

and

$$[M^1, M^1]_t \le (\Delta\sigma^*_{\max})^2 \int_0^t |\theta(s)|_2^2 \,\mathrm{d}s, \tag{92}$$

where

$$(\sigma^*_{\max})^2 = \sup |\sigma_k|^2 \tag{93}$$

and

$$(\Delta\sigma^*_{\max})^2 = \sup |k^2\sigma_k|^2. \tag{94}$$

Here the k are the standard deviations of the kth components of the noise. This is not to be confused with the subscript k we use next to indicate an index in time. Note that we have obtained estimates on the L^2-norm of θ. We thus proceed by defining the following events

$$B_k = \left\{ \sup_{s\in[0,k]} |M_s| \ge \frac{K_1}{4}(k+1)^\delta \right\} . \tag{95}$$

We proceed by making an estimate of the probability of this event.

$$\begin{aligned}
\mathbb{P}\{B_k\} &\le \frac{4^2}{(K_1)^2(k+1)^{2\delta}}\mathbb{E}[|M_s|^2] \quad (\text{Doob}'\text{sinequalityappliedto } B_k)\\
&\le \frac{4^2}{(K_1)^2(k+1)^{2\delta}}\mathbb{E}\left[\sup_{s\in[0,k]}|M_s|^2\right]\\
&\le \frac{4^2}{(K_1)^2(k+1)^{2\delta}}\mathbb{E}[([M,M]_k)] \quad (\text{Burkholder–Davis–Gundy inequality})\\
&\le \frac{4^2}{(K_1)^2(k+1)^{2\delta}}\mathbb{E}\left[(\sigma^*_{\max})^2\int_0^t|\theta(s)|_2^2\right]\mathrm{d}s \quad (\text{estimateon } \mathbb{E}[([M,M|]_k)])\\
&\le \frac{4^2}{(K_1)^2(k+1)^{2\delta}}(\sigma^*_{\max})^2\int_0^k\mathbb{E}[|\theta(s)|_2^2]\,\mathrm{d}s\\
&\le \frac{4^2}{(K_1)^2(k+1)^{2\delta}}(\sigma^*_{\max})^2 Ck \quad (\text{estimateon } \mathbb{E}[|\theta(s)|_2^2]).
\end{aligned}$$

Therefore we have that

$$\mathbb{P}\{B_k\} \le \frac{Ck}{(k+1)^{2\delta}}. \tag{96}$$

We note that

$$\mathbb{P}\left\{M_t \le \frac{K_1}{4}(t)^\delta\right\} = 1 - \mathbb{P}\left\{\bigcup_k B_k\right\} \ge 1 - \sum_k \mathbb{P}\{B_k\}. \tag{97}$$

We see that these sums are finite for sufficiently large δ. In particular, $\delta > 1$ suffices. We note

$$\mathbb{P}\left\{M_t \le \frac{K_1}{4}(t)^\delta\right\} \ge 1 - \sum_k \mathbb{P}\{B_k\}. \tag{98}$$

We can make the sum $\sum_k \mathbb{P}\{B_k\}$ arbitrarily small by increasing K_1, since

$$\sum_k \mathbb{P}\{B_k\} \le \frac{1}{K_1^2}\sum_k \frac{Ck}{(k+1)^{2\delta}} \le \frac{1}{K_1^2}\sum_k \frac{1}{k^p} \le \frac{C}{K_1^2}. \tag{99}$$

Here $p > 1$. Thus, given $a \in (0,1)$, if we choose $K_1 = \sqrt{\frac{2C}{a}}$, we obtain

$$\mathbb{P}\left\{M_t > \frac{K_1}{4}t^\delta\right\} \le \frac{a}{2}. \tag{100}$$

The same argument applies to the process M_t^1. Similar estimates can be made on $\mathbb{P}\{A_k\}$ where

$$A_k = \left\{ \sup_{s\in[0,k]} |M_s^1| \geq \frac{K_1}{4}(k+1)^\delta \right\}. \tag{101}$$

Thus after performing a similar analysis as above, we can choose K_1 large enough to have

$$\mathbb{P}\left\{ M_t^1 > \frac{K_1}{4}(t)^\delta \right\} \leq \frac{a}{2}, \tag{102}$$

and then combining (100) and (102) yields

$$\mathbb{P}\left\{ M_t + M_t^1 \leq \frac{K_1}{2}(t+1)^\delta \right\} \geq 1 - a, \tag{103}$$

for $t \geq 0$, as was required. This completes the proof of the lemma. □

We next prove the following lemma.

Lemma 4.4. *Let μ_p be a measure induced on $C((-\infty, 0], L^2(X))$ by any given stationary measure μ. Fix any $K_0 \geq 0$ and $\delta > \frac{1}{2}$. Then for μ_p-almost every trajectory $\theta(s)$ in $C((-\infty, 0], L^2(X))$, there exists a constant T_1 such that for $s \leq 0$ we have that*

$$|\theta(s)|_2^2 \leq \epsilon_0 + K_0 \min(T_1, |s|)^\delta. \tag{104}$$

Proof. The basic energy estimates give us

$$|\theta(t)|_2^2 \leq |\theta_0|_2^2 - C_1 \int_{t_0}^t |\theta(s)|_2^2 \,\mathrm{d}s + \epsilon_0(t - t_0) + \int_{t_0}^t \langle \theta(s), \mathrm{d}\mathbb{W}(s) \rangle_2 . \tag{105}$$

Define the quantity

$$F_k(s) = -C_1 \int_{-k}^s |\theta(t)|_2^2 \,\mathrm{d}t + \int_{-k}^s \langle \theta(t), \mathrm{d}\mathbb{W}(t) \rangle_2 . \tag{106}$$

By the above definition, we have that, for any $k \geq 1$,

$$\sup_{s\in[-k,-k+1]} |\theta(s)|_2^2 \leq |\theta(-k)|_2^2 + \epsilon_0 + \sup_{s\in[-k,-k+1]} F_k(s). \tag{107}$$

We now define the event

$$A_k = \left\{ \theta(s) : \sup_{s\in[-k,-k+1]} |\theta(t)|_2^2 \leq \epsilon_0 + K_0 |k-1|^\delta \right\}. \tag{108}$$

Let

$$U_T = \bigcap_{k \geq T} A_k. \tag{109}$$

Then

$$U_T^c = \left(\bigcap_{k \geq T} A_k \right)^c = \bigcup_{k \geq T} A_k^c, \tag{110}$$

and we have that

$$\mu_p(U_T^c) = \mu_p \left(\bigcup_{k \geq T} A_k^c \right) \leq \sum_{k \geq T} \mu_p(A_k^c). \tag{111}$$

It follows from the definition of a measure that

$$\mu_p(A_k^c) \leq \mu_p \left\{ (\theta(s)) : |\theta(-k)|_2^2 \geq \frac{K_0}{2} |K-1|^\delta \right\}$$
$$+ \mu_p \left\{ (\theta(s)) : \sup_{s \in [-k,-k+1]} F_k(s) \geq \frac{K_0}{2} |K-1|^\delta \right\}. \tag{112}$$

We will now estimate each of the quantities on the right-hand side of the above inequality. We proceed with the first one. Chebyshev's inequality yields

$$\mu_p \left\{ (\theta(s)) : |\theta(-k)|_2^2 \geq \frac{K_0}{2} |K-1|^\delta \right\} \leq \frac{16}{K_0^2 |k-1|^{2\delta}} (\mathbb{E} |\theta(-k)|_2^2). \tag{113}$$

We now use the earlier derived energy estimates to yield

$$\mu_p \left\{ (\theta(s)) : |\theta(-k)|_2^2 \geq \frac{K_0}{2} |K-1|^\delta \right\} \leq \frac{C}{|k-1|^{2\delta}}. \tag{114}$$

Here C absorbs the uniform bounds of the energy estimates derived earlier. If we choose $\delta > \frac{1}{2}$, then

$$\sum_k \frac{C}{|k-1|^{2\delta}} < \infty. \tag{115}$$

Now we have shown that

$$[M_k, M_k] \leq (\sigma_{\max}^*)^2 \int_{-k}^{s} |\theta(t)|_2^2 \, dt. \tag{116}$$

We have that

$$F_k(s) \leq C_1 M_k - \frac{1}{(\sigma^*_{\max})^2}[M_k, M_k]. \tag{117}$$

Recall the exponential martingale inequality for any positive constants α and β:

$$\mathbb{P}\left\{ \sup_{s\in[-k,0]} M_k(s) - \frac{\alpha}{2}[M_k, M_k] \geq \beta \right\} \leq e^{-\alpha\beta}. \tag{118}$$

It therefore follows that

$$\begin{aligned}\mu_p &\left\{(\theta(s)) : \sup_{s\in[-k,-k+1]} F_k(s) \geq \tfrac{K_0}{2}|k-1|^{2\delta}\right\} \\ &\leq e^{-\frac{2K_0}{(\sigma_{\max})^2}|k-1|^{\delta}} + e^{-\frac{2K_0}{(\sigma^*_{\max})^2}|k-1|^{\delta}} \\ &\leq C_1 e^{-C_2|k-1|^{\delta}}.\end{aligned} \tag{119}$$

Again, clearly for $\delta > \frac{1}{2}$, we have that

$$\sum_k C_1 e^{-C_2|k-1|^{\delta}} < \infty. \tag{120}$$

Hence, we arrive at

$$\sum_{k\geq T} \mu_p(A_k^c) \leq \sum_k \frac{C}{|k-1|^{2\delta}} + \sum_k C_1 e^{-C_2|k-1|^{\delta}} < \infty. \tag{121}$$

This implies that

$$\sum_{k\geq T} \mu_p(A_k^c) < \infty, \tag{122}$$

and by the Borel–Cantelli lemma we have that

$$\mu_p(\limsup_{k\to\infty} A_k^c) = 0. \tag{123}$$

This tells us that for large values of k the complement of A_k would μ_p-almost never occur. Hence, A_k would μ_p-almost certainly occur. Hence, there must exist a time T_1 such that

$$|\theta(s)|_2^2 \leq \epsilon_0 + K_0 \min(T_1, |s|)^{\delta}. \tag{124}$$

Therefore the lemma is proved. □

5 Control of High Modes

We now state a lemma which gives an estimate on the difference of high modes of two different solutions to (22). This lemma is crucial for a proof of the third condition that follows subsequently.

Lemma 5.1. *Suppose there exist two solutions to the stochastic infinite Darcy–Prandtl number model:*

$$\theta_1(t) = l(t) + h_1(t) \tag{125}$$

and

$$\theta_2(t) = l(t) + h_2(t). \tag{126}$$

Then there exists a positive constant C such that when N is chosen so that

$$-\gamma = -(1-\alpha)N^2 + \beta < 0, \tag{127}$$

then $\theta_1 = \theta_2$*, i.e.,* $h_1(t) = h_2(t)$*. Furthermore, given a solution* $\theta(t)$*, any* h_0*, and* $t \leq 0$*, the limit exists:*

$$\lim_{t_0 \to -\infty} \Phi_{t_0,t}(l, h_0) = h(t). \tag{128}$$

Note that the α and β referred to in the Lemma above are introduced in condition 2, via (37) in [10]. The Φ is the same as introduced earlier.

Proof. Let $\rho(t) = h_1(t) - h_2(t)$. Then subtracting the requisite equations yields

$$\frac{d\rho}{dt} = -A\rho + P_h[R(\theta_1 - \theta_2)]. \tag{129}$$

We multiply the above equation by ρ, integrate by parts, and use condition 2 to yield

$$\begin{aligned}\frac{d|\rho|_2^2}{dt} &= -\langle A\rho, \rho\rangle + \langle [R(\theta_1 - \theta_2)], \rho\rangle \\ &\leq -\langle A\rho, \rho\rangle + \frac{1}{4}\langle A\rho, \rho\rangle + (C + K(\theta_1)|\rho|_2^2) \\ &\leq \frac{-3N^2}{4}|\rho|_2^2 + (C + K(\theta_1))|\rho|_2^2.\end{aligned}$$

Then there exists T_2 depending on t and θ_1 such that for $t_0 \leq T_2$, we have that

$$-\left(\frac{-3N^2}{4}\right)(t - t_0) + \int_{t_0}^{t} (C + K(\theta_1))\,, ds \leq -\gamma(t - t_0). \tag{130}$$

Hence, by Gronwall's inequality, we have

$$
\begin{aligned}
|\rho|_2^2 &\leq \exp\left\{-(\frac{-3N^2}{4})(t-t_0) + \int_{t_0}^t (C + K(\theta_1))\,\mathrm{d}s\right\} \\
&\leq e^{-\gamma(t-t_0)}|\rho(t_0)|_2^2 \\
&\leq e^{-\gamma(t-t_0)}[\epsilon_0 + K_0(t_0)^\delta].
\end{aligned}
$$

It follows then that for any time $t_0 \leq \min(T_1, T_2)$, we have, as $t_0 \to -\infty$, exponential decay when $N^2 \geq \frac{4}{3(t-t_0)}\int_{t_0}^t (C + K(\theta_1))\,\mathrm{d}s$.

For the second part of the lemma, let the high mode of the given solution $\theta(t)$ be h_1, and the solution to the high-mode equation starting from t_0 and h_0 be h_2. Then we have

$$
|\rho|_2^2 \leq |h(t_0) - h_0|_2^2 \exp\left\{-\left(\frac{3N^2}{4}\right) + \int_{t_0}^t K(\theta_1(s))\,\mathrm{d}s\right\}. \tag{131}
$$

By the argument made before, $\rho(t)$ decays to 0 as $t_0 \to -\infty$, and the limit equals $h(t)$. This proves the lemma. □

5.1 *Proof of Third Condition*

We are now in a position to verify the third condition, via (39) from [10]. We fix $L^0 \in P$ and $\bar{h}(0)$, which is an initial value for the high mode at time 0. Let $L^s = S_s^\omega L^0$ define S_s^ω, and $l(s) = L^t(s)$ for $s \leq t$. Then we know that with probability 1, $h(s) = \Phi_s(L^s)$ where $\theta(s) = (l(s), h(s))$, by Lemma 4.3. Fix a constant C_0 such that $|\theta(0)|_2^2 + |\nabla\theta(0)|_2^2 \leq C_0$. For any positive C, we define

$$
D(C) = \left\{f \in C([0,\infty), L_l^2) : |\theta|_2^2 + |\nabla\theta|_2^2 + \int_0^t (|\nabla\theta|_2^2 + |\Delta\theta|_2^2)\,\mathrm{d}s \leq C_1 + C_2\,t + C\,t^\delta\right\}. \tag{132}
$$

Here $\theta^* = f(s) + \Phi_s(f, \Phi_0(L_0))$.

Now we project $\theta(t)$ onto H_l, and by Lemma 5.1 we have that for any $a \in (0,1)$, there exists a C such that

$$
\mathbb{P}\left\{\omega : S_s^\omega L^0 \in D(C)\right\} > 1 - a > 0. \tag{133}
$$

Therefore if we set $\bar{h}(s) = \Phi_s(L^s, \bar{h}(0))$ and $\rho(s) = h(s) - \bar{h}(s)$, then $\theta = l + h = l + \bar{h} + \rho$. We state the following Lemma.

Lemma 5.2. *For the set $D(C)$ as defined in* (132), *the following estimate holds*

$$\sup_{\{\omega : S_t^{\omega} L^0 \in D(C)\}} \int_0^{\infty} |D(l(t), h(t), \bar{h}(t))|^2 \, dt < \infty, \tag{134}$$

where $D(l, h, \bar{h})$ is defined in Eq. (44).

Proof. We have

$$\begin{aligned}
&|D(l(t), h(t), \bar{h}(t)|_2^2 \\
&= \sup_{\{w \in L^2, |w|_2 \le 1\}} |Ra_D \left\langle P_l(P(\mathbf{k}(l+h) \cdot \nabla(l+h)) - P(\mathbf{k}(l+\bar{h}) \cdot \nabla(l+\bar{h}))), w \right\rangle|^2 \\
&\le \sup_{\{w \in L^2, |w|_2 \le 1\}} |Ra_D \ \langle P_l(P(\mathbf{k}(l+h) \cdot \nabla \rho)), w \rangle \ |^2 \\
&\quad + \sup_{\{w \in L^2, |w|_2 \le 1\}} | \ \langle Ra_D \ P_l(P(\mathbf{k} \nabla (l+h) \rho)), w \rangle \ |^2 \\
&\le \sup_{\{w \in L^2, |w|_2 \le 1\}} |Ra_D \ \langle P_l(P(\mathbf{k}(l+h) \cdot \rho)), \nabla w \rangle \ |^2 \\
&\quad + \sup_{\{w \in L^2, |w|_2 \le 1\}} | \ \langle Ra_D \ P_l(P(\mathbf{k}(l+h)\rho)), \nabla w \rangle \ |^2 \\
&\le \sup_{\{w \in L^2, |w|_2 \le 1\}} Ra_D^2 \ |P_l(\nabla w)|_{\infty}^2 |P(\mathbf{k}(l+h))|_2^2 |\rho|_2^2 \\
&\quad + \sup_{\{w \in L^2, |w|_2 \le 1\}} Ra_D^2 \ |P_l(\nabla w)|_{\infty}^2 |P(\mathbf{k}(l+h))|_2^2 |\rho|_2^2 \\
&\le \sup_{\{w \in L^2, |w|_2 \le 1\}} Ra_D^2 (|w|_{H^3}^2) |\theta|_2^2 |\rho|_2^2 \\
&\le C(N) \ Ra_D^2 |\theta|_2^2 |\rho|_2^2.
\end{aligned}$$

This follows via integration by parts and the embeddings

$$H^2(X) \hookrightarrow L^{\infty}(X) \hookrightarrow L^2(X), \tag{135}$$

and

$$H^3(X) \hookrightarrow W^{1,\infty}(X). \tag{136}$$

Note that if $L^t \in D(C)$, we can use the estimates on $|\theta|_2^2$ in Lemma 4.2, and on $|\rho|_2^2$ in Lemma 5.1 to yield

$$|D(l(t), h(t), \bar{h}(t)|_2^2 \le |\rho(0)|_2^2 C(N) e^{(-N^2 + \frac{1}{2} + C)t} (C_1 + C_2 t). \tag{137}$$

If N is chosen such that $N^2 > \frac{1}{2} + C$ as $t \to \infty$, the right-hand side decays exponentially. This yields

$$\begin{aligned}
&\sup_{\{\omega : S_t^{\omega} L^0 \in D(C)\}} \int_0^{\infty} |D(l(t), h(t), \bar{h}(t))|_2^2 \,\mathrm{d}t \\
&\qquad \le \sup_{\{\omega : S_t^{\omega} L^0 \in D(C)\}} \int_0^{\infty} t^{\delta} C(N) e^{(-N^2+\frac{1}{2}+C)t} (C_1 + C_2 t) \,\mathrm{d}t \\
&\qquad < K \\
&\qquad < \infty.
\end{aligned}$$

This proves the Lemma. □

Thus condition 3 via (39) from [10] is verified.

5.2 *Estimation of High Modes*

The following Lemma will be useful in proving the last condition from [10]. This gives an estimate on the high modes of a solution to the stochastic infinite Darcy–Prandtl number model.

Lemma 5.3. *If $h(t)$ is the high-mode component of the solution to the stochastic Darcy–Boussinesq equation with low-mode forcing $l \in C([0,t], L_l^2)$, then $|h(t)|_2^2$ is bounded by a constant that depends only on $|h(0)|_2^2$ and $\int_0^t |l(s)|_2^2 \,\mathrm{d}s$.*

Proof. We start with the equations for the high modes of the system, multiply them by the high-mode component, and integrate the result by parts to yield

$$\begin{aligned}
\frac{\mathrm{d}dt}{|} h|_2^2 &\le -2|\nabla h|_2^2 - 2Ra_D P_h(P(\mathbf{k}(l+h)) \cdot \nabla(l+h)h + 2Ra_D\, \gamma'(z) P_h[l+h]h \\
&\le -2|\nabla h|_2^2 - 2Ra_D \,\langle P(\mathbf{k}(l+h) \cdot \nabla l, h\rangle - \langle P(\mathbf{k}(l+h) \cdot \nabla h, h\rangle + C\,|h|_2^2 \\
&\le -2|\nabla h|_2^2 - 2Ra_D \int_H (l l_z h + h l_z h) \,\mathrm{d}\mathbf{x} + 2Ra_D |h|_2^2 \\
&\le Ra_D |l_z|_\infty |l|_2^2 + Ra_D |l_z|_\infty |h|_2^2 + 2Ra_D |l_z|_\infty |h|_2^2 + C\,|h|_2^2 \\
&\le Ra_D |\Delta l_z|_2 |l|_2^2 + Ra_D |\Delta l_z|_2 |h|_2^2 + 2Ra_D |\Delta l_z|_2 |h|_2^2 + C\,|h|_2^2 \\
&\le Ra_D C(N) |l|_2^2 + (3C(N) Ra_D + C)|h|_2^2 \\
&\le Ra_D C(N) |l|_2^2 + C(N) Ra_D |h|_2^2.
\end{aligned}$$

These follow via integration by parts, the embedding $H^2(X) \hookrightarrow L^\infty(X)$, and the uniform estimate on $|\theta|_2$. Thus, via the Gronwall inequality, we obtain

$$|h|_2^2 \le |h(0)|_2^2 + \int_0^t C(N)Ra_D|l|_2^2 \exp\left\{C\int_0^s |l(\tau)|_2^2\,\mathrm{d}\tau + C(N)Ra_D s\right\}\,\mathrm{d}s. \quad (138)$$

This completes the proof of the Lemma. □

5.3 *Proof of the Fourth Condition*

Let us assume that $|l(0)|_2 \le M$. We define, for $f \in C([0,\infty), L_l^2)$ and $0 \le s \le T$, $\theta(s) = f(s) + \Phi_s(f, \Phi_0(L^0))$. Set

$$D_T(C) = \left\{f \in C([0,\infty), L_l^2) : \int_0^t |\theta|_2^2\,\mathrm{d}s < CT, 0 \le t \le T\right\}. \quad (139)$$

We now state the following Lemma.

Lemma 5.4. *For the set $D_T(C)$ defined in* (139), *the following estimate holds:*

$$\sup_{\{L^t \in D_T\}} \int_0^t |G(l(s), \Phi_s(L^s, \Phi_0(L^0))|_2^2\,\mathrm{d}s < \infty. \quad (140)$$

Proof. We know $h(s) = \Phi_s(L^s, \Phi_0(L^0))$ with probability 1, from Lemma 5.1. Thus we have

$$\begin{aligned}
&|G(l(s), \Phi_s(L^s, \phi_0(L^0)))|_2 \\
&= \sup_{\{w\in L^2, |w|_2\le 1\}} \langle P_l(-P(\mathbf{k}(l+h))\cdot\nabla(l+h)) - Ra_D\, P_l(l+h), w\rangle \\
&\le \sup_{\{w\in L^2, |w|_2\le 1\}} \frac{1}{2}\langle P_l(P(\nabla(l+h)))^2, w\rangle + \sup_{\{w\in L^2, |w|_2\le 1\}} \langle Ra_D\, P_l(l+h), w\rangle \\
&\le \sup_{\{w\in L^2, |w|_2\le 1\}} |P_l\nabla w|_\infty \langle |l+h|, |l+h|\rangle + Ra_D\, |P_l\nabla w|_2 |l|_2 \\
&\le \sup_{\{w\in L^2, |w|_2\le 1\}} |\nabla w|_\infty (|l+h|_2^2 + Ra_D\, |l|_2^2) \\
&\le C(N)(|l|_2^2 + |h|_2^2).
\end{aligned}$$

Here $C(N)$ absorbs various constants. Therefore we have that for $l \in D_T(C^*)$

$$\begin{aligned}
|h|_2^2 &\le |h(0)|_2^2 + \int_0^t C(N)Ra_D|l|_2^2 \exp\left\{C\int_0^s |l(\tau)|_2^2\,\mathrm{d}\tau + (C(N)Ra_D)s\right\}\mathrm{d}s \\
&\le |h(0)|_2^2 + \int_0^T C(N)Ra_D|l|_2^2 \exp\left\{C\int_0^s |l(\tau)|_2^2\,\mathrm{d}\tau + (C(N)Ra_D)s\right\}\mathrm{d}s.
\end{aligned}$$

This is less than some C by Lemma 5.3, as we are on a finite time interval $[0, T]$. Here C depends only on $|h_0|_2$ and the C and T used to define $D_T(C)$. Therefore

$$
\begin{aligned}
\sup_{L^t \in D_T} \int_0^T & |G(l(s), \Phi_s(L^s, \Phi_0(L^0))|_2^2 \, \mathrm{d}s \\
&\leq \int_0^T C(N)(|l|_2^2 + |h|_2^2) \, \mathrm{d}s \\
&\leq C(N)CT + \int_0^T C(N)Ra_D |l|_2^2 \exp\left\{ C \int_0^t |l(s)|_2^2 \, \mathrm{d}s + (C(N)Ra_D)t \right\} \mathrm{d}t \\
&\leq C_{12}\, T + C_{13} e^{C_{14} T} \\
&\leq K \\
&< \infty.
\end{aligned}
$$

Thus the Lemma is proved. □

Condition 4 via (40) from [10] now follows. Since we have verified that the stochastic infinite Darcy–Prandtl number model satisfies all the conditions as postulated in [10], we can apply the results that appear therein, to yield the following result:

Theorem 2. *The stochastic infinite Darcy–Prandtl number model, as defined via* (22)*, possesses a unique invariant measure for*

$$N^2 > \frac{4C}{3}, \tag{141}$$

where C is the uniform constant appearing in Lemma 4.2.

6 Conclusion

In conclusion, we have shown the uniqueness of invariant measure for the stochastic infinite Darcy–Prandtl number model, under stochastic forcing of low modes. Various questions remain open at this point, for example, the uniqueness of an invariant measure for the stochastic Darcy–Boussinesq system. In this case, we would have to add noise to both the velocity and temperature equations. This is conceivably a harder question to tackle as the Darcy–Boussinesq system is only weakly dissipative.

We have shown in [16] that the stationary statistical properties for the deterministic Darcy–Boussinesq system are upper semicontinuous after lifting in the singular limit, i.e., for $\mu_\epsilon \in IM_\epsilon, 0 \leq \epsilon \leq \epsilon_0$, there exists a weakly convergent subsequence, denoted μ_ϵ, and $\mu_0 \in IM_0$ such that

$$\mu_\epsilon \rightharpoonup L(\mu_0). \tag{142}$$

Here μ_ϵ is an invariant measure for the Darcy–Boussinesq system and μ_0 is an invariant measure for the infinite Darcy–Prandtl number model. It would be of further interest then to derive uniqueness conditions on the μ_ϵ, and μ_0 as well.

Another difficult question to consider would be the zero-noise limit of both the stochastic infinite Darcy–Prandtl number model and stochastic Darcy–Boussinesq system. In particular, since we have shown uniqueness for the stochastic infinite Darcy–Prandtl number model, we could consider a "δ" model of the stochastic infinite Darcy–Prandtl number model, where δ is a simple multiplicative parameter which enters the system as

$$d\theta = (\Delta\theta - P(\mathbf{k}\theta)\cdot\nabla\theta + Ra_D P(\theta))\,dt + \delta\,d\mathbb{W}, \tag{143}$$

$$\theta|_{z=0} = 0, \quad \theta|_{z=1} = 0. \tag{144}$$

Hence, we would have a unique invariant measure for each fixed δ. What can be said about the limit of these as δ goes to zero, the "zero-noise limit"? Some work has been done in this regard for axiom A systems [27]. However, these are difficult to verify pragmatically for most physical systems we deal with.

Acknowledgements The first author would like to sincerely acknowledge the suggestions and guidance provided by his PhD advisor Dr. Xiaoming Wang. These greatly helped in completing the current work.

References

1. Adams, R.A.: Sobolev Spaces. Academic, New York (1975)
2. Charbeneau, R.J.: Groundwater Hydraulics and Pollutant Transport. Prentice Hall, Upper Saddle River (2000)
3. Da Prato, G., Zabczyk, J.: Ergodicity for Infinite Dimensional Systems. London Mathematical Society Lecture Note Series, vol. 229. Cambridge University Press, Cambridge (1996)
4. Doering, C., Constantin, P.: Heat transfer in convective turbulence. Nonlinearity **9**, 1049–1060 (1996)
5. Doering, C., Constantin, P.: Bounds for heat transport in a porous layer. J. Fluid Mech. **376**, 263–296 (1998)
6. Foias, C., Manley, O., Rosa, R., Temam, R.: Navier-Stokes Equations and Turbulence. Encyclopedia of Mathematics and Its Applications, vol. 83. Cambridge University Press, Cambridge (2001)
7. Horton, C.W., Rogers, F.T.: Convection currents in a porous medium. J. Appl. Phys. **16**, 367–370 (1945)
8. Lapwood, E.R.: Convection of a fluid in a porous medium. Math. Proc. Camb. Philos. Soc. **44**, 508–521 (1948)
9. Lewis, M.A.: Water in Earth Science Mapping for Planning, Development and Conservation. McCall, J., Marker, B. (eds.). Graham and Trotman, London (1989)
10. Liu, D., Weinan, E.: Gibbsian dynamics and invariant measures for stochastic PDE. J. Stat. Phys. **108**, 1125–1156 (2002)
11. Ly, H.V., Titi, E.S.: Global Gevrey regularity for the Benard convection in a porous medium with zero Darcy–Prandtl number. J. Nonlinear Sci. **9**, 333–362 (1999)

12. Mattingly, J.: The stochastically forced Navier-Stokes equations: energy estimates and phase space contractions. Ph.d thesis, Princeton University (1998)
13. Mattingly, J., Weinan, E., Sinai, Y.: Gibbsian dynamics and ergodicity for the stochastically forced Navier-Stokes equations. Commun. Math. Phys. **224**, 83–106 (2001)
14. Nield, D., Bejan, A.: Convection in Porous Media. 2nd edn. Springer, New York (1999)
15. Parshad, R.D.: Asymptotic behaviour of convection in porous media. PhD Thesis, Florida State University (2009)
16. Parshad, R.D.: Asymptotic behaviour of the Darcy–Boussinesq system at large Darcy–Prandtl number. Discrete Contin. Dyn. Syst A. Special issue 18. **26**(4), 1441–1469 (2010)
17. Saad, M.: Ensembles inertiels pour un modele de convection naturelle dissipatif, en milieu poreux. C.R. Acad. Sci Paris **Serie I 316**, 1277–1280 (1993)
18. Saling, J.: Radioactive Waste Management. Taylor and Francis, New York (2001)
19. Speight, J.: Enhanced Oil Recovery Handbook. Gulf Publishing Company, Houston (2009)
20. Temam, R.: Infinite-Dimensional Dynamical Systems in Mechanics and Physics, 2nd edn. Applied Mathematical Sciences, vol. 68. Springer, New York (1997)
21. Vafai, K.: Handbook of Porous Media. Marcel Dekker, New York (2000)
22. Wang, X.: Infinite Prandtl number limit of Rayleigh-Benard convection. Commun. Pure Appl. Math. **57**, 1265–1282 (2004)
23. Wang, X.: Asymptotic behavior of global attractors to the Boussinesq system for Rayleigh-Benard convection at large Prandtl number. Commun. Pure Appl. Math. **60**, 1293–1318 (2007)
24. Wang, X.: Upper semi-continuity of stationary statistical properties of dissipative systems. Discrete Contin. Dyn. Syst. A **23**, 521–540 (2009) [special issue dedicated to Prof. Li Ta-Tsien]
25. Wang, X.: Lecture notes on elementary statistical theories with applications to fluid systems. 2007 Shanghai Mathematics Summer School in Fudan University. To be published by Higher Education Press, Beijing (2009)
26. Wu, M., Lee, J.: Ergodicity for the dissipative Boussinesq equations with random forcing. J. Stat. Phys. **117**, 929–957 (2004)
27. Young, L.S.: What are SRB measures and which dynamical systems have them? J. Stat. Phys. **108**, 733–754 (2002)

Pricing Barrier Options Using Integral Transforms

Edgard Ngounda, Kailash C. Patidar, and Edson Pindza

Abstract Barrier options are a class of exotic options that are traded in over-the-counter markets worldwide. These options are particularly attractive for their lower cost compared to vanilla options. However, the closed form analytical solutions for the partial differential equations modeling these options are not easy to obtain and therefore one usually seeks numerical approaches to find them. In this paper, we consider two types of exotic options, namely a single barrier European down-and-out call and a double barrier European knock-out call options. Like some other standard and nonstandard options, these barrier options also have non-smooth payoffs at the exercise price. This non-smooth payoff is the main cause of the reduction in accuracy when the classical numerical methods, for example, lattice method, Monte Carlo method, or other methods based on finite difference and finite elements are used to solve such problems. In fact, the same happens when one uses the spectral method which is known to preserve the exponential accuracy. In order to retain this high-order accuracy, in this paper we propose a spectral decomposition method which approximates the unknown solution by rational interpolants on each sub-domain. The resulting semi-discrete problem is solved by a contour integral method. Our numerical results affirm that the proposed approach is very robust and gives very reliable results.

Keywords Barriers options • Laplace transform • Spectral methods • Domain decomposition

AMS Subject Classification (2010): 65M70; 65R10; 91G60

E. Ngounda • K.C. Patidar (✉) • E. Pindza
Department of Mathematics and Applied Mathematics, University of the Western Cape, Private Bag X17, Bellville 7535, South Africa
e-mail: kpatidar@uwc.ac.za

B. Toni (ed.), *New Trends and Advanced Methods in Interdisciplinary Mathematical Sciences*, STEAM-H: Science, Technology, Engineering, Agriculture, Mathematics & Health, DOI 10.1007/978-3-319-55612-3_10

1 Introduction

The market for barrier options has been expanding rapidly over the last two decades [20]. They are characterized by the dependence of their payoffs on the path of the underlying asset throughout their lifetime. When the asset price reaches a specified barrier level, these options are either exercisable (activated) in which case they are called knock-in options or they expire (extinguish) in which case they are called knock-out options.

Both knock-in and knock-out options are divided into down and up options depending on the level of the asset price compared to the barrier level. On one hand, knock-out options are either up-and-out or down-and-out. The up-and-out options can be exercised unless the asset price reaches the barrier level from below and the down-and-out options are exercisable until the asset price reaches the barrier level from above. Knock-in options are also classified as the up-and-in and down-and-in knock-in options. The up-and-in options are exercisable if the asset reaches the barrier level from below whereas the down-and-in options can be exercised only if the asset price reaches the barrier from above the barrier level.

An important issue of pricing barrier options is whether the barrier crossing is monitored in continuous time or in discrete time. A continuously monitored option is an option which is monitored constantly between the current and the final time T at maturity of the option. A discretely monitored option is only monitored at discrete time $t \leq t_1 < t_2 <, \ldots, < t_N < T$. On the market, barrier options are mostly discretely monitored. Analytical solutions of these options are expressed in terms of an m-dimensional integral over the m-dimensional multivariate normal distribution. However, the computation of the multivariate normal distribution is compromised when $m > 5$ (where m is the number of monitored points) [16], and thus one has to resort to numerical methods for practical implementation [4, 8, 12]. We prefer to solve the options that are monitored continuously under the Black–Scholes framework because they give their holders more flexibility in exercising.

As far as numerical methods for barrier options valuation are concerned, lattice methods have been used quite extensively despite their low level of accuracy. However, the implementation of these methods presents a computational challenge. Numerous techniques have been developed to improve their computational cost [11], still the improvement is not fully satisfying. Recently, Monte Carlo methods have emerged as a viable alternative to lattice methods. In this context, Barraquand and Martineau [1] obtained some promising results for multi-asset American options.

Finite-difference methods appear to be cost effective and flexible, compared to lattice and Monte Carlo methods, for pricing barrier options. In [3], the authors considered an explicit finite-difference approach. They constructed a grid which lies right on the barrier and applied interpolation to find the values of the option. Following on the same principle, Figlewski et al. [9] used an adaptive method to refine the mesh in regions of interest (around a barrier) and obtained very efficient results.

In [22], Zvan et al. took advantage of the higher order implicit methods to successfully price a variety of barrier option problems. Their approach used a fine grid around the location of the barrier. Fang and Oosterlee [6] considered the Fourier-based numerical method for American and barrier options. They obtained exponential results. This approach was extended in [7] to Bermudan and Barrier options under the Heston stochastic volatility model. The two-dimensional pricing problem was dealt with by a combination of a Fourier cosine series expansion in one dimension and high-order quadrature rules in the other dimension. Error analysis and experiments confirmed a fast error convergence.

A radial basis approach with Crank-Nicolson's discretization in time was proposed in [10] for pricing barrier, European, and Asian options. Recently a high-order accurate implicit finite-difference method to price various types of barrier options was applied in [14]. This approach was used for both the discretely and continuously monitored options. The scheme was also applied to the analysis of Greeks such as Δ and Γ of the option.

In [14], Ndogmo and Ntwiga used a fourth-order θ-method for the continuously and discretely monitored options. To refine grid points in regions of interest, they used a coordinate transformation. The strength of their approach lies on a probability-based optimal determination of the boundary conditions, along with the option values. The authors in [5] developed a robust method for both continuously and discretely monitored barrier options. The method is based on the method-of-line and can evaluate the solution of the option price problem and its Greeks very accurately. Furthermore, this approach efficiently handles standard barrier options and barrier options with early exercise feature.

In this paper, we propose the use of a multi-domain spectral method and the Laplace transformed method. The multi-domain method uses spectral method directly in each sub-domain. Appropriate matching conditions are imposed at the interface of the sub-domains to ensure the continuity of the solution and that of its first derivative. After the spatial discretization, the resulting semi-discrete problem is solved by a contour integral method.

The remainder of the paper is structured as follows. Section 2 outlines the model problem. In Section 3, we discuss the basic idea of spectral domain decomposition method and its implementation to the model problem. In Section 4, we describe the application of the Laplace transform to solve the semi-discrete problem. Numerical results are presented in Section 5. In Section 6, we draw some concluding remarks and present the scope for future research.

2 Description of Model Problem

We assume that the asset satisfies the following stochastic differential equation

$$\frac{dS}{S} = \mu dt + \sigma dW_t, \tag{1}$$

where μ is a constant representing the drift rate, σ is the volatility of the underlying asset, and dW_t is a Wiener process with mean zero and variance of dt. Under the Itô process, (1) gives the following Black–Scholes partial differential equation for the valuation of an option V:

$$\frac{\partial V}{\partial t} + \frac{1}{2}\sigma^2 S^2 \frac{\partial^2 V}{\partial S^2} + rS\frac{\partial V}{\partial S} - rV = 0. \tag{2}$$

Here, t is the current time at which the option $V(S, t)$ is valued before the expiration of the option at time T. For numerical computation, we set $\tau = T - t$ and rewrite (2) to a convenient form as follows

$$\frac{\partial V}{\partial \tau} = \frac{1}{2}\sigma^2 S^2 \frac{\partial^2 V}{\partial S^2} + rS\frac{\partial V}{\partial S} - rV, \tag{3}$$

with the boundary, initial, and barrier conditions given as follows:

- For single barrier down-and-out call, the boundary conditions are

$$V(0, \tau) = 0, \qquad V(S_{\max}, \tau) = S_{\max} - Ke^{-r\tau}, \tag{4}$$

whereas the barrier constraint is

$$V(S, \tau) = \begin{cases} 0, & S \le X \quad \text{and} \quad 0 \le \tau \le T, \\ V(S, \tau), & \text{otherwise.} \end{cases} \tag{5}$$

The initial condition is given by

$$V(S, 0) = \begin{cases} 0, & S \le X, \\ S - K, & \text{otherwise.} \end{cases} \tag{6}$$

Here, X is barrier level, K is the strike price, and $S_{\max}$ is chosen sufficiently large.

- For double barrier knock-out call, the boundary conditions are

$$V(0, \tau) = 0, \qquad V(S_{\max}, \tau) = 0, \tag{7}$$

and barrier constraint is given by

$$V(S, \tau) = \begin{cases} 0, & S \le X_1 \quad \text{or} \quad S \ge X_2 \quad \text{and} \quad 0 \le \tau \le T, \\ V(S, \tau), & \text{otherwise.} \end{cases} \tag{8}$$

The initial condition is given by

$$V(S,0) = \begin{cases} 0, & S \leq X_1, \\ S - E, & X_1 \leq S \leq X_2, \\ 0, & S \geq X_2. \end{cases} \tag{9}$$

Here, X_1 and X_2 represent the lower and upper level barriers, respectively, K is the strike price, and $S_{\max}$ is chosen sufficiently large.

We use the spectral domain decomposition approach for discretization of (3) in the next section.

3 Spectral Domain Decomposition Method

The exponential convergence rate of spectral method is strictly dependent on the smoothness of the unknown function. However, option pricing problems involve PDEs with initial conditions that are non-smooth. Therefore, direct application of a spectral method to the solution of the Black–Scholes equation leads to a low-order (slowly converging) approximation.

While dealing with problems having non-smooth initial data, Tee and Trefethen [19] proposed a new spectral method based on rational interpolants. In their paper, they considered problems whose solutions exhibit localized regions of rapid variation, such as sharp spikes and thin layers, but are still analytic on the computational interval. Unlike this work, here we consider decomposing the domain rather than using a grid stretching in order to maintain the exponential accuracy.

Instead of spectral methods based on polynomial interpolants, we shall consider spectral method based on rational interpolants for their improved stability properties compared to their polynomial counterparts [2]. To begin with, let us note that the rational approximation for a function u at the Chebyshev points ξ_k, $k = 0, 1 \ldots, N$ is given by

$$u_N(\xi) = \frac{\displaystyle\sum_{k=0}^{N} \frac{w_k}{\xi - \xi_k} u(\xi_k)}{\displaystyle\sum_{k=0}^{N} \frac{w_k}{\xi - \xi_k}}, \tag{10}$$

where the w_k; $k = 0, 1, \ldots, N$ are the barycentric weights defined by $w_0 = 1/2$, $w_N = (-1)^N/2$, and $w_k = (-1)^k$; $k = 1, \ldots, N-1$. For the spectral polynomial method, the mth order differentiation matrix associated with the rational interpolant (10) is given by

$$u_N^{(m)}(\xi_k) = \sum_{k=0}^{N} \frac{d^m}{d\xi^m} \left(\frac{\displaystyle\sum_{k=0}^{N} \frac{w_k}{\xi - \xi_k} u(\xi_k)}{\displaystyle\sum_{k=0}^{N} \frac{w_k}{\xi - \xi_k}} \right) \tag{11}$$

$$= \sum_{k=0}^{N} D_{jk}^{(m)} u(\xi_k), \tag{12}$$

where $D_{jk}^{(m)}$ are the entries of the differentiation matrix of order m. Formula to construct $D_N^{(m)}$ are given by Schneider and Werner in [17] for $m = 1$ and $m = 2$ and later generalized for any order by Tee in [19]. The first- and second-order differentiation matrices are given by the following formulas

$$D_{jk}^{(1)} = \begin{cases} \dfrac{w_k}{w_j(\xi_j - \xi_k)}, & j \neq k \\ -\displaystyle\sum_{i \neq k} D_{ji}^{(1)}, & j = k, \end{cases} \tag{13}$$

and

$$D_{jk}^{(2)} = \begin{cases} 2D_{jk}^{(1)} \left(D_{jj}^{(1)} - \dfrac{1}{\xi_j - \xi_k} \right), & j \neq k, \\ -\displaystyle\sum_{i \neq k} D_{ji}^{(2)}, & j = k. \end{cases} \tag{14}$$

The above expression will be used when we approximate the first and second derivative terms in (3).

From our earlier discussion in this section, we know that the direct application of spectral method to the Black–Scholes problem leads to low-order convergence results. This poor convergence is attributed due to the non-smoothness of the initial condition at the strike price. This motivates our attempt to recover the spectral accuracy. To this end, we consider decomposing the domain into different sub-domains. We split the domain at the point of discontinuity and apply the Chebyshev spatial discretization in each sub-domain to reduce the problem to a set of coupled ordinary differential equations in time. Since the function is analytic on each subinterval including the point of discontinuity (strike K), spectral accuracy can be obtained provided that appropriate matching conditions are set across this point of discontinuity.

To begin with this, we consider the domain decomposition approach for (3). Since the initial condition is non-smooth at the strike price K, we split the interval

$[0, S_{\max}]$ into two subintervals, $\vartheta = [0, K]$ and $\chi = [K, S_{\max}]$, of lengths $d^\vartheta = K$ and $d^\chi = S_{\max} - K$, respectively. To apply the Chebyshev discretization, we map each of these subintervals to the reference interval $[-1, 1]$ by using the linear transformation

$$S^{\vartheta,\chi}(\xi) = \begin{cases} \frac{d^\vartheta}{2}(\xi + 1), & S \in [0, K], \\ \frac{d^\chi}{2}(\xi + 1) + K, & S \in [K, S_{\max}]. \end{cases} \tag{15}$$

The linear transformation to the derivative of $S^{\vartheta,\chi}(\xi)$ leads to

$$dS^{\vartheta,\chi}(\xi) = \begin{cases} \frac{d^\vartheta}{2} d\xi, & S \in [0, K], \\ \frac{d^\chi}{2} d\xi, & S \in [K, S_{\max}], \end{cases} \tag{16}$$

which implies

$$\frac{\partial}{\partial S} \equiv \frac{\partial}{\partial \xi}\frac{\partial \xi}{\partial S} \equiv \begin{cases} \frac{2}{d^\vartheta}\frac{\partial}{\partial \xi}, & S \in [0, K], \\ \frac{2}{d^\chi}\frac{\partial}{\partial \xi}, & S \in [K, S_{\max}]. \end{cases} \tag{17}$$

From Eqs. (16) and (17), we can rewrite (3) into each sub-domain as follows

$$\frac{\partial V^\vartheta}{\partial \tau} = \frac{2}{(d^\vartheta)^2}\sigma^2 S^\vartheta(\xi)^2 \frac{\partial^2 V^\vartheta}{\partial \xi^2} + \frac{2r}{d^\vartheta} S^\vartheta(\xi)\frac{\partial V^\vartheta}{\partial \xi} - rV^\vartheta, \tag{18}$$

and

$$\frac{\partial V^\chi}{\partial \tau} = \frac{2}{(d^\chi)^2}\sigma^2 S^\chi(\xi)^2 \frac{\partial^2 V^\chi}{\partial \xi^2} + \frac{2r}{d^\chi} S^\chi(\xi)\frac{\partial V^\chi}{\partial \xi} - rV^\chi. \tag{19}$$

Using matrix notations, we have

$$\dot{V}^\vartheta = \frac{2}{(d^\vartheta)^2}\sigma^2 (W^\vartheta)^2 D^{(2)} V^\vartheta + \frac{2r}{d^\vartheta} W^\vartheta D^{(1)} V^\vartheta - rV^\vartheta, \tag{20}$$

and

$$\dot{V}^\chi = \frac{2}{(d^\chi)^2}\sigma^2 (W^\chi)^2 D^{(2)} V^\chi \frac{2r}{d^\chi} W^\chi D^{(1)} V^\chi - rV^\chi, \tag{21}$$

where '$\cdot$' denotes the differentiation w.r.t. τ, and $W^{\vartheta,\chi} = \operatorname{diag}\left((2/d^{\vartheta,\chi}) S^{\vartheta,\chi}(\xi)\right)$.

Defining

$$B^{\vartheta} := \frac{2}{(d^{\vartheta})^2}\sigma^2 (W^{\vartheta})^2 D^{(2)}, \tag{22}$$

$$C^{\vartheta} := \frac{2r}{d^{\vartheta}} W^{\vartheta} D^{(1)}, \tag{23}$$

we can rewrite (20) and (21) as

$$\dot{V}^{\vartheta} = (B^{\vartheta} V^{\vartheta}) + (C^{\vartheta} V^{\vartheta}) - rV^{\vartheta}, \tag{24}$$

and

$$\dot{V}^{\chi} = (B^{\chi} V^{\chi}) + (C^{\chi} V^{\chi}) - rV^{\chi}. \tag{25}$$

To impose the boundary conditions, we substitute the first and last equations from the first and second sub-domains, respectively, by the corresponding boundary conditions in (7). Two additional equations are enforced to ensure that the boundary conditions are satisfied. To ensure the continuity of the solution and that of its first derivative at the interface, we impose the following matching conditions across the point of discontinuity, i.e., at the strike price K. Defining

$$V^{\diamond} = V_N^{\vartheta} = V_0^{\chi}, \tag{26}$$

and

$$\frac{\partial V_N^{\vartheta}}{\partial \xi} = \frac{\partial V_0^{\chi}}{\partial \xi}, \tag{27}$$

we see that from Eqs. (24)–(26) implies

$$2\dot{V}^{\diamond} = \left(\left(B^{\vartheta} + C^{\vartheta} - rI^{\vartheta}\right) V^{\vartheta}\right)_N + ((B^{\chi} + C^{\chi} - rI^{\chi})\, V^{\chi})_0 , \tag{28}$$

where I^{ϑ} and I^{χ} are identity matrices on ϑ and χ, respectively.

On the other hand, Eq. (27) yields

$$\frac{2}{d^{\vartheta}}\left(D^{(1)} V^{\vartheta}\right)_N - \frac{2}{d^{\chi}}\left(D^{(2)} V^{\chi}\right)_0 = 0. \tag{29}$$

The solution vector V is obtained from the system of $2N + 2$ equations from (24), (25), (28), (29), and the boundary conditions. We combine the approximations from the two sub-domains with boundary and matching conditions included into the following global system.

$$\dot{V} = AV. \tag{30}$$

Here, V is the vector

$$V = \left(V_0^{\vartheta}, \ldots, V_N^{\vartheta}, V_0^{\chi}, \ldots, V_N^{\chi}\right),$$

and

$$A = B + C - rI,$$

where I is the identity, B and C are block diagonal matrices of $B^{\vartheta,\chi}$ and $C^{\vartheta,\chi}$, respectively. After the discretization, the resulting semi-discrete problem is solved using Laplace method as discussed in the next section.

4 Solving Semi-discrete Problem by Applying Laplace Transform

Following our approach, presented in our recent work [15], on a Talbot contour [18] of hyperbolic type, we can analyze the error by decomposing it into three parts: Discretization error, Truncation error, and Conditioning error; based on which we can derive optimal contour parameters. We have also shown in that work that the optimal error while using the Laplace transform method is exponentially small. In what follows, we discuss its application to solve the problem under consideration.

A direct application of the Laplace transform to (30) leads to

$$(zI - A)\widehat{V} = V_0, \tag{31}$$

where I is the identity matrix and $\widehat{V}$ the Laplace transform of $V(\cdot, t)$ defined by

$$\widehat{V}(\cdot, z) = \int_0^{\infty} V(\cdot, t)e^{-zt}dt. \tag{32}$$

The inverse is evaluated on a contour Γ known as the Bromwich contour as

$$V(\cdot, t) = \frac{1}{2\pi i}\int_{\Gamma} e^{zt}\widehat{V}(\cdot, z)dz, \qquad t > 0. \tag{33}$$

The contour Γ is chosen such that it encloses all the singularities of $\widehat{V}(., z)$.

$$z = z(\ell), \quad -\infty < \ell < \infty,$$

with the property that $\operatorname{Re} z \to -\infty$ as $\ell \to \pm\infty$.

On the contour $z(\ell) = \widetilde{\mu}\,(1 + \sin(i\ell - \alpha))\,, \quad \ell \in \mathbb{R}$, the inversion formula (33) can be rewritten as

$$V(\cdot, t) = \frac{1}{2\pi i}\int_{-\infty}^{\infty} e^{z(\ell)t}\widehat{V}\left(\cdot, z(\ell)\right) z'(\ell)d\ell, \tag{34}$$

where

$$z'(\ell) = \widehat{\mu} i \cos(i\ell - \alpha).$$

For $h > 0$ such that $\ell_k = kh$, where k is an integer, the trapezoidal rule can then be expressed by

$$V(\cdot, t) \approx \frac{h}{2\pi i} \sum_{k=-\infty}^{\infty} e^{z(\ell_k)t} \widehat{V}(\cdot, z(\ell_k))\, z'(\ell_k). \tag{35}$$

In practice, the infinite sum has to be truncated at a finite integer M. Furthermore, because of the symmetry of the contour, we can rewrite (35) as

$$V_M(\cdot, t) \approx \text{Re}\left\{ \frac{h}{\pi i} \sum_{k=0}^{M}{}^{\circ} e^{z(\ell_k)t} \widehat{V}(\cdot, z(\ell_k))\, z'(\ell_k) \right\}. \tag{36}$$

where '∘' indicates that the first term is divided by 2. The benefit of using (36) is that it reduces by half the summation (35) and subsequently the number of linear system to be evaluated in (31).

Now to analyze the error, we note that in our approximation, we encounter three types of errors: (1) The application of the trapezoidal rule (35) to the unbounded integral (34) introduces discretization error; (2) a truncation of the infinite series (35) produces a truncation error, and (3) the evaluation of (35) done in floating point environment introduces a roundoff error. Estimates on these errors are as follows:

4.1 Discretization Error

The discretization error is the difference between the continuous formula (34) and the corresponding trapezoidal formula (35), i.e.,

$$E_d = \frac{1}{2\pi i} \int_{-\infty}^{\infty} e^{z(\ell)t} \widehat{V}(\cdot, z(\ell))\, z'(\ell) d\ell - \frac{h}{2\pi i} \sum_{k=-\infty}^{\infty} e^{z(\ell_k)t} \widehat{V}(\cdot, z(\ell_k))\, z'(\ell_k). \tag{37}$$

Following estimate on the discrete error (37) can be obtained by using contour integral to represent E_d (see [15] for further details):

$$|E_d| \leq \frac{2e^{-2\pi d/h} C}{1 - e^{-2\pi d/h}} \to 0 \qquad \text{as} \qquad h \to 0. \tag{38}$$

4.2 Truncation Error

The truncation error is given by

$$\begin{aligned} E_t &= \frac{h}{2\pi i}\sum_{k=-\infty}^{\infty} e^{z(\ell_k)t}\widehat{V}\left(\cdot, z(\ell_k)\right) z'\left(\ell_k\right) - \frac{h}{2\pi i}\sum_{k=-\infty}^{\infty} e^{z(\ell_k)t}\widehat{V}\left(\cdot, z(\ell_k)\right) z'\left(\ell_k\right) \\ &= \frac{h}{2\pi i}\left(\sum_{k=-\infty}^{N-1} e^{z(\ell_k)t}\widehat{V}\left(\cdot, z(\ell_k)\right) z'\left(\ell_k\right) + \sum_{k=N+1}^{\infty} e^{z(\ell_k)t}\widehat{V}\left(\cdot, z(\ell_k)\right) z'\left(\ell_k\right)\right) \\ &= \frac{h}{\pi i}\sum_{k=N+1}^{\infty} e^{z(\ell_k)t}\widehat{V}\left(z(\ell_k)\right) z'\left(\ell_k\right). \end{aligned} \tag{39}$$

Because of the exponential factor $e^{z(\ell_k)t}$, the terms in the sum decrease exponentially as $k \to \infty$.

4.3 Conditioning Error

We note that in (35), the approximation $\tilde{f}(t)$ requires the evaluation of the transformed $F(z_k) = F(z(\ell_k))$, for $k = -M, -M+1, \ldots, M-1, M$. In reality, these evaluations are affected by roundoff errors which means that the actual approximation that takes place is $\tilde{F}(z_k) = \widetilde{V}(\cdot, z_k) + \rho_k,\ k = -M, -M+1, \ldots, M-1, M$, where $\rho_k > 0$ is a small value such that $|\rho_k| \le \epsilon$, with ϵ given by the machine precision. As a result, (35) implies

$$\begin{aligned} \check{f}_M(t) &= \frac{h}{2\pi i}\sum_{k=-M}^{M} e^{z_k t}\left(\widetilde{V}(\cdot, z_k) + \rho_k\right) z_k' \\ &= V_M(t) + \frac{h}{2\pi i}\sum_{k=-M}^{M} e^{z_k t}\rho_k z_k'. \end{aligned}$$

Setting

$$f_\rho(t) = \frac{h}{2\pi i}\sum_{k=-M}^{M} e^{z_k t}\rho_k z_k'.$$

and using the techniques discussed in [15], we can see that

$$\left\| f_\rho(t)\right\| \le \frac{e^{\widetilde{\mu}t}}{\pi e \varrho \sin\alpha}\frac{\rho}{t}\left[L\left((1-\varrho)\widetilde{\mu}t\sin\alpha\right)\right]. \tag{40}$$

We note that (40) is independent of $\widehat{V}(\cdot, z)$ and propagates moderately with respect to $\widetilde{\mu}t$.

In summary, we can see that the total error is fully controllable as long as we choose the optimal values of the associated parameters. The derivation of these parameters is described below.

4.4 Derivation of Optimal Contour Parameters

To derive optimal parameters of the contour when solving barrier options pricing problems, we note that the computational effort in the evaluation of $V_M(\cdot, t)$ in (35) comes from the evaluation of the Laplace transform $\widetilde{V}(\cdot, z)$ in (31) for each $z(\ell_k)$. Furthermore, we note that this evaluation of $\widetilde{V}(\cdot, t)$ is independent of time t and thus can be carried out once and subsequently use the same evaluation to approximate $V_M(\cdot, t)$ at different time level over an interval $[t_0, \Lambda t_0]$ for an integer Λ. Note also that various methods for finding an optimal contour were developed in [13, 21]. In [21], the following convergence estimate for the family of hyperbolic contours was derived

$$E_d = \mathcal{O}(e^{-2\pi(\pi/2-\alpha)/h}), \quad E_{-d} = \mathcal{O}(e^{\widetilde{\mu}t-2\pi\alpha/h}), \quad h \to 0. \tag{41}$$

Hence the total discretization error on the strip $(-d, d)$ is given by

$$E_r = E_d + E_{-d}.$$

Moreover, the truncation error that we found satisfies

$$E_t = \mathcal{O}(e^{\widetilde{\mu}t(1-\sin\alpha\cosh(hM))}), \quad M \to \infty. \tag{42}$$

To obtain an optimal contour parameter, we argue as follows: first note that only the second equation in (41) and Eq. (42) are time dependent. On one hand, the error E_t decreases when t increases and thus is maximum at t_0. To see this, we consider $\alpha \in (\pi/4, \pi/2)$ whichimpliesthattheinequality $1/\sqrt{2} < \sin\alpha < 1$ holds. For $M \to \infty$, multiplication of both sides of the inequality by $\cosh(hM)$ yields

$$\frac{\cosh(hM)}{\sqrt{2}} < \sin\alpha\cosh(hM) < \cosh(hM).$$

Since $\cosh(hM)/\sqrt{2} > 1$ for sufficiently large M, we get

$$1 - \sin\alpha\cosh(hM) \leq 0, \quad \text{for a fixed } h \neq 0 \quad \text{and} \quad M \to \infty.$$

On the other hand, the discretization error E_{-d} increases with t and so is maximum at $t = \Lambda t_0$.

To estimate the optimal parameters of the contour on $[t_0, \Lambda t_0]$, an asymptotic balance of the three errors at their maximum, i.e., E_t, E_d, E_{-d} is required. To this end, we set

$$-2\pi\,(\pi/2-\alpha)\,/h = \widetilde{\mu}\Lambda t_0 - 2\pi\alpha/h = \widetilde{\mu}t_0\,(1-\sin\alpha\cosh(hM))\,. \tag{43}$$

We solve these equations for $\widetilde{\mu}$ and h. From the first equation, we have

$$\widetilde{\mu}h = \frac{4\alpha\pi - \pi^2}{\Lambda t_0}. \tag{44}$$

The last equation in (43) together with (44) yields

$$\cosh(hM) = \frac{(\pi-2\alpha)\,\Lambda - \pi + 4\alpha}{(4\alpha-\pi)\sin\alpha},$$

From this, it follows that

$$A(\alpha) = hM = \cosh^{-1}\left(\frac{(\pi-2\alpha)\,\Lambda - \pi + 4\alpha}{(4\alpha-\pi)\sin\alpha}\right). \tag{45}$$

Therefore, we obtain

$$h = \frac{A(\alpha)}{M} \tag{46}$$

and

$$\widetilde{\mu} = \frac{4\alpha\pi - \pi^2}{h\Lambda t_0} = \frac{4\alpha\pi - \pi^2}{A(\alpha)}\frac{M}{\Lambda t_0}. \tag{47}$$

The contour parameters (47) and (46) are fixed and time independent. As a result, the corresponding contour is also fixed over the interval $[t_0, \Lambda t_0]$. From the parameters derived above, the error is

$$E_M = \mathcal{O}(e^{-B(\alpha)M}), \tag{48}$$

where

$$B(\alpha) = \frac{\pi^2 - 2\pi\alpha}{\cosh^{-1}\left(\frac{(\pi-2\alpha)\Lambda+4\alpha-\pi}{(4\alpha-\pi)\sin\alpha}\right)}. \tag{49}$$

The optimal error is obtained when $B(\alpha)$ attains its maximum for each value of Λ. In our computation, we choose $\Lambda = 50$ and obtained optimal parameters as listed in Table 1. Below we discuss the numerical results obtained by using the proposed approach.

Table 1 Optimal parameters of the contour for $\Lambda = 50$

Λ	α	$A(\alpha)$	$\mu\Lambda t_0/M$	$B(\alpha)$
50	0.9381	5.5582	0.3452	0.7152

Table 2 Values of the European single barrier down-and-out options using the CN, ILT-FDM, and our ILT-SDDM

S	Exact sol.	CN	ILT-FDM	ILT-SDDM	Error Error (CN)	Error (ILT-FDM)	(ILT-SDDM)
7	0.0000	0.0000	0.0000	0.0000	0.000	0.000	0.000
8	0.0000	0.0000	0.0000	0.0000	0.000	0.000	0.000
9	0.0000	0.0000	0.0000	0.0000	0.000	0.000	0.000
11	1.294595	1.293735	1.294598	1.294595	8.6E-4	1.9E-5	2.2E-13
13	3.116046	3.116054	3.116043	3.116046	7.0E-5	1.8E-5	1.8E-13
15	5.100318	5.100347	5.100310	5.100318	2.8E-5	5.1E-5	3.8E-13
17	7.099529	7.099532	7.099529	7.099529	3.3E-5	5.6E-6	6.7E-13
19	9.099502	9.099502	9.099502	9.099502	8.8E-5	3.3E-7	5.8E-13

5 Numerical Results

In this section, we present some numerical results which show the performance of our approach for single and double barrier options. As an example of a single barrier option, we consider the European down-and-out options whereas for double barrier option, we consider the knock-out call options.

In Table 2, we present the values of a down-and-out call option obtained by using the inverse Laplace transform combined with the finite-difference method (ILT-FDM), Crank-Nicolson method (CN), and our Laplace transform method combined with spectral decomposition method (ILT-SDDM). Furthermore, we list the errors in the solutions obtained by these three methods. The first column gives the values of the asset whereas the second column contains the values of the exact solution. Next three columns represent the solutions obtained by above three methods and then the remaining columns show the errors. The other parameters used in the simulations are $K = 10, \sigma = 0.2, r = 0.02, \tau = 0.5$, and $X = 9$. We see that the errors obtained by the proposed approach are very small.

Figure 1 (top) shows the graphs of the exact and numerical solution of the European down-and-out option price and that of the computed solution using the ILT-SDDM. We have also computed the Δ and Γ as shown in the bottom left and bottom right plots of this figure. The convergence results obtained by CN, ILT-FDM, and ILT-SDDM methods are presented in Fig. 2 (top). In this figure, we also present errors for Δ (bottom left) and Γ (bottom right) obtained by using our ILT-SDDM. We use the same parameters as those used for the results presented in Table 2.

In Table 3, we show the results obtained for two barriers X_1 and X_2. In this experiment, we solve the problem for asset price ranging from $S_{\min} = 90$ to $S_{\max} = 115$. The barriers are applied at $X_1 = 95$ and $X_2 = 110$. Other parameters are $K = 100, \sigma = 0.25, r = 0.05$, and $\tau = 0.25$. The table includes results obtained

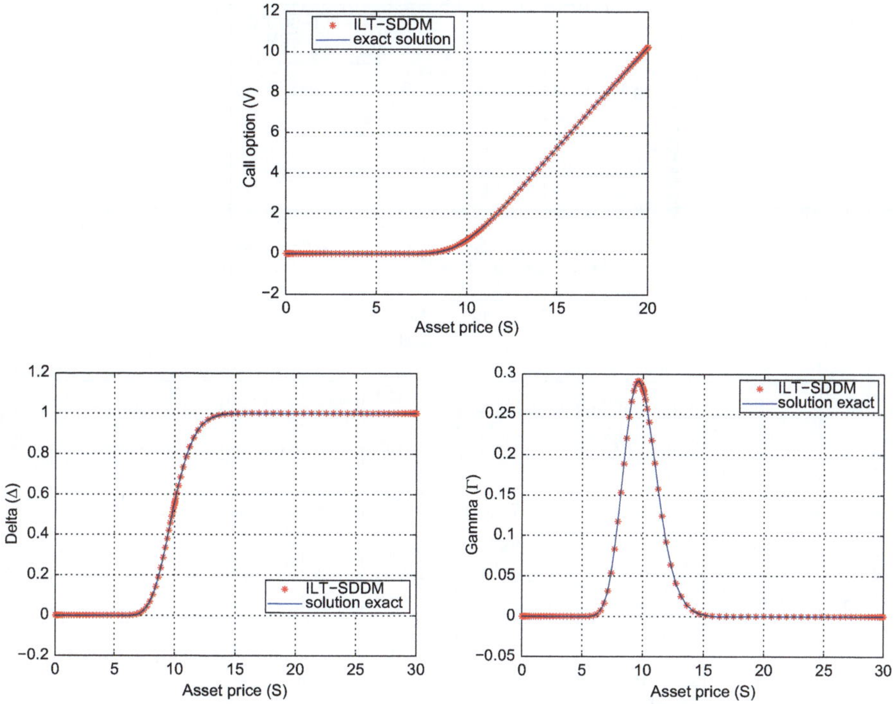

Fig. 1 Values of the European single barrier down-out call option (*top*) and its Δ (*bottom left*) and Γ (*bottom right*) using $K = 10$; $\sigma = 0.20$; $r = 0.05$; $T = 0.50$, and $X = 9$

by using our ILT-SDDM, and by using Crank-Nicolson (CN) and Crank-Nicolson improved (CN-improved) methods presented in [20]. We also list the errors in the solutions obtained by these methods. As reference solutions at the barriers, we use the values 0.0969796 at X_1 and 0.081481 at X_2 as obtained in [20]. The results confirm that the proposed approach outperforms the other methods mention here.

Figure 3 shows the graphs that we obtained by using the ILT-SDDM to price double barrier options for the set of parameters used for the computation of results presented in Table 3. The top figure shows the value of the option where the bottom figures show its Δ (left) and Γ (right).

It is clear from all the tabular results and graphs that the proposed method is very competitive.

6 Concluding Remarks and Scope for Future Research

In this paper, we presented a robust numerical method to price a class of barrier options. These are traded in over-the-counter markets worldwide. These options are particularly attractive for their lower cost compared to vanilla options. However, the

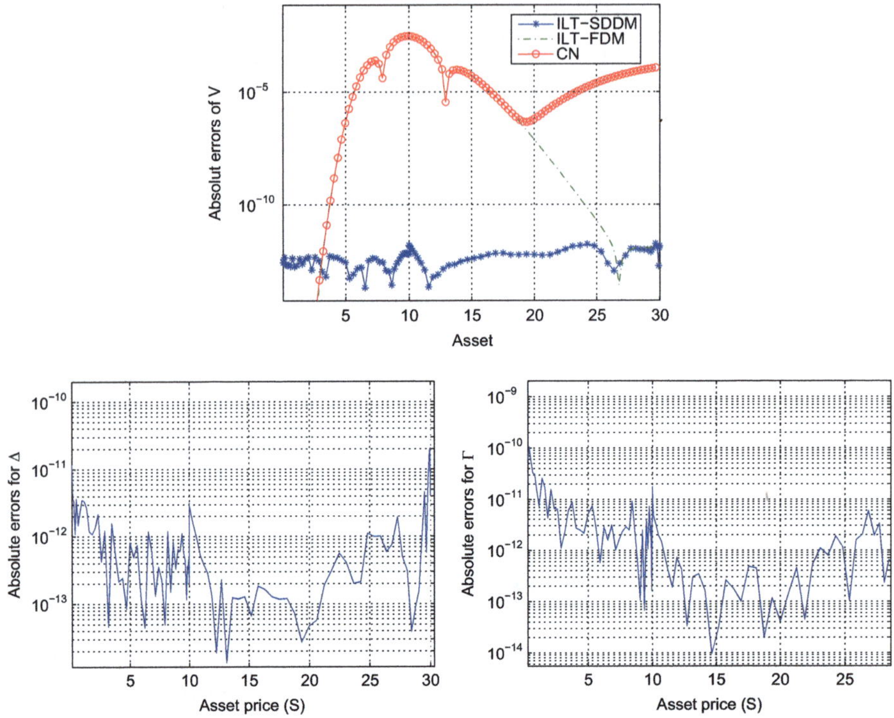

Fig. 2 Convergence of CN method, ILT-FDM, and ILT-SDDM for the single barrier European down-out call options (*top*), Δ (*bottom left*) and Γ (*bottom right*) using $K = 10$; $\sigma = 0.20$; $r = 0.05$; $T = 0.50$, and $X = 9$. We took $N = 50$ in each of the two sub-domains

closed form analytical solutions for the partial differential equations modeling these options are not easy to obtain and therefore one usually seeks numerical approaches to find them. We described the spectral domain decomposition and the Laplace transform methods for valuing single and double barriers option prices. As can be seen from Tables 2, 3 and Figs. 2, 3, the proposed approach gave very accurate results for the down-and-out call and the double barrier options.

With regard to the extension of the proposed approach for multi-asset problems, we would like to point out that there are significant technicalities involved with derivation of optimal contour parameters as whole of the analysis has to be extended for multiple dimensions.

Acknowledgements The two authors Edgard Ngounda and Edson Pindza acknowledge the Agence Nationale des Bourses du Gabon for the financial support. Kailash C. Patidar's research was supported by the South African National Research Foundation.

Table 3 Values of the European double barrier knock-out options using the CN, CN-improved, and our ILT-SDDM

Valueat	Reference sol.	CN [20]	CN-improved [20]	ILT-SDDM	Error (CN)	Error (CN-improved)	Error (ILT-SDDM)
X_1	0.096979	−0.094017	0.097002	0.096979	1.90E-1	3.10E-5	9.20E-8
X_2	0.081481	−0.909154	0.081503	0.081481	9.90E-1	2.89E-5	8.70E-8

Parameter values are $K = 100$; $\sigma = 0.25$; $r = 0.05$; $T = 0.5$; $X_1 = 95$, and $X_2 = 110$

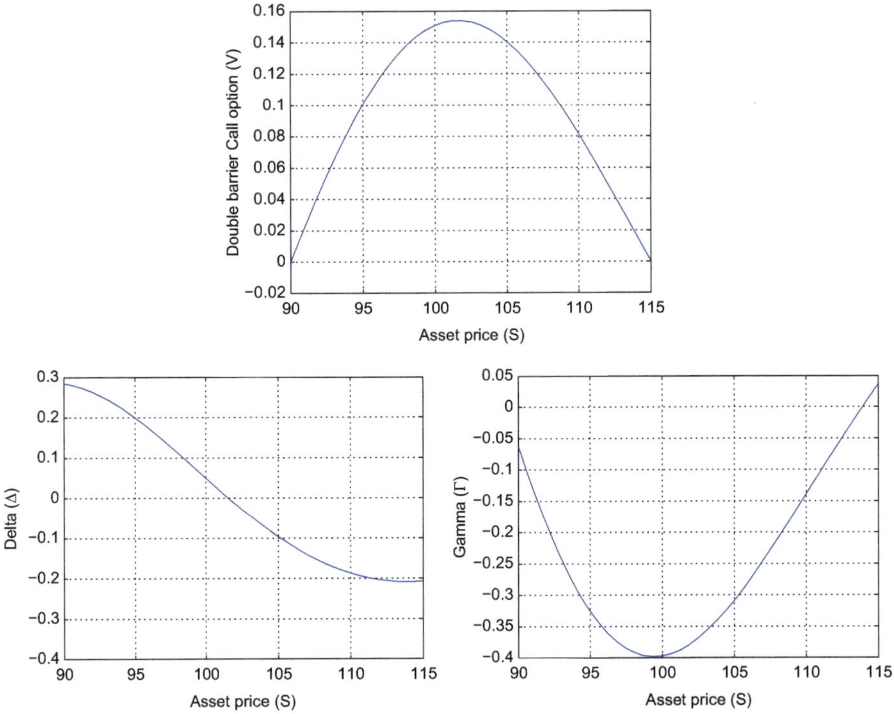

Fig. 3 Values of the European double barrier knock-out call option (*top*) and its Δ (*bottom left*) and Γ (*bottom right*) using $K = 100$; $\sigma = 0.25$; $r = 0.05$; $T = 0.5$; $X_1 = 95$, and $X_2 = 110$

References

1. Barraquand, J., Martineau, D.: Numerical valuation of high dimensional multivariate American securities. J. Financ. Quant. Anal. **30**, 383–405 (1995)
2. Berrut, J.P., Mittelmann, H.D.: Rational interpolation through the optimal attachment of poles to the interpolating polynomial. Numer. Algorithms **23**(4), 315–328 (2000)
3. Boyle, P.P., Tian, Y.S.: An explicit finite difference approach to the pricing of barrier options. Appl. Math. Financ. **5**, 17–43 (1998)
4. Broadie, M., Glasserman, P., Kou, S.G.: A continuity correction for discrete barrier options. Math. Financ. **7**, 325–349 (1997)
5. Chiarella, C., Kang, B., Meyer, G.: The evaluation of barrier option prices under stochastic volatility. Comput. Math. Appl. **64**(6), 2034–2048 (2012)
6. Fang, F., Oosterlee, C.W.: Pricing early-exercise and discrete barrier options by Fourier-Cosine series expansions. Numer. Math. **114**(1), 27–62 (2009)
7. Fang, F., Oosterlee, C.W.: A Fourier-based valuation method for Bermudan and barrier options under Heston′s model. SIAM J. Financ. Math. **2**, 439–463 (2011)
8. Feng, L., Linetsky, V.: Pricing discretely monitored barrier options and defaultable bonds in Lévy process models: a fast Hilbert transform approach. Math. Financ. **18**(3), 337–384 (2008)
9. Figlewski, S., Gao, B.: The adaptive mesh model: a new approach to efficient option pricing. J. Financ. Econ. **53**, 313–351 (1999)

10. Goto, Y., Fei, Z., Kan, S., Kita, E.: Options valuation by using radial basis function approximation. Eng. Anal. Bound. Elem. **31**, 836–843 (2007)
11. Haug, E.: Closed form valuation of American barrier options. Int. J. Theor. Appl. Financ. **4**, 355–359 (2001)
12. Kou, S.G., Wang, H.: First passage times of a jump-diffusion process. Adv. Appl. Probab. **35**, 504–531 (2003)
13. López-Fernández, M., Palencia, C.: On the numerical inversion of the Laplace transform of certain holomorphic mappings. Appl. Numer. Math. **51**(2–3), 289–303 (2004)
14. Ndogmo, J.C., Ntwiga, D.B.: High-order accurate implicit methods for barrier option pricing. Appl. Math. Comput. **218**, 2210–2224 (2011)
15. Ngounda, E., Patidar, K.C., Pindza, E.: Contour integral method for European options with jumps. Commun. Nonlinear Sci. Numer. Simul. **18**, 478–492 (2013)
16. Petralla, G., Kou, S.: Numerical pricing of discrete barrier and lookback options via Laplace transforms. J. Comput. Financ. **8**(1), 1–37 (2004)
17. Schneider, C., Werner, W.: Some new aspect of rational interpolation. Math. Comput. **47**(175), 285–299 (1986)
18. Talbot, A.: The accurate numerical inversion of Laplace transforms. IMA J. Appl. Math. **23**(1), 97–120 (1979)
19. Tee, W.: A rational collocation with adaptively transformed Chebyshev grid points. SIAM J. Sci. Comput. **28**(5), 1798–1811 (2006)
20. Wade, B.A., Khaliq, A.Q.M., Yousuf, M., Vigo-Aguiar, J., Deininger, R.: On smooth of the Crank-Nicolson scheme and higher order scheme for pricing barrier options. J. Comput. Appl. Math. **204**, 144–158 (2007)
21. Weideman, J.A.C., Trefethen, L.N.: Parabolic and hyperbolic contours for computing the Bromwich integral. Math. Comput. **76**(259), 1341–1356 (2007)
22. Zvan, R., Vetzal, K.R., Forsyth, P.A.: PDE methods for pricing barrier options. J. Econ. Dyn. Control **24**, 1563–1590 (2000)

Philosophy of Adelic Physics

Matti Pitkänen

Abstract The p-adic aspects of Topological Geometrodynamics (TGD) will be discussed. Introduction gives a short summary about classical and quantum TGD. This is needed since the p-adic ideas are inspired by TGD based view about physics.

p-Adic mass calculations relying on p-adic generalization of thermodynamics and super-symplectic and super-conformal symmetries are summarized. Number theoretical existence constrains lead to highly non-trivial and successful physical predictions. The notion of canonical identification mapping p-adic mass squared to real mass squared emerges, and is expected to be a key player of adelic physics allowing to map various invariants from p-adics to reals and vice versa.

A view about p-adicization and adelization of real number based physics is proposed. The proposal is a fusion of real physics and various p-adic physics to single coherent whole achieved by a generalization of number concept by fusing reals and extensions of p-adic numbers induced by given extension of rationals to a larger structure and having the extension of rationals as their intersection.

The existence of p-adic variants of definite integral, Fourier analysis, Hilbert space, and Riemann geometry is far from obvious and various constraints lead to the idea of number theoretic universality (NTU) and finite measurement resolution realized in terms of number theory. An attractive manner to overcome the problems in case of symmetric spaces relies on the replacement of angle variables and their hyperbolic analogs with their exponentials identified as roots of unity and roots of e existing in finite-dimensional algebraic extension of p-adic numbers. Only group invariants—typically squares of distances and norms—are mapped by canonical identification from p-adic to real realm and various phases are mapped to themselves as number theoretically universal entities.

Also the understanding of the correspondence between real and p-adic physics at various levels—space-time level, imbedding space level, and level of "world of classical worlds" (WCW)—is a challenge. The gigantic isometry group of WCW and the maximal isometry group of imbedding space give hopes about a resolution of the problems. Strong form of holography (SH) allows a non-local correspondence

M. Pitkänen (✉)
Karkinkatu 3 I 3, 03600 Karkkila, Finland
e-mail: matpitka6@gmail.com; http://tgdtheory.com/public_html/

B. Toni (ed.), *New Trends and Advanced Methods in Interdisciplinary Mathematical Sciences*, STEAM-H: Science, Technology, Engineering, Agriculture, Mathematics & Health, DOI 10.1007/978-3-319-55612-3_11

between real and p-adic space-time surfaces induced by algebraic continuation from common string world sheets and partonic 2-surfaces. Also local correspondence seems intuitively plausible and is based on number theoretic discretization as intersection of real and p-adic surfaces providing automatically finite "cognitive" resolution. The existence of p-adic variants of Kähler geometry of WCW is a challenge, and NTU might allow to realize it.

I will also sum up the role of p-adic physics in TGD inspired theory of consciousness. Negentropic entanglement (NE) characterized by number theoretical entanglement negentropy (NEN) plays a key role. Negentropy Maximization Principle (NMP) forces the generation of NE. The interpretation is in terms of evolution as increase of negentropy resources.

Keywords Adelic physics • Space-time • Planck constants • P-adicization • Topological Geometrodynamics

1 Introduction

I have developed during last 39 years a proposal for unifying fundamental interactions which I call "Topological Geometrodynamics" (TGD). During last 20 years TGD has expanded to a theory of consciousness and quantum biology and also p-adic and adelic physics have emerged as one thread in the number theoretical vision about TGD.

Since quantum TGD and physical arguments have served as basic guidelines in the development of p-adic ideas, the best manner to introduce the subject of p-adic physics is by describing first TGD briefly.

In this article I will consider the p-adic aspects of TGD—the first thread of the number theoretic vision—as I see them at this moment.

1. I will describe p-adic mass calculations based on p-adic generalization of thermodynamics and super-conformal invariance [27, 34] with number theoretical existence constrains leading to highly non-trivial and successful physical predictions. Here the notion of canonical identification mapping p-adic mass squared to real mass squared emerges and is expected to be key player of adelic physics and allow to map various invariants from p-adics to reals and vice versa.
2. I will propose the formulation of p-adicization of real physics and adelization meaning the fusion of real physics and various p-adic physics to single coherent whole by a generalization of number concept fusing reals and p-adics to larger structure having algebraic extension of rationals as a kind of intersection.

   The existence of p-adic variants of definite integral, Fourier analysis, Hilbert space, and Riemann geometry is far from obvious, and various constraints lead to the idea of NTU and finite measurement resolution realized in terms of number theory. Maybe the only manner to overcome the problems relies on the idea that various angles and their hyperbolic analogs are replaced with their exponentials and identified as roots of unity and roots of e existing in finite-dimensional algebraic extension of p-adic numbers. Only group invariants—typically squares of distances and norms—are mapped by canonical identification from p-adic to real realm and various phases are mapped to themselves as number theoretically universal entities.

Another challenge is the correspondence between real and p-adic physics at various levels: space-time level, imbedding space level, and WCW level. Here the enormous symmetries of WCW and those of imbedding space are in crucial role. Strong form of holography (SH) allows a correspondence between real and p-adic space-time surfaces induced by algebraic continuation from string world sheets and partonic 2-surface, which can be said to be common to real and p-adic space-time surfaces.

3. In the last section I will describe the role of p-adic physics in TGD inspired theory of consciousness. The key notion is negentropic entanglement (NE) characterized in terms of number theoretic entanglement negentropy (NEN). Negentropy Maximization Principle (NMP) would force the growth of NE. The interpretation would be in terms of evolution as increase of negentropy resources—Akashic records as one might poetically say. The newest finding is that NMP in statistical sense follows from the mere fact that the dimension of extension of rationals defining adeles increases unavoidably in statistical sense—separate NMP would not be necessary.

In the sequel I will use some shorthand notations for key principles and key notions. Quantum field theory (QFT); relativity principle (RP); equivalence principle (EP); general coordinate invariance (GCI); world of classical worlds (WCW); Strong form of GCI (SGCI); strong form of holography (SH); preferred extremal (PE); zero energy ontology (ZEO); quantum criticality (QC); hyper-finite factor of type II_1 (HFF); number theoretical universality (NTU); canonical identification (CI); Negentropy Maximization Principle (NMP); negentropic entanglement (NE); number theoretical entanglement negentropy (NEN) are the most often occurring acronyms.

2 TGD Briefly

This section gives a brief summary of classical and quantum TGD, which to my opinion is necessary for understanding the number theoretic vision.

2.1 Space-Time as 4-Surface

TGD forces a new view about space-time as 4-surface of 8-D imbedding space. This view is extremely simple locally but by its many-sheetedness topologically much more complex than GRT space-time.

2.1.1 Energy Problem of GRT as Starting Point

The physical motivation for TGD was what I have christened the energy problem of general relativity [61, 68].

1. The notion of energy is ill-defined because the basic symmetries of empty space-time are lost in the presence of gravity. The presence of matter curves empty Minkowski space M^4 so that its isometries realized as transformations leaving the distances between points and thus shapes of 4-D objects invariant are lost. Noether's theorem states that symmetries and conservation laws correspond to each other. Hence conservation laws are lost and conserved quantities are ill-defined. Usually this is not seen a practical problem since gravitation is so weak interaction.
2. The proposed way out of the problem is based on the assumption that space-times are imbeddable as 4-surfaces to some 8-dimensional space $H = M^4 \times S$ by replacing the points of 4-D empty Minkowski space with 4-D very small internal space S. The space S is unique from the requirement that the theory has the symmetries of standard model: $S = CP_2$, where CP_2 is complex projective space with four real dimensions [61]. Isometries of space-time are replaced with those of imbedding space. Noether's theorem predicts the classical conserved charges for given general coordinate invariant (GCI) action principle.

Also now the curvature of space-time codes for gravitation. Equivalence principle (EP) and general coordinate invariance (GCI) of GRT augmented with relativity principle (RT) of SRT remain the basic principles. Now, however, the number of solutions to field equations—preferred extremals (PEs)—is dramatically smaller than in Einstein's theory [25, 67].

1. An unexpected bonus was geometrization classical fields of standard model for $S = CP_2$. Also the space-time counterparts for field quanta emerge naturally but this requires a profound generalization of the notion of space-time: the topological inhomogeneities of space-time surface are identified as particles. This means a further huge reduction for dynamical field like variables at the level of single space-time sheet. By general coordinate invariance (GCI) only four imbedding space coordinates appear as variables analogous to classical fields: in a typical GUT their number is hundreds.
2. CP_2 also codes for the standard model quantum numbers in its geometry in the sense that electromagnetic charge and weak isospin emerge from CP_2 geometry: the corresponding symmetries are not isometries so that electroweak symmetry

breaking is coded already at this level. Color quantum numbers correspond to the isometries of CP_2 defining an unbroken symmetry: this also conforms with empirical facts. The color of TGD, however, differs from that in standard model in several aspects and LHC has began to exhibit these differences via the unexpected behavior of what was believed to be quark gluon plasma [37]. The conservation of baryon and lepton numbers follows as a prediction. Leptons and quarks correspond to opposite chiralities for imbedding space spinors.

3. What remains to be explained in standard model is family replication phenomenon for leptons and quarks. Both quarks and leptons appear as three families identical apart from having different masses. The conjecture is that fermion families correspond to different topologies for 2-D surfaces characterized by genus telling the number g (genus) of handles attached to sphere to obtain the surface: sphere, torus, etc. The 2-surfaces are identified as "partonic 2-surfaces" whose orbits are light-like 3-surface at which the signature of the induced metric of space-time surface transforms from Minkowskian to Euclidian. The partonic orbits replace the lines of Feynman diagrams in TGD Universe in accordance with the replacement of point-like particle with 2-surface.

   Only the three lowest genera are observed experimentally. A possible explanation is in terms of conformal symmetries: the genera $g \leq 2$ allow always Z_2 as a subgroup of conformal symmetries (hyper-ellipticity) whereas higher genera in general do not. The handles of partonic 2-surfaces could form analogs of unbound many-particle states for $g > 2$ with a continuous spectrum of mass squared but for $g = 2$ form a bound state by hyper-ellipticity [27].

4. Later further arguments in favor of $H = M^4 \times CP_2$ have emerged. One of them relates to twistorialization and twistor lift of TGD [49, 69, 70]. 4-D Minkowski space is unique space-time with Minkowskian signature of metric in the sense that it allows twistor structure. This is a problem in attempts to introduce twistors to General Relativity Theory (GRT) and a serious obstacle in the quantization based on twistor Grassmann approach, which has demonstrate its enormous power in the quantization of gauge theories. In TGD framework one can ask whether one could lift also the twistor structure to the level of H. M^4 has twistor structure and so does also CP_2: which is the only Euclidian 4-manifold allowing twistor space, which is also a Kähler manifold! This led to the notion of twistor lift of TGD inducing rather recent breakthrough in the understanding of TGD.

TGD can be also seen as a generalization of hadronic string model—not yet superstring model since this model became fashionable 2 years after the thesis about TGD [20]. Later it has become clear that string like objects, which look like strings but are actually 3-D, are basic stuff of TGD Universe and appear in all scales [29, 67]. Also strictly 2-D string world sheets popped up in the formulation of quantum TGD (analogy with branes) [53, 65] so that one can say that string model in 4-D space-time is part of TGD.

Concluding, TGD generalizes standard model symmetries and provides an incredibly simple proposal for a dynamics: only four classical field variables and in fermionic sector only quark and lepton like spinor fields. The basic objection against TGD looks rather obvious in the light of after-wisdom. One loses linear superposition of fields, which holds in good approximation in ordinary field theories, which are almost linear. The solution of the problem to be discussed later relies on the notion many-sheeted space-time [68].

2.1.2 Many-Sheeted Space-Time

The replacement of the abstract manifold geometry of general relativity with the geometry of 4-surfaces brings in the shape of surface as seen from the perspective of 8-D space-time as additional degrees of freedom giving excellent hopes of realizing the dream of Einstein about geometrization of fundamental interactions.

The work with the generic solutions of the field equations assignable to almost any variational principle satisfying GCI led soon to the realization that the topological structure of space-time in this framework is much more richer than in GRT.

1. Space-time decomposes into space-time sheets of finite size. This led to the identification of physical objects that we perceive around us as space-time sheets. The original identification of space-time sheet was as a surface of H with outer boundary. For instance, the outer boundary of the table would be where that particular space-time sheet ends (what "ends" means is not, however, quite obvious!). We would directly see the complex topology of many-sheeted space-time! Besides sheets also string like objects and elementary particle like objects appear so that TGD can be regarded also as a generalization of string models obtained by replacing strings with 3-D surfaces.

   It turned that boundaries are probably excluded by boundary conditions. Rather, two sheets with boundaries must be glued along their boundaries together to get double covering. Sphere can be seen as simplest example of this kind of covering: northern and southern hemispheres are glued along equator together.
2. The original vision was that elementary particles are topological inhomogeneities glued to these space-time sheets using topological sum contacts. This means drilling a hole to both sheets and connecting with a very short cylinder. Two-dimensional illustration should give the idea. In this conceptual framework material structures and shapes would not be due to some mysterious substance in slightly curved space-time but reduce to space-time topology just as energy-momentum currents reduce to space-time curvature in GRT.

   This view has gradually evolved to much more detailed picture. Elementary particles have wormhole contacts as basic building bricks. Wormhole contact is very small region with *Euclidian (!)* signature of the induced metric connecting two Minkowskian space-time sheets with light-like boundaries carrying spinor fields and there particle quantum numbers. Wormhole contact carries magnetic monopole flux through it and there must be second wormhole contact in order to have closed lines of magnetic flux. Particle world lines are replaced with 3-D light-like surfaces—orbits of partonic 2-surfaces—at which the signature of the induced metric changes.

One might describe particle as a pair of magnetic monopoles with opposite charges. With some natural assumptions the explanation for the family replication phenomenon in terms of the genus g of the partonic 2-surface is not affected. Bosons emerge as fermion anti-fermion pairs with fermion and anti-fermion at the opposite throats of the wormhole contact. In principle family replication phenomenon should have bosonic analog. This picture assigns to particles strings connecting the two wormhole throats at each space-time sheet so that string model mathematics becomes part of TGD.

The notion of classical field differs in TGD framework in many respects from that in Maxwellian theory.

1. In TGD framework fields do not obey linear superposition and all classical fields are expressible in terms of four imbedding space coordinates in given region of space-time surface. Superposition for classical fields is replaced with *superposition of their effects* [57, 61]—in full accordance with operationalism. Particle can topologically condense simultaneously to several space-time sheets by generating topological sum contacts (not stable like the wormhole contacts carrying magnetic monopole flux and defining building bricks of particles). Particle "experiences" the superposition of the effects of the classical fields at various space-time sheets rather than the superposition of the fields.

   It is also natural to expect that at macroscopic length scales the physics of classical fields (to be distinguished from that for field quanta) can be explained using only four primary field like variables. Electromagnetic gauge potential has only four components and classical electromagnetic fields give an excellent description of physics. This relates directly to electroweak symmetry breaking in color confinement which in standard model imply the effective absence of weak and color gauge fields in macroscopic scales. TGD, however, predicts that copies of hadronic physics and electroweak physics could exist in arbitrary long scales [36] and there are indications that just this makes living matter so different as compared to inanimate matter.
2. The notion of induced gauge field means that one induces electroweak gauge potentials defining the so-called spinor connection at space-time surface (induction of bundle structure). Induction boils down locally to a projection of the imbedding space vectors representing the spinor connection. The classical fields at the imbedding space level are non-dynamical and fixed and extremely simple: one can say that one has generalization of constant electric field and magnetic fields in CP_2. The dynamics of the 3-surface, however, implies that induced fields can form arbitrarily complex field patterns. This is essentially dynamics of shadows.

   Induced gauge fields are not equivalent with ordinary free gauge fields. For instance, the attempt to represent constant magnetic or electric field as a space-time surface has a limited success. Only a finite portion of space-time carrying this field allows realization as 4-surface. I call this topological field quantization. The magnetization of electric and magnetic fluxes is the outcome. Also gravitational field patterns allowing imbedding are very restricted: one implication is that topological with over-critical mass density are not globally imbeddable. This would explain why the mass density in cosmology can be at most critical. This solves one of the mysteries of GRT based cosmology [45].

Quite generally, the field patterns are extremely restricted: not only due to imbeddability constraint but also due to the fact that by SH only very restricted set of space-time surfaces can appear solutions of field equations: I speak of preferred extremals (PEs) [25, 67, 68]. One might speak about archetypes at the level of physics: they are in quite strict sense analogies of Bohr orbits in atomic physics: this implies by the realization of GCI. This kind of simplicity does not conform with what we observed. The way out is many-sheeted space-time. Although fields do not superpose, particles experience the superposition of effects from the archetypal field configurations (superposition is replaced with set theoretic union).

3. The important implication is that one can assign to each material system a field identity since electromagnetic and other fields decompose to topological field quanta. Examples are magnetic and electric flux tubes and flux sheets and topological light rays representing light propagating along tube like structure without dispersion and dissipation making em ideal tool for communications [41]. One can speak about field body or magnetic body of the system.

   Field body indeed becomes the key notion distinguishing TGD inspired model of quantum biology from competitors but having applications also in particle physics since also leptons and quarks possess field bodies. The is evidence for the Lamb shift anomaly of muonic hydrogen [9] and the color magnetic body of u quark whose size is somewhat larger than the Bohr radius could explain the anomaly [37]. The magnetic flux tubes of magnetic body carry monopole fluxes existing without generating currents. In cosmology the flux tubes assignable to the remnants of cosmic strings make possible long range magnetic fields in all scales impossible in standard cosmology. Also super-conductivity is proposed to rely on dark $h_{eff} = n \times h$ Cooper pairs at pairs of flux tubes carrying monopole flux.

GRT and gauge theory limit of TGD is obtained as an approximation.

1. GRT/gauge theory type description is an approximation obtained by lumping together the space-time sheets to single region of M^4, with gravitational fields and gauge potentials as sums of corresponding induced field quantities at space-time surface geometrized in terms of geometry of H. Gravitational field corresponds to the deviation of the induced metric from Minkowski metric using M^4 coordinates for space-time surface so that the description applies only in long length scale limit.

   Space-time surface has both Minkowskian and Euclidian regions. Euclidian regions are identified in terms of what I call generalized scattering/twistor diagrams. The 3-D boundaries between Euclidian and Minkowskian regions have degenerate induced 4-metric and I call them light-like orbits of partonic 2-surfaces or light-like wormhole throats analogous to blackhole horizons. The interiors of blackholes are replaced with the Euclidian regions and every physical system is characterized by this kind of region.

Lumping of sheets together implies that global conservation laws cannot hold exactly true for the resulting GRT type space-time. Equivalence principle (EP) as Einstein's equations stating conservation laws locally follows as a local remnant of Poincare invariance.

2. Euclidian regions are identified as slightly deformed pieces of CP_2 connecting two Minkowskian space-time regions. Partonic 2-surfaces defining their boundaries are connected to each other by magnetic flux tubes carrying monopole flux.

   Wormhole contacts connect two Minkowskian space-time sheets already at elementary particle level, and appear in pairs by the conservation of the monopole flux. Flux tube can be visualized as a highly flattened square traversing along and between the space-time sheets involved. Flux tubes are accompanied by fermionic strings carrying fermion number. Fermionic strings give rise to string world sheets carrying vanishing induced em charged weak fields (otherwise em charge would not be well-defined for spinor modes). String theory in space-time surface becomes part of TGD. Fermions at the ends of strings can get entangled and entanglement can carry information.

3. Strong form of GCI (SGCI) states that light-like orbits of partonic 2-surfaces on one hand and space-like 3-surfaces at the ends of causal diamonds on the other hand provide equivalent descriptions of physics. The outcome is that partonic 2-surfaces and string world sheets at the ends of CD can be regarded as basic dynamical objects.

   Strong form of holography (SH) states the correspondence between quantum description based on these 2-surfaces and 4-D classical space-time description, and generalizes AdS/CFT correspondence. One has huge super-symplectic symmetry algebra acting as isometries of WCW with conformal structure [28, 53, 65], conformal algebra of light-cone boundary extending the ordinary conformal algebra, and ordinary Kac-Moody and conformal symmetries of string world sheets. This explains why 10-D space-time can be replaced with ordinary space-time and 4-D Minkowski space can be replaced with partonic 2-surfaces and string world sheets. This holography looks very much like the one we are accustomed with!

2.2 Zero Energy Ontology

In standard ontology of quantum physics physical states are assumed to have positive energy. In zero energy ontology (ZEO) [54] physical states decompose to pairs of positive and negative energy states such that the net values of the conserved quantum numbers vanish. The interpretation of these states in ordinary ontology would be as transitions between initial and final states, physical events.

2.2.1 ZEO and Positive Energy Ontology

ZEO is consistent with the crossing symmetry of QFTs meaning that the final states of the quantum scattering event can be described formally as negative energy states. As long as one can restrict the consideration to either positive or negative energy part of the state ZEO is consistent with positive energy ontology. This is the case when the observer characterized by a particular CD studies the physics in the time scale of much larger CD containing observer's CD as a sub-CD. When the time scale sub-CD of the studied system is much shorter that the time scale of sub-CD characterizing the observer, the interpretation of states associated with sub-CD is in terms of quantum fluctuations.

ZEO solves the problem, which emerges in any theory assuming symmetries giving rise to conservation laws. The problem is that the theory itself is not able to characterize the values of conserved quantum numbers of the initial state of say cosmology. In ZEO this problem disappears since in principle any zero energy state is obtained from any other state by a sequence of quantum jumps without breaking of conservation laws. The fact that energy is not conserved in GRT based cosmologies can be also understood since each CD is characterized by its own conserved quantities. As a matter of fact, one must speak about average values of conserved quantities since one can have a quantum superposition of zero energy states with the quantum numbers of the positive energy part varying over some range.

At the level of principle the implications are quite dramatic. In quantum jump as recreation replacing the quantum Universe with a new one it is possible to create entire sub-Universes from vacuum without breaking the fundamental conservation laws. From the point of view of consciousness theory the important implication is that "free will" is consistent with the laws of physics. This makes obsolete the basic arguments in favor of materialistic and deterministic world view.

2.2.2 Zero Energy States Are Located Inside Causal Diamond

By quantum classical correspondence zero energy states must have space-time and imbedding space correlates.

1. Positive and negative energy parts of zero energy state reside at future and past light-like boundaries of causal diamond (CD) identified as intersection of future and past directed light-cones and visualizable as double cone. The analog of CD in cosmology is big bang followed by big crunch. Penrose diagrams provide an excellent 2-D visualization of the notion. CDs form a fractal hierarchy containing CDs within CDs. Disjoint CDs are possible and CDs could also intersect.

   The interpretation of CD in TGD inspired theory of consciousness is an imbedding space correlate for perceptive field of conscious entity: the contents of conscious experience are about the region defined by CD. At the level of space-time sheets the experience come from space-time sheets in the interior of CD. Whether the sheets can be assumed to continue outside CD is still unclear.

Quantum measurement theory must be modified in ZEO since state function reduction can happen at both boundaries of CD and the reduced states at opposite boundaries are related by time reversal. One can also have quantum superposition of CDs changing between reductions at active boundary followed by localization in the moduli space of CDs with the tip of passive boundary fixed. Quantum measurement theory generalizes to a theory of consciousness with continuous entity identified as a sequence of state function reductions at active (changing) boundary of CD [23].

2. By number theoretical universality (NTU) the temporal distances between the tips of the intersecting light-cones are assumed to come as integer multiples $T = m \times T_0$ of a fundamental time scale T_0 defined by CP_2 size R as $T_0 = R/c$. p-Adic length scale hypothesis [39, 66] motivates the stronger hypothesis that the distances tend to come as octaves of T_0: $T = 2^n T_0$. One prediction is that in the case of electron this time scale is 0.1 s defining the fundamental biorhythm. Also in the case u and d quarks the time scales correspond to biologically important time scales given by 10 ms for u quark and by 2.5 ms for d quark [24]. This means a direct coupling between microscopic and macroscopic scales.

2.3 *Quantum Physics as Physics of Classical Spinor Fields in WCW*

The notions of Kähler geometry of "world of classical worlds" (WCW) and WCW spinor structure are inspired by the vision about the geometrization of the entire quantum theory.

2.3.1 Motivations for WCW

The notion of "world of classical worlds" (WCW) [28, 33, 65] was forced by the failure of both path integral approach and canonical quantization in TGD framework. The idea is that the Kähler function defining WCW Kähler geometry is determined by the real part of an action S determining space-time dynamics and receiving contributions from both Minkowskian and Euclidian regions of space-time surface X^4 (note that $\sqrt{g_4}$ is proportional to imaginary unit in Minkowskian regions).

1. If S is space-time volume both canonical quantization and path integral would make sense at least formally since in principle one could solve the time derivatives of four imbedding space coordinates as functions of canonical momentum densities (general coordinate invariance allows to eliminate four coordinates). The calculation of path integral is, however, more or less hopeless challenge in practice.

2. A mere space-time volume as action is, however, not physically attractive. This was thought to leave under consideration only Kähler action S_K—Maxwell action for the induced Kähler form expressible in terms of gauge potential defined by the induced Kähler gauge potential of CP_2. This action has, however, a huge vacuum degeneracy. Any space-time surface with at most 2-D CP_2 projection, which is Lagrangian sub-manifold of CP_2, is vacuum extremal. Symplectic transformations acting like U(1) gauge transformations generate new vacuum extremals. They, however, fail to act as symmetries of non-vacuum extremals so that gauge invariance is not in question: the deviation of the induced metric from flat metric is the reason for the failure. This degeneracy is assumed to give rise to what might be called 4-D spin glass degeneracy meaning that the landscape for the maxima of Kähler function is fractal.
3. Canonical quantization fails because by the extreme non-linearity of the action principle making it is impossible to solve time derivatives explicitly in terms of canonical momentum densities. The problem is especially acute for the canonical imbedding of empty Minkowski space to $M^4 \times CP_2$. The action is vanishing up to fourth order in imbedding space coordinates so that canonical momentum densities vanish identically and there is no hope of defining propagator in path integral approach. The mechanical analog would be criticality around where the potential reduces to $V \propto x^4$. Quantum criticality is indeed a basic aspect of TGD Universe.

The hope held for a long time was that WCW geometry allowing to get rid of path integral would solve the problems. One could, however, worry about vacuum degeneracy implying that WCW metric becomes extremely degenerate for vacuum extremals and also holography becomes extremely non-unique for them. Also the expected failure of perturbative approach around M^4 is troublesome.

2.3.2 WCW and Twistor Lift of TGD

During last year this picture has indeed changed, thanks to what might be called twistor lift of TGD [49, 69, 70] inspired by twistor Grassmann approach to supersymmetric gauge theories [7]. Remarkably, twistor lift would provide automatically the fundamental couplings of standard model and GRT and also the scale assigned to GUTs as CP_2 radius. PEs would be both extremals of Kähler action and minimal surfaces.

1. The basic observation is E^4, and its Euclidian compactification S^4 and CP_2 are completely unique in that they allow twistor space with Kähler structure [3]. This was discovered by Hitchin at roughly the same time as I discovered TGD! This generalizes to M^4 having a generalization of ordinary Kähler structure to what I have called Hamilton–Jacobi structure by decomposition $M^4 = M^2 \times E^2$, where M^2 allows hypercomplex structure [49, 69]. One can consider also integral distributions of tangent decompositions $M^4 = M^2(x) \times E^2(x)$, depending on position. The twistor space has a double fibration by S^2 with base spaces

identifiable as M^4 and conformal compactification of M^4 for which metric is defined only up to conformal scaling. The first fibration $M^4 \times S^2$ with a well-defined metric would correspond to the classical TGD.

2. Both Newton's constant G and cosmological constant Λ emerge from twistor lift in M^4 factor. The radius of S^2 is identified in terms of Planck length $l_P = \sqrt{G}$. For CP_2 factor, the radius corresponds to the radius of CP_2 geodesic sphere. 4-D Kähler action can be lifted to 6-D Kähler action only for $M^4 \times CP_2$ so that TGD would be completely unique both mathematically and physically. The twistor space of CP_2 is flag-manifold $SU(3)/U(1) \times U(1)$ having interpretation as the space for the choices of quantization axis of color isospin and hypercharge. This choice could correspond to a selection of Eguchi–Hanson complex coordinates for CP_2 by fixing their phase angles in which isospin and hypercharge rotations induce shifts.
3. The physically motivated conjecture is that the PEs can be lifted to their 6-D twistor bundles with S^2 serving as a fiber, that one induce the twistor structure and the outcome is equal to the twistor structure of space-time surface, and that this condition is at least part of the PE property. This would correspond to the solution of massless wave equations in terms of twistors in the original twistor approach of Penrose [13]. The analog of spontaneous compactification would lead to 4-D action equal to Kähler action plus volume term. One could of course postulate this action directly without mentioning twistors at all.

   The coefficient of the volume term would correspond to dark energy density characterized by cosmological constant Λ being extremely small in cosmological scales. It removes vacuum degeneracy although the situation remains highly non-perturbative. This can be combined with the earlier conjecture that cosmological constant Λ behaves as $\Lambda \propto 1/p$ under p-adic coupling constant evolution so that Λ would be large in primordial cosmology.
4. The generic extremals of space-time action would depend on coupling parameters, which does not fit with the number theoretic vision inspiring speculations that space-time surface can be seen as quaternionic sub-manifolds of 8-D octonionic space-time [48], satisfying quaternion analyticity [69], or a 4-D generalization of holomorphy. By SH the extremals are, however, "preferred." What could this imply?

   Intriguingly, all known non-vacuum extremals and also CP_2 type vacuum extremals having null-geodesic as M^4 projection are extremals of both Kähler action and volume term separately! The dynamics for volume term and Kähler action effectively decouple and coupling constants do not appear at all in field equations. The twistor lift would only select minimal surface amongst vacuum extremals, modify the Kähler function of WCW identifiable as exponent for the real part of action, and provide a profound mathematical and physical motivation for cosmological constant Λ remaining mysterious GRT framework. One could even hope that preferred extremals are nothing but minimal surface extremals of Kähler action with the vanishing conditions for some sub-algebra of super-symplectic algebra satisfied automatically!

The analog of decoupling of Kähler action and volume term should take place also for induced spinors. This is expected if mere analyticity properties make spinor modes solutions of modified Dirac equations. This is true in 2-D case Hamilton–Jacobi structure should guarantee this in 4-D case [53, 69].

PEs depend on coupling parameters only via boundary conditions stating the vanishing of Noether charges for a sub-algebra of super-symplectic algebra and its commutator with entire algebra. Also the conservation conditions at 3-D light-like surfaces at which the signature of metric changes imply dependence on coupling parameters. These conditions allow the transfer of classical charges between Minkowskian and Euclidian regions necessary to understand momentum exchange between particles and environment classically only if Kähler couplings strength is complex—otherwise there is no exchange of conserved quantities since their real *resp.* imaginary at the two sides [30]. Interestingly, also in twistor Grassmann approach the massless poles in propagators are complex.

This picture conforms with the conjecture that discrete p-adic evolution of the Kähler coupling strength in subset of primes near prime powers of two corresponds to complex zeros of zeta [30]. This conforms also with the conjectured discreteness of p-adic coupling constant evolution by phase transitions changing the values of coupling parameters. One implication is that all loop corrections in functional integral vanish.

5. In path integral approach quantum TGD would be extremely non-perturbative around extremals for which Kähler action vanishes. Same is true also in WCW approach. The cure would be provided by the hierarchy of Planck constants $h_{eff}/h = n$, which effectively scales Λ down to Λ/n. n would be the number sheets of the M^4 covering defined by the space-time surface: the action of Galois group for the number theoretic discretization of space-time surface could give rise to this covering. The finiteness of the volume term in turn forces ZEO: the volume of space-time surface is indeed finite due to the finite size of CD.

Consider now the delicacies of this picture.

1. Should assign also to M^4 the analog of symplectic structure giving an additional contribution to the induced Kähler form? The symmetry between M^4 and CP_2 suggests this, and this term could be highly relevant for the understanding of the observed CP breaking and matter antimatter asymmetry [79]. Poincare invariance is not lost since the needed moduli space for M^4 Kähler forms would be the moduli space of CDs forced by ZEO in any case, and M^4 Kähler form would serve as the correlate for fixing rest system and spin quantization axis in quantum measurement.
2. Also induced spinor fields are present. The well-definedness of electromagnetic charge for the spinor modes forces in the generic case the localization of the modes of induced spinor fields at string world sheets (and possibly to partonic 2-surfaces) at which the induced charged weak gauge fields and possibly also neutral Z^0 gauge field vanish. The analogy with branes and super-symmetry force to consider two options.

Option I: The *fundamental* action principle for space-time surfaces contains besides 4-D action also 2-D action assignable to string world sheets, whose topological part (magnetic flux) gives rise to a coupling term to Kähler gauge potentials assignable to the 1-D boundaries of string world sheets containing also geodesic length part. Super-symplectic symmetry demands that modified Dirac action has 1-, 2-, and 4-D parts: spinor modes would exist at both string boundaries, string world sheets, and space-time interior. A possible interpretation for the interior modes would be as generators of space-time super-symmetries [56].

This option is not quite in the spirit of SH and string tension appears as an additional parameter. Also the conservation of em charge forces 2-D string world sheets carrying vanishing induced W fields and this is in conflict with the existence of 4-D spinor modes unless they satisfy the same condition. This looks strange.

Option II: Stringy action and its fermionic counterpart are effective actions only and justified by SH. In this case there are no problems of interpretation. SH requires only that the induced spinor fields at string world sheets determine them in the interior much like the values of analytic function at curve determine it in an open set of complex plane. At the level of quantum theory the scattering amplitudes should be determined by the data at string world sheets. If induced W fields at string world sheets are vanishing, the mixing of different charge states in the interior of X^4 would not make itself visible at the level of scattering amplitudes! In this case 4-D spinor modes do not define space-time super-symmetries.

3. Why the string world sheets coding for effective action should carry vanishing weak gauge fields? If M^4 has the analog of Kähler structure [79], one can speak about Lagrangian sub-manifolds in the sense that the sum of the symplectic forms of M^4 and CP_2 projected to Lagrangian sub-manifold vanishes. Could the induced spinor fields for effective action be localized to generalized Lagrangian sub-manifolds? This would allow both string world sheets and 4-D space-time surfaces but SH would select 2-D Lagrangian manifolds. At the level of effective action the theory would be incredibly simple.

Induced spinor fields at string world sheets could obey the "dynamics of avoidance" in the sense that *both* the induced weak gauge fields W, Z^0 and induced Kähler form (to achieve this U(1) gauge potential must be sum of M^4 and CP_2 parts) would vanish for the regions carrying induced spinor fields. They would couple only to the *induced em field (!)* given by the R_{12} part of CP_2 spinor curvature [21] for $D = 2, 4$. For $D = 1$ at boundaries of string world sheets the coupling to gauge potentials would be non-trivial since gauge potentials need *not* vanish there. Spinorial dynamics would be extremely simple and would conform with the vision about symmetry breaking of weak group to electromagnetic gauge group.

The projections of canonical currents of Kähler action to string world sheets would vanish, and the projections of the 4-D modified gamma matrices would define just the induced 2-D metric. If the induced metric of space-time surface reduces to an orthogonal direct sum of string world sheet metric and metric acting in normal space, the flow defined by 4-D canonical momentum currents is parallel to string world sheet. These conditions could define the "boundary" conditions at string world sheets for SH.

This admittedly speculative picture has revolutionized the understanding of both classical and quantum TGD during last year [68–70]. In particular, the construction of single-sheeted PEs as minimal surfaces allows a kind of lego like engineering of more complex PEs [74]. The minimal surface equations generalize Laplace equation of Newton's gravitational theory to non-linear massless d'Alembert equation with gravitational self-coupling. One obtains the analog of Schwartschild solution and radiative solutions describing also gravitational radiation [68]. An open question is whether classical theory makes sense if also the analog of Kähler form in M^4 is allowed.

2.3.3 Identification of WCW

The notion of WCW [28, 33, 65] was inspired by the super-space approach of Wheeler in which three-geometries are the basic geometric entities.

1. In TGD framework 3-surfaces take this role. Einstein's program for geometrizing classical physics is generalized to a geometrization of entire quantum physics. Hermitian conjugation corresponds to complex conjugation in infinite-dimensional context so that WCW must have Kähler geometry. The geometrization of fermionic statistics/oscillator operators is in terms of gamma matrices of WCW expressible as linear combinations of oscillator operators for second quantized induced spinor field. Formally purely classical spinor modes of WCW represent many-fermion states as functionals of 3-surface. One can also interpret gamma matrices as generators of super-conformal symmetries in accordance with the fact that also SUSY involves Clifford algebra.

   In ZEO the entanglement coefficients between positive and negative energy parts of zero energy states determine the S-matrix so that S-matrix would be coded by the modes of WCW spinor fields. Twistor approach to TGD [69] suggests that the S-matrix reduces completely to the symmetries defined by the multi-local (locus corresponds to partonic 2-surface) generators of the Yangian associated with the super-symplectic algebra.
2. ZEO forces to identify 3-surfaces as pairs of 3-surfaces with members at the opposite boundaries of CD. SH reduces them to a collection of partonic 2-surfaces at boundaries of CD plus number theoretic discretization in space-time interior. Basic geometric objects are pairs of initial and final states (coordinates for both in mechanical analogy) rather than initial states with initial value conditions (coordinates and velocities in mechanical analogy) and initial value problem transforms to boundary value problem. Processes rather than states become the basic elements of ontology: this has far reaching consequences in biology and neuroscience.

3. The realization of GCI requires that the definition of WCW Kähler function assigns to a "physically" 3-surface a unique 4-surface for 4-D general coordinate transformations to act: "physically" could mean "apart from transformations acting as gauge transformations" not affecting the action and conserved classical charges. The outcome is holography.
4. Strong form of holography (SH) would emerge as follows. The condition that light-like 3-surfaces defining boundaries between Euclidian and Minkowskian regions are basic geometric entities equivalent with pairs of space-like 3-surfaces at the ends of given causal diamond CD implies SH: partonic 2-surfaces and their 4-D tangent space data should code the physics. One could also speak about almost/effective two-dimensionality. Tangent space data could in turn be coded by string world sheets. Number theoretical discretization of space-time interior with preferred coordinates in the extension of rationals could give meaning for "almost."
5. Kähler metric is expressible both in terms of second derivatives of Kähler function K [33] and as anticommutators of WCW gamma matrices expressible as linear combinations of fermionic oscillator operators. This suggests a close relationship between space-time dynamics and spinor dynamics.

   Super-symplectic symmetry between the action defining space-time surfaces (Kähler action plus volume term) and modified Dirac action would realize this relationship. This is achieved if the modified gamma matrices are defined by the canonical momentum currents of 2-D action associated with string world sheets. These currents are parallel to the string world sheets. This implies the analog of AdS/CFT correspondence requiring only that induced spinor modes at string world sheets determine them in space-time interior (this is like analytic continuation). The localization of spinor modes at string world sheets is *not* required as I believed first.

The geometry of loop spaces developed by Freed [2] serves as a model in the construction of WCW Kähler geometry [65].

1. The existence of loop space Riemann connection requires maximal isometry group identifiable as Kac-Moody group so that Killing vector fields span the entire tangent space of the loop space.
2. In TGD framework the properties of Kähler action lead to the idea that WCW is union of homogenous or even symmetric spaces of symplectic algebra acting at the boundary of $\delta CD \subset \delta CD_+ \cup \delta CD_-$, $\delta CD_\pm \subset \delta M^4_\pm \times CP_2$. ZEO requires that the conserved quantum numbers for physical states are opposite for the positive and negative energy parts of the states at the two opposite boundary parts of CD. The symmetric spaces G/H in the union are labelled by zero modes, which do not appear in the line element as differentials but only as parameters of the metric. Conserved Noether charges of isometries and symplectic invariants of examples of zero modes as also the super-symplectic Noether charges invariant under complex conjugation of WCW coordinates.
3. Homogenous spaces of the symplectic group G are obtained by dividing a subgroup H. An especially attractive option is suggested by the fractal structure

of the symplectic algebra containing an infinite hierarchy of sub-algebras G_n for which conformal weights are $n > 0$-multiples of those for G. For this option $H = G_n$ is isomorphic to G and one could have infinite hierarchies of inclusions analogous to the hierarchy of inclusions of hyper-finite factors of type II_1 (HFFs). PE property requires almost two-dimensionality and elimination of huge number of degrees of freedom. The natural condition is that the Noether charges of G_n vanish at the ends of CD. A stronger condition is that also the Noether charges for $[G, G_n]$ vanish. This implies effective normal algebra property and G/G_n acts effectively like group.

The inclusion of HFFs would define measurement resolution with included factor acting like gauge algebra. Measurement resolution would be naturally determined by the number theoretic discretization of the space-time surface so that physics as geometry and number theory visions would meet each other.

4. This inclusion hierarchy can be identified in terms of quantum criticality (QC). The transitions $n \to kn$ increasing the value of $n > 0$ reduce QC since pure gauge symmetries are reduced, and new physical super-symplectic degrees of freedom emerge. QC also requires that Kähler couplings strength analogous to temperature is analogous to critical temperature so that the quantum theory is uniquely defined if their is only one critical temperature. Spectrum for α_K seems more plausible and the possibility that Kähler coupling strength depends on the level of the number theoretical hierarchy defined by the allowed extensions of rationals can be considered [30].

2.3.4 WCW Spinor Structure

The basic idea is geometrization of quantum states by identifying them as modes of WCW spinor fields [53, 65]. This requires definition of WCW spinors and WCW spinor structure, WCW gamma matrices and Dirac operator, etc.

The starting point is the definition of WCW gamma matrices using a representation analogous to the usual vielbein representation as linear combinations of flat space gamma matrices. The conceptual leap is the observation that there is no need to assume that the counterparts of flat space gamma matrices have vectorial quantum numbers. Instead, they are identified as fermionic oscillator operators for second quantized free induced spinor fields at space-time surface.

This allows geometrization of the fermionic statistics since WVW spinors for a given 3-surface are analogous to fermionic Fock states. One can also say that spinor structure follows as a square root of metric and also that the spinor basis defines a geometric correlate of Boolean mind [26]. The dependence of WCW spinor field on 3-surface represents the bosonic degrees of freedom not reducible to many-fermion states. For instance, most of hadron mass would be associated with these degrees of freedom.

Quantum TGD involves Dirac equations at space-time level, imbedding space level, and level of WCW. The dynamics of the induced spinor fields is related by super-symmetry to the action defining space-time surfaces as preferred extremals [53, 65].

1. The gamma matrices in the equation—modified gamma matrices—are determined by contractions of the canonical momentum currents of Kähler action with the imbedding space gamma matrices. The localization at string world sheets for which only induced neutral weak fields or only em field are non-vanishing is accompanied by the integrability condition that various conserved currents run along string world sheets: one can speak of sub-flow.
2. Modified Dirac equation can be solved exactly just like in the case of string models using holomorphy and the properties of complexified modified gamma matrices. This is expected to be true also in 4-D case by Hamilton–Jacobi structure. If the dynamics of avoidance is realized the modified Dirac equation would be essentially free Dirac equation and holomorphy would allow to solve it.

At the level of WCW one obtains also the analog of massless Dirac equation as the analog of Super-Virasoro conditions of Super-Virasoro algebra.

1. The fermionic counterparts of super-conformal gauge conditions assignable with sub-algebra G_n of super-symplectic conformal symmetry associated with the both light-cone boundary (light-like radial coordinate), with conformal symmetries of light-cone boundary, and with string world sheets.
2. The ground states of super-symplectic representations satisfy massless imbedding space Dirac equation in imbedding space so that Dirac equations in WCW, in imbedding space, and at string world sheets are involved. In twistorialization also massless M^8 Dirac equation emerges in the tangent space M^8 of imbedding space assignable to the partonic 2-surfaces and generalizes the 4-D light-likeness with its 8-D counterpart applying to states with M^4 mass. Here octonionic representation of imbedding space gamma matrices emerges naturally and allows to speak about 8-D analogs of Pauli's sigma matrices [49].

2.4 *Quantum Criticality, Measurement Resolution, and Hierarchy of Planck Constants*

The notions of quantum criticality (QC), finite measurement resolution, and hierarchy of Planck constants proposed to give rise to dark matter as phases of ordinary matter are central for TGD [31, 52, 62].

These notions relate closely to the strong form of holography (SH) implied by strong form of general coordinate invariance (SGCI). In adelic physics all this would relate closely to the hierarchy of extensions of rationals serving as a correlate for number theoretical evolution.

2.4.1 Finite Measurement Resolution and Fractal Inclusion Hierarchy of Super-Symplectic Algebras

The notion of finite measurement resolution realized in terms of inclusions of hyper-finite factors of type II_1 (HFFs) is one of the ideas that emerged for about decade ago [52]. The vision is that the action of the included algebra generates states not distinguishable from the original state in the measurement resolution used.

The fractal hierarchy of isomorphic sub-algebras of super-symplectic algebra—call it g—defines an excellent candidate for the realization of finite measurement resolution. Similar hierarchies can be assigned also for the extended super-conformal algebra assignable with light-like boundaries of CD and with Kac-Moody and conformal algebras assignable to string world sheets.

An interesting possibility is that the conformal weights assignable to infinitesimal scaling operator of the light-like radial coordinate of light-cone boundary correspond to zeros of Riemann zeta [66, 72]. A kind of dual spectrum would correspond to conformal weights that correspond to logarithms for powers of primes. One can identify the conformal weight as negative of the pole of fermionic zeta $z_F = \zeta(s)/\zeta(2s)$ natural in TGD framework. The real part of conformal weight for the generators is $h_R = -1/4$ for "non-trivial" poles and positive integer $h = n > 0$ for "trivial" poles. There is also a pole for $h = -1$. Hence one obtains tachyonic ground states, which must be assumed also in p-adic mass calculations [34].

Also the generators of Yangian algebra [49] integrating the algebras assignable to various partonic 2-surfaces to a multi-local algebra are labelled by a non-negative integer n analogous to conformal weight and telling the number of partonic 2-surfaces involved with the action of the generator. Also this algebra has similar fractal hierarchy of sub-algebras so that the considerations that follow might apply also to it. Now that number of partonic 2-surface would play the role of measurement resolution.

As noticed, there are also other algebras, which allow conformal hierarchy if one can restrict the conformal weights to be non-negative. The first of them generates generalized conformal transformations of light-cone boundary depending on light-like radial coordinate as parameter: also now radial conformal weights for generators can have zeros of zeta as spectrum. As a special case one obtains infinite-dimensional group of isometries of light-cone boundary. Second one corresponds to ordinary conformal and Kac-Moody symmetries for induced spinor fields acting on string world sheets. Also here similar hierarchy of sub-algebras can be considered. In the following argument one restricts to super-symplectic algebra assumed to act as isometries of WCW.

Consider now how the finite measurement resolution could be realized as an infinite hierarchy of super-symplectic gauge symmetry breakings. The physical picture relies on quantum criticality of TGD Universe. The levels of the hierarchy labelled by positive integer n and a ball at the top of ball at... serve as a convenient metaphor.

1. The sub-algebra g_n for which conformal weights of generators (whose commutators give the sub-algebra) are positive integer multiples for those of the entire algebra g defines the algebra acting as pure gauge algebra defining a sub-group of symplectic group. The action of g_n as gauge algebra would mean that it affects on degrees of freedom below the measurement resolution. One can assign to this algebra a coset space G/G_n of the entire symplectic group G and of subgroup G_n. This coset space would describe the dynamical degrees of freedom. If the subgroup were a normal subgroup, the coset space would be a group. This is not the case now since the commutator $[g, g_n]$ of the entire algebra with the sub-algebra does not belong to g_n.

   However, if one poses stronger—physically very attractive—gauge conditions stating that not only g_n but also the commutator algebra $[g, g_n]$ annihilates the physical states and that corresponding classical Noether charges vanish, one obtains effectively a normal subgroup and one has good hopes that coset space acts effectively as group, which is finite-dimensional as far as conformal weights are considered.
2. $n > 0$ is essential for obtaining effective normal algebra property. Without this assumption the commutator $[g, g_n]$ would be entire g. If the spectrum of super-symplectic conformal weights is integer valued it is not obvious why one should pose the restriction $n \geq 1$.
3. In this framework pure conformal invariance could reduce to a finite-dimensional gauge symmetry. A possible interpretation would be in terms of Mc-Kay correspondence [4] assigning to the inclusions of HFFs labelled by integer $n \geq 3$ a hierarchy of simply laced Lie-groups. Since the included algebra would naturally correspond to degrees of freedom not visible in the resolution used, the interpretation as a dynamical gauge group is suggestive. The dynamical gauge group could correspond to n-dimensional Cartan algebra acting in conformal degrees of freedom identifiable as a simply laced Lie group. This would assign a infinite hierarchy of dynamical gauge symmetries to the broken conformal gauge invariance acting as symmetries of dark matter. This still leaves infinite number of degrees of freedom assignable to the imbedding space Hamiltonians and spectrum generated by zeros of zeta but this might have interpretation in terms of gauging so that additional vanishing conditions for Noether charges are suggestive.

2.4.2 Dark Matter as Large Phases with Large Gravitational Planck Constant $h_{eff} = h_{gr}$

D. Da Rocha and Laurent Nottale [11] have proposed that Schrödinger equation with Planck constant $\hbar$ replaced with what might be called gravitational Planck constant $\hbar_{gr} = \frac{GmM}{v_0}$ ($\hbar = c = 1$). v_0 is a velocity parameter having the value $v_0 = 144.7 \pm .7$ km/s giving $v_0/c = 4.6 \times 10^{-4}$. This is rather near to the peak orbital velocity of stars in galactic halos. Also subharmonics and harmonics of v_0 seem to appear. The support for the hypothesis coming from empirical data is impressive [42, 44].

1. The proposal is that a Schrödinger equation results from a fractal hydrodynamics. Many-sheeted space-time, however, suggests that astrophysical systems are at some levels of the hierarchy of space-time sheets macroscopic quantum systems and that only the generalizations of Bohr orbits are involved. The space-time sheets in question would carry dark matter.
2. Nottale's hypothesis would predict a gigantic value of $\hbar_{gr}$. Equivalence principle and the independence of gravitational Compton length $\Lambda_{gr} = \hbar_{gr}/m = GM/v_0 = 2r_S/v_0$ (typically astrophysical scale) on mass m imply however that one can restrict the values of mass m to masses of microscopic objects so that $\hbar_{gr}$ would be much smaller. Large $\hbar_{gr}$ could provide a solution of the blackhole collapse (IR catastrophe) problem encountered at the classical level. The resolution of the problem inspired by TGD inspired theory of living matter is that it is the dark matter at larger space-time sheets, which is quantum coherent in the required time scale [44].

   One could criticize the hypothesis since it treats the masses M and m asymmetrically: this is only apparently true [62].
3. It is natural to assign the values of Planck constants postulated by Nottale to the space-time sheets mediating gravitational interaction and identifiable as magnetic flux tubes (quanta). The cross section of the flux tube corresponds to a sphere $S^2_i \subset CP_2$, $i = I, II$ [70]. S^2_I is homologically non-trivial carrying Kähler magnetic monopole flux. S^2_{II} is homologically trivial carrying vanishing Kähler magnetic flux but non-vanishing electroweak flux [70].

   The flux tubes of type I have both Kähler magnetic energy and dark energy due to the volume action. Flux tubes of type II would have only the volume energy. Both flux tubes could be remnants of cosmic string phase of primordial cosmology. The energy of these flux quanta would be correlated for galactic dark matter and volume action and also magnetic tension would give rise to negative "pressure" forcing accelerated cosmological expansion. This leads to a rather detailed vision about the evolution of stars and galaxies identified as bubbles of ordinary and dark matter inside flux tubes identified also as dark energy.
4. Both theoretical consistency and certain experimental findings from astrophysics [10, 14] and biology [58, 59] suggest the identification $h_{eff} = n \times h = h_{gr}$. The large value of h_{gr} can be seen as a manner to reduce the string tension of fermionic strings so that gravitational (in fact all!) bound states can be described in terms of strings connecting the partonic 2-surfaces defining particles (analogous to AdS/CFT description) [65]. The values $h_{eff}/h = n$ can be interpreted in terms of a hierarchy of breakings of super-conformal symmetry in which the super-conformal generators act as gauge symmetries only for a sub-algebras with conformal weights coming as multiples of n. Macroscopic quantum coherence in astrophysical scales is implied. If also modified Dirac action is present, part of the interior degrees of freedom associated with the fermionic part of conformal algebra become physical.

   Fermionic oscillator operators could generate super-symmetries and sparticles could correspond to dark matter with $h_{eff}/h = n > 1$. One implication would be that at least part if not all gravitons would be dark and be observed only through their decays to an ordinary high frequency graviton ($E = hf_{high} = h_{eff}f_{low}$) or to a bunch of n low energy gravitons.

2.4.3 Hierarchies of Quantum Criticalities, Planck Constants, and Dark Matters

Quantum criticality is one of the cornerstone assumptions of TGD. In the original approach the value of Kähler coupling strength α_K together with CP_2 radius R fixed quantum TGD is analogous to critical temperature. Twistor lift [70] brings in additional coupling constant Λ obeying p-adic coupling constant evolution and Planck length l_G, which like CP_2 radius would not obey coupling constant evolution (as also G). The values of these parameters should be fixed by quantum criticality. What else does quantum criticality mean is, however, far from obvious, and I have pondered the notion repeatedly both from the point of view of mathematical description and phenomenology [33, 53, 65].

1. Criticality is characterized by long range correlations and sensitivity to external perturbations and living systems define an excellent example of critical systems—even in the scale of populations since without sensitivity and long range correlations cultural evolution and society would not be possible. For a physicist with the conceptual tools of existing theoretical physics the recent information society in which the actions of people at different side of globe are highly correlated should look like a miracle.
2. The hierarchy of Planck constants with dark matter identified as phases of ordinary matter with non-standard value $h_{eff} = n \times h$ of Planck constant is one of the "almost-predictions" of TGD is definitely something essentially new physics. The phase transition transforming ordinary matter to dark matter in this sense generates long range quantal correlations and even macroscopic quantum coherence.

   Finding a universal mechanism generating dark matter has been a key challenge during last 10 years. Could quantum criticality having classical or perhaps even thermodynamical criticality as its correlate be always accompanied by the generation of dark matter? If this were the case, the recipe would be stupefyingly simple: create a critical system! Dark matter would be everywhere and we would have observed its effects for centuries! Magnetic flux tubes (possibly carrying monopole flux) define the space-time correlates for long range correlations at criticality and would carry the dark matter. They are indeed key players in TGD inspired quantum biology.
3. Change of symmetry is assigned with criticality as also conformal symmetry (in 2-D case). In TGD framework conformal symmetry is extended and infinite hierarchy of breakings of conformal symmetry so that a sub-algebras of various conformal algebras with conformal weights coming as integer multiples of integer n defining h_{eff} would occur.
4. Phase separation is what typically occurs at criticality and one should understand also this. The strengthening of this hypothesis with the assumption $h_{eff} = h_{gr}$, where $h_{gr} = GMm/v_0$ is the gravitational Planck constant originally introduced by Nottale [62, 64]. In the formula v_0 has dimensions of velocity, and will be proposed to be determined by a condition relating the size of the system with mass M to the radius within which the wave function of particle m with $h_{eff} = h_{gr}$ is localized in the gravitational field of M.

The condition $h_{eff} = h_{gr}$ implies that the integer n in h_{eff} is proportional to the mass of the particle. The implication is that particles with different masses reside at flux tubes with different Planck constant and separation of phases indeed occurs.

5. What is remarkable is that neither gravitational Compton length nor cyclotron energy spectrum depends on the mass of the particle. This universality could play key role in living matter. One can assign Planck constant also to other interactions such as electromagnetic interaction so that one would have $h_{em} = Z_1Z_2e^2/v_0$. The phase transition could take place when the perturbation series based on the coupling strength $\alpha = Z_1Z_2e^2/\hbar$ ceases to converge. In the new phase perturbation series would converge since the coupling strength is proportional to $1/h_{eff}$. Hence criticality and separation into phases serve as criteria as one tries to see whether the earlier proposals for the mechanisms giving rise to large h_{eff} phases make sense. One can also check whether the systems to which large h_{eff} has been assigned are indeed critical.

One example of criticality is super-fluidity. Superfluids exhibit rather mysterious looking effects such as fountain effect [6] and what looks like quantum coherence of superfluid containers, which should be classically isolated. These findings serve as a motivation for the proposal that genuine superfluid portion of superfluid corresponds to a large h_{eff} phase near criticality at least and that also in other phase transition like phenomena a phase transition to dark phase occurs near the vicinity [62].

But how does quantum criticality relate to number theory and adelic physics? $h_{eff}/h = n$ has been identified as the number of sheets of space-time surface identified as a covering space of some kind. Number theoretic discretization defining the "spine" for a monadic space-time surface. Pitkänen [77] defines also a covering space with Galois group for an extension of rationals acting as covering group. Could n be identifiable as the order for a sub-group of Galois group? If this is the case, the proposed rule for h_{eff} changing phase transitions stating that the reduction of n occurs to its factor would translate to spontaneous symmetry breaking for Galois group and spontaneous—symmetry breakings indeed accompany phase transitions.

2.4.4 TGD Variant of AdS/CFT Duality

AdS/CFT duality [8] has provided a powerful approach in the attempts to understand the non-perturbative aspects of super-string theories. The duality states that conformal field theory in n-dimensional Minkowski space M^n identifiable as a boundary of $n+1$-dimensional space AdS_{n+1} is dual to a string theory in $AdS_{n+1} \times S^{9-n}$.

As a mathematical discovery AdS/CFT duality is extremely interesting but it seems that it need not have much to do with physics as such. From TGD point of view the reason is obvious: the notion of conformal invariance is quite too limited. In TGD framework conformal invariance is extended to a super-symplectic symmetry in $\delta M^4_\pm \times CP_2$, whose Lie-algebra has the structure of conformal algebra. Also ordinary super-conformal symmetries associated with string world sheets are present as well as generalization of 2-D conformal symmetries to their analogs at light-cone boundary and light-like orbits of partonic 2-surfaces. In this framework AdS/CFT duality is expected to be modified.

The matrix elements $G_{K\overline{L}}$ of Kähler metric of WCW can be expressed in two manners. As contractions of the derivatives $\partial_K \partial_{\overline{L}} K$ of the Kähler function of WCW with isometry generators or as anticommutators $\{\Gamma_K, \Gamma_{\overline{L}}\}$ of WCW gamma matrices identified as super-symplectic Noether super charges assignable to fermionic strings connecting partonic 2-surfaces. Kähler function is identified as real part of the action: if coupling parameters are real it reduces to the action for the Euclidian space-time regions with 4-D CP_2 projection and otherwise contains contributions from both Minkowskian and Euclidian regions. The action defines the modified gamma matrices appearing in modified Dirac action as contractions of canonical momentum currents with imbedding space gamma matrices.

This observation suggests that there is a super-symmetry between action and modified Dirac action. The problem is that induced spinor fields naive of SH and also well-definedness of em charge demand the localization of induced spinor modes at 2-D string world sheets. This simply cannot be true. On the other hand, SH only requires that the data about induced spinor fields and space-time surface at the string world sheets is enough to construct the modes in space-time interior.

This leaves two options if one assumes that SH is exact (recall, however, that the number theoretic interpretation for the hierarchy of Planck constants suggests that the number theoretic spin of monadic space-time surface represents additional discrete data needed besides that assignable to string world sheets to describe dark matter). As found in the Section 2.3.2, there are two options.

Option I: The analog of brane hierarchy is realized at the level of fundamental action. There is a separate fundamental 2-D action assignable with string world sheets—area and topological magnetic flux term—as also world line action assignable to the boundaries of string world sheets. By previous argument string tension should be determined by the value of the cosmological constant Λ obeying p-adic coupling constant evolution rather than by G: otherwise there is no hope about gravitationally bound states above Planck scale. String tension would appear as an additional fundamental coupling parameter (perhaps fixed by quantum criticality). This option does not quite conform with the spirit of SH.

Option II: 4-D space-time action and corresponding modified Dirac action defining fundamental actions are expressible as effective actions assignable to string world sheets and their boundaries. String world sheet effective action could be expressible as string area for the effective metric defined by the anticommutators of modified gamma matrices at string world sheet. If the sum of the induced Kähler forms of M^4 and CP_2 vanishes at string world sheets the effective metric would be the induced 2-D metric: this together with the observed CP breaking could provide a justification for the introduction of the analog of Kähler form in M^4. String tension would be dynamical rather than determined by l_P and depend on Λ, l_P, R, and α_K. This representation of Kähler action would be one aspect of the analog of AdS/CFT duality in TGD framework.

Both options would allow to understand how strings connecting partonic 2-surfaces give rise to the formation of gravitationally bound states. Bound states of macroscopic size are possible only if one allows hierarchy of Planck constants and this is required also by the (extremely) small value of Λ (in cosmic scales).

Consider the concrete realizations for this vision.

1. SGCI requires effective two-dimensionality. In given UV and IR resolutions partonic 2-surfaces and string world sheets are assignable to a finite hierarchy of CDs inside CDs with given CD characterized by a discrete scale coming as an integer multiple of a fundamental scale (essentially CP_2 size). Λ would closely relate to the size scale of CD. String world sheets have boundaries consisting of either light-like curves in induced metric at light-like wormhole throats and space-like curves at the ends of CD whose M^4 projections are light-like. These braids carrying fermionic quantum numbers intersect partonic 2-surfaces at discrete points.
2. This implies a rather concrete analogy with $AdS_5 \times S_5$ duality, which describes gluons as open strings. In zero energy ontology (ZEO) string world sheets are indeed a fundamental notion and the natural conjecture is that these surfaces are minimal surfaces, whose area by quantum classical correspondence depends on the quantum numbers of the external particles.

2.4.5 String Tension of Gravitational Flux Tubes

For Planckian cosmic strings only quantum gravitational bound states of length of order Planck length are possible. There must be a mechanism reducing the string tension. The *effective* string tension assignable to magnetic flux tubes must be inversely proportional to $1/h_{eff}^2$, $h_{eff} = n \times h = h_{gr} = 2\pi GMm/v_0$ in order to obtain gravitationally bound states in macroscopic length scales identified as structures for which partonic 2-surfaces are connected by flux tubes accompanied by fermionic strings.

The reason is that the size scale of (quantum) gravitationally bound states of masses M and m is given by gravitational Compton length $\Lambda_{gr} = GM/v_0$ [44, 62, 64] assignable to the gravitational flux tubes connecting the masses M and m. If the string tension is of order Λ_{gr}^2 this is achieved since the typical length of string would be Λ_{gr}. Gravitational string tension must be therefore of order $T_{gr} \sim 1/\Lambda_{gr}^2$. How could this be achieved? One can imagine several options and here only the option based on the assumptions

1. Twistor lift makes sense.
2. Fundamental action is 4-D for both space-time and fermionic degrees of freedom and 2-D string world sheet action is an effective action realizing SH. Note effective action makes also possible braid statistics, which does not make sense at fundamental level.

3. Also M^4 carries the analog of Kähler form and the sum of induced Kähler forms from M^4 and CP_2 vanishes at string world sheets and also weak gauge fields vanishes at string world sheets leaving only em field.

is considered since it avoids all the objections that I have been able to invent.

For the twistor lift of TGD [70] predicting cosmological constant Λ depending on p-adic length scale $\Lambda \propto 1/p$ the gravitational strings would be naturally homologically trivial cosmic strings. These vacuum extremals of Kähler action transform to minimal surface extremals with string tension given by $\rho_{vac}S$, where ρ_{vac} the density of dark energy assignable to the volume term of the action and S the transverse area of the flux tube. One should have $\rho_{vac}S = 8\pi\Lambda S/G = 1/\Lambda_{gr}^2$ so that one would have

$$8\pi\Lambda S = \frac{G}{\Lambda_{gr}^2} \ .$$

Λ for flux tubes (characterizing the size of CDs containing them) would depend on the gravitational coupling Mm.

2.5 Number Theoretical Vision

Physics as infinite-D spinor geometry of WCW and physics as generalized number theory are the two basic vision about TGD. The number theoretical vision involves three threads [46–48].

1. The first thread [47] involves the notion of number theoretical universality NTU: quantum TGD should make sense in both real and p-adic number fields (and their algebraic extensions induced by extensions of rationals). p-Adic number fields are needed to understand the space-time correlates of cognition and intentionality [32, 39, 40].

   p-Adic mass calculations lead to the notion of a p-adic length scale hierarchy quantifying the notion of the many-sheeted space-time [32, 39]. One of the first applications was the calculation of elementary particle masses [34]. The basic predictions are only weakly model independent since only p-adic thermodynamics for Super-Virasoro algebra are involved. Not only the fundamental mass scales would reduce to number theory but also particle masses are predicted correctly under rather mild assumptions and are exponentially sensitive to the p-adic length scale predicted by p-adic length scale hypothesis. Also predictions such as the possibility of neutrinos to have several mass scales were made on the basis of number theoretical arguments and have found experimental support [27, 34].
2. Second thread [48] is inspired by the dimensions $D = 1, 2, 4, 8$ of the basic objects of TGD and assumes that classical number fields are in a crucial role in TGD. 8-D imbedding space would have octonionic structure and space-time

surfaces would have associative (quaternionic) tangent space or normal space. String world sheets could correspond to commutative surfaces. Also the notion of $M^8 - H$-duality is part of this thread and states that quaternionic 4-surfaces of M^8 containing preferred M^2 in its tangent space can be mapped to PEs in H by assigning to the tangent space CP_2 point parametrizing it. M^2 could be replaced by integrable distribution of $M^2(x)$. If PEs are also quaternionic one has also $H-H$ duality allowing to iterate the map so that PEs form a category. Also quaternion analyticity of PEs is a highly attractive hypothesis [49]. For instance, it might be possible to interpret string world sheets and partonic 2-surfaces appearing in strong form of holography (SH) as co-dimension 2 surfaces analogous to poles of analytic function in complex plane. Light-like 3-surfaces might be seen as analogs of cuts. The coding of analytic function by its singularities could be seen as analog of SH.

3. The third thread [46] corresponds to infinite primes and leads to several speculations. The construction of infinite primes is structurally analogous to a repeated second quantization of a supersymmetric arithmetic quantum field theory with free particle states characterized by primes. The many-sheeted structure of TGD space-time could reflect directly the structure of infinite prime coding it. Space-time point would become infinitely structured in various p-adic senses but not in real sense (that is cognitively) so that the vision of Leibniz about monads reflecting the external world in their structure is realized in terms of algebraic holography. Space-time becomes algebraic hologram and realizes also Brahman = Atman idea of Eastern philosophies.

3 p-Adic Mass Calculations and p-Adic Thermodynamics

p-Adic mass calculations carried for the first time around 1995 were the stimulus eventually leading to the number theoretical vision as a kind dual for the geometric vision about TGD. In this section I will roughly describe the calculations [27, 34] and the questions and challenges raised by them.

3.1 p-Adic Numbers

Like real numbers, p-adic numbers (http://tinyurl.com/hmgqtoh) can be regarded as completions of the rational numbers to a larger number field [32]. Each prime p defines a p-adic number field allowing the counterparts of the usual arithmetic operations.

1. The basic difference between real and p-adic numbers is that p-adic topology is ultra-metric. Ultrametricity means that the distance function $d(x,y)$ (the counterpart of $|x-y|$ in the real context) satisfies the inequality

$$d(x,z) \leq Max\{d(x,y), d(y,z)\} \ ,$$

(Max(a, b) denotes maximum of a and b) rather than the usual triangle inequality

$$d(x,z) \leq d(x,y) + d(y,z) \ .$$

2. The topology defined by p-adic numbers is compact-open. Hence the generalization of manifold obtained by gluing together n-balls fails because smallest open n-balls are just points and one has totally disconnected topology.
3. p-Adic numbers are not well-ordered like real numbers. Therefore one cannot assign orientation to the p-adic number line. This in turn leads to difficulties with attempts to define definite integrals and the notion of differential form although indefinite integral is well-defined. These difficulties serve as important guidelines in the attempts to understand what p-adic physics is and also how to fuse real and various p-adic physics to a larger structure.
4. p-Adic numbers allow an expansion in powers of p analogous to the decimal expansion

$$x = \sum_{n \geq 0} x_n p^n \ ,$$

and the number of terms in the expansion can be infinite so that p-adic number need not be finite as a real number. The norm of the p-adic number (counterpart of $|x|$ for real numbers) is defined as

$$N_p(x) = \sum_{n \geq 0} x_n p^n = p^{-n_0} \ ,$$

and depends only very weakly on p-adic number. The ultra-metric distance function can be defined as $d_p(x,y) = N_p(x-y)$.
5. p-Adic numbers allow a generalization of the differential calculus. The basic rules of the p-adic differential calculus are the same as those of the ordinary differential calculus. There is, however, one important new element: the set of the functions having vanishing p-adic derivative consists of the so-called pseudo-constants, which are analogs of real valued piecewise constant functions. In the real case only constant functions have vanishing derivative. This implies that p-adic differential equations are non-deterministic. This non-determinism is identified as a counterpart of the non-determinism of cognition and imagination [40].

3.2 *Model of Elementary Particle*

p-Adic mass calculations [27, 34] rely heavily on a topological model for elementary particle and it is appropriate to describe it before going to the summary of calculations.

3.2.1 Family Replication Phenomenon Topologically

One of the basic ideas of TGD approach to particle physics has been genus-generation correspondence: boundary components of the 3-surface should be carriers of elementary particle numbers and the observed particle families should correspond to various boundary topologies.

With the advent of zero energy ontology (ZEO) this picture has changed somewhat.

1. The wormhole throats identified as light-like 3-surfaces at which the induced metric of the space-time surface changes its signature from Minkowskian to Euclidian correspond to the light-like orbits of partonic 2-surfaces. One cannot of course exclude the possibility that also boundary components allow to satisfy boundary conditions without assuming vacuum extremal property of nearby space-time surface.

   The intersections of the wormhole throats with the light-like boundaries of causal diamonds (CDs) identified as intersections of future and past directed light cones ($CD \times CP_2$ is actually in question but I will speak about CDs) define special partonic 2-surfaces and the conformal moduli of these partonic 2-surfaces appear in the elementary particle vacuum functionals [27] naturally. A modification of the original simple picture came from the proposed identification of physical particles as bound states of two wormhole contacts connected by tubes carrying monopole fluxes.
2. For generalized scattering diagrams stringy trouser vertices are replaced with vertices at which the ends of light-like wormhole throats meet. This vertex is the analog of 3-vertex for Feynman diagrams in particle physics lengths scales and for the biological replication (DNA and even cell) in macroscopic length scales.

   In this picture the interpretation of the analog of trouser vertex is in terms of propagation of same particle along two different paths. This interpretation is mathematically natural since vertices correspond to 2-manifolds rather than singular 2-manifolds, which are just splitting to two disjoint components. Second complication comes from the weak form of electric-magnetic duality forcing to identify physical particles as weak strings with magnetic monopoles at their ends and one should understand also the possible complications caused by this generalization.

These modifications force to consider several options concerning the identification of light fermions and bosons and one can end up with a unique identification only by making some assumptions. Masslessness of all wormhole throats—also those appearing in internal lines—and dynamical $SU(3)$ symmetry for particle generations are attractive general enough assumptions of this kind. Bosons and their possible spartners would correspond to wormhole contacts with fermion and anti-fermion at the throats of the contact. The expectation was that free fermions and their possible spartners correspond to CP_2 type vacuum extremals with single wormhole throat. It turned, however, that dynamical $SU(3)$ symmetry forces to identify massive (and possibly topologically condensed) fermions as pairs of (g, g)

type wormhole contacts. The existence of higher boson families would mean breaking of quark and lepton universality and there are indications for this kind of anomaly [36] .

3.2.2 The Notion of Elementary Particle Vacuum Functional

Obviously one must know something about the dependence of the elementary particle state functionals on the geometric properties of the boundary component and in the sequel an attempt to construct what might be called elementary particle vacuum functionals (EPVFs) is made. The basic assumptions underlying the construction are the following ones [27]:

1. EPVFs depend on the geometric properties of the two-surface X^2 representing elementary particle.
2. EPVFs possess extended Diff invariance: all 2-surfaces on the orbit of the 2-surface X^2 correspond to the same value of the vacuum functional. This condition is satisfied if vacuum functionals have as their argument, not X^2 as such, but some 2- surface Y^2 belonging to the unique orbit of X^2 (determined by the principle selecting PE as a generalized Bohr orbit [25, 33, 67]) and determined in general coordinate invariant manner.
3. ZEO allows to select uniquely the partonic 2-surface as the intersection of the wormhole throat at which the signature of the induced four-metric changes with either the upper or lower boundary of $CD \times CP_2$. This is essential since otherwise one could not specify the vacuum functional uniquely.
4. Vacuum functionals possess conformal invariance and therefore for a given genus depend on a finite number of variables specifying the conformal equivalence class of Y^2.
5. Vacuum functionals satisfy the cluster decomposition property: when the surface Y^2 degenerates to a union of two disjoint surfaces (particle decay in string model inspired picture), vacuum functional decomposes into a product of the vacuum functionals associated with disjoint surfaces.
6. EPVFs are stable against the decay $g \rightarrow g_1 + g_2$ and one particle decay $g \rightarrow g-1$. This process corresponds to genuine particle decay only for stringy diagrams. For generalized scattering diagrams the interpretation is in terms of propagation along two different paths simultaneously.

In [27] the construction of EPVFs is described in detail. This requires some basic concepts related to the description of the space of the conformal equivalence classes of Riemann surfaces and the concept of hyper-ellipticity. Since theta functions will play a central role in the construction of the vacuum functionals, also their basic properties are needed. Also possible explanations for the experimental absence of the higher fermion families are considered. Concerning p-adic mass calculations, the key question is how to construct p-adic variants of EPVFs.

3.3 p-Adic Mass Calculations

3.3.1 p-Adic Thermodynamics

Consider first the basic ideas of p-adic thermodynamics.

1. p-Adic valued mass squared is identified as thermal mass in p-adic thermodynamics. Boltzmann weights $exp(-E/T)$ do not make sense if one just replaces exponent function with the p-adic variant of its Taylor series. The reason is that $exp(x)$ has p-adic norm equal to 1 for all acceptable values of the argument x (having p-adic norm smaller than one) so that partition function does not have the usual exponential convergence property. Nothing however prevents from consider Boltzmann weights as powers p^n making sense for integer values of n. Here the p-adic norm approaches zero for $n \to +\infty$: thus the correspondences $e^{-E/T} \leftrightarrow p^{E/T_p}$.

   The values of E/T_p must be quantized to integers. This is guaranteed if E is integer valued in suitable unit of energy and $1/T_p$ has integer valued spectrum using same unit for T_p. Super-conformal invariance guarantees integer valued spectrum of E, which in the recent case corresponds to mass squared. These number theoretical conditions are very powerful and lead to the quantization of also thermal mass squared for given p-adic prime p.
2. The p-adic mass squared is mapped to real number by canonical identification $I : \sum x_n p^n \to \sum x_n p^{-n}$ or its variant for rationals. Canonical identification is continuous and maps powers of p^n to their inverses. One modification of canonical identification maps rationals m/n in their representation in which m and n have no common divisors to $I(m)/I(n)$. The predictions of calculations depend in some cases on which variant one uses but rational option looks the most reasonable choice.
3. p-Adic length scale hypothesis states that preferred p-adic primes correspond to powers of 2: $p \simeq 2^k$, but smaller than 2^k. The values of k form with $p = 2^k - 1$ is prime—Mersenne prime—are especially favored. The nearer the prime p to 2^k, the more favored p is physically. One justification for the hypothesis is that preferred primes have been selected by an evolutionary process.
4. It turns out that p-adic temperature is $T_p = 1$ for fermions. For gauge bosons $T_p \leq 1/2$ seems to be necessary assumption for gauge bosons implying that the contribution to mass squared is very small so that super-symplectic contribution assignable to the wormhole magnetic flux tube dominates for weak bosons. For canonical identification $m/n \to I(m)/I(n)$ second order contribution to fermionic mass squared is very small.
5. The large values of p-adic prime p guarantee that the p-adic thermodynamics converges extremely rapidly. For $m/n \to I(m)/I(n)$ already the second order contribution is extremely small since the expansion for the real mass squared is in terms of $1/p$ and for electron with $p = M_{127}$ one has $p \sim 10^{38}$. Hence the calculations are essentially exact and errors are those of the model. It is quite possible that calculations could be done exactly using exact expressions for the

super-symplectic partition functions generalized to p-adic context. The success of the p-adic mass calculations is especially remarkable because p-adic length scale hypothesis $p \simeq 2^k$ predicts exponential sensitivity of the particle mass scale on k.

3.3.2 Symmetries

The number theoretical existence of p-adic thermodynamics requires powerful symmetries to guarantee integer valued spectrum for the thermalized contribution to the mass squared.

1. Super-conformal symmetry with integer valued conformal weights for Virasoro scaling generator L_0 is essential because it predicts in string models that mass squared is apart from ground state contribution integer valued in suitable units. In TGD framework fermionic string world sheets are characterized by super-conformal symmetry. This gives the p-adic thermodynamics assumed in the calculations. One could, however, assign Super-Virasoro algebra also to super-symplectic algebra having its analog as sub-algebra with positive integer conformal weights. Same applies to the extended conformal algebra of light-cone boundary.
2. TGD, however, predicts also generalization of conformal symmetry associated with light-cone boundary involving ordinary complex conformal weights and the conformal weight associated with the light-like radial coordinate. For the latter conformal weights for the generators of supersymmetry might be given by $h = -s_n/2$. s_n zero of zeta or pole $h = -s = -1$ of zeta.

   Also super-symplectic symmetries would have similar radial spectrum of conformal weights. Conformal confinement requiring that the conformal weights of states are real implies that the spectrum of conformal weights for physical states consists of non-negative integers as for ordinary super-conformal invariance.

   It is not clear whether thermalization occurs in these degrees of freedom except perhaps for trivial conformal weights. These degrees of freedom need not therefore contribute to thermal masses of leptons and quarks but would give dominating contribution to hadron masses and weak boson masses. The negative conformal weights predicted by $h = -s/2$ hypothesis predict that ground state weight is negative for super-symplectic representations and must be compensated for massless states.

   The assumption that ground state conformal weight is negative and thus tachyonic is essential in case of p-adic mass calculations [34], and only for massless particles (graviton, photon, gluons) it vanishes or is of order $O(1/p)$. This could be achieved if the ground state of super-symplectic representation has $h = 0$.
3. Modular invariance [27] assignable to partonic 2-surfaces is a further assumption similar to that made in string models. This invariance means that for a given genus the dynamical degrees of freedom of the partonic 2-surface correspond to finite-dimensional space of Teichmueller parameters. For genus $g = 0$ this space is trivial.

Also modular invariance for string world sheets can be considered. By SH the information needed in mass calculations should be assignable to partonic 2-surfaces: the assumption is that one can assign this information to single partonic 2-surface. Stringy contribution would be seen only in scattering amplitudes.

This might be true only effectively: the recent view about elementary particles is that they are pairs of wormhole contacts connected by flux tubes defining a closed monopole flux and wormhole throats of contact have same genus for light states. Furthermore the quantum numbers of particle are associated with single throat for fermions and with opposite throats of single contact for bosons. The second wormhole contact would carry neutralizing weak charges to realize the finite range of weak interactions as "weak confinement."

The number of genera is infinite and one must understand why only three quark and lepton generations are observed. An attractive explanation is in terms of symmetry. For the three lowest genera the partonic 2-surfaces are always hyper-elliptic and have thus global conformal Z_2 symmetry. For higher genera this is not true always and EPVFs constructed from the assumption of modular invariance vanish for the hyper-elliptic surfaces. This suggests that the higher genera are very massive or can be interpreted as many-particle states of handles, which are not bound states but have continuous mass squared.

3.3.3 Contributions to Mass Squared

There are several contributions to the p-adic thermal mass squared come from the degrees of freedom, which are thermalized.

Super-conformal degrees of freedom associated with string world sheets are certainly thermalized. p-Adic mass calculations strongly suggest that the number of super-conformal tensor factors is $N = 5$ but also $N = 4$ and $N = 6$ can be considered marginally.

I have considered several identifications of tensor factors and not found a compelling alternative. If one assumes that super-symplectic degrees of freedom do not contribute to the thermal mass, string world sheets should explain masses of elementary fermions. Here charged lepton masses are the test bench. On the other hand, if super-symplectic degrees of freedom contribute one obtains additional tensor factor assignable to $h = -s/2$, s trivial zero of zeta. Only one tensor factor emerges since Hamiltonians correspond to the products of functions of the coordinates of light-cone boundary and CP_2.

1. $SU(2)_L \times U(1)$ gives two tensor factors. $SU(3)$ gives one tensor factor. The two transversal degrees of freedom for string world sheet suggest 2 degrees of freedom corresponding to Abelian group E^2. Rotations, however, transform these degrees to each other so that 1 tensor factor should emerge. This gives four tensor factors. Could it correspond to the degrees of freedom parallel to string at its end assignable to wormhole throat? Could normal vibrations of partonic 2-surface? This would $N = 5$ tensor factors. Another possibility is that the fifth tensor factor comes from super-symplectic Super-Virasoro algebra defined by trivial conformal weights.

2. Super-symplectic contributions need not be present for ordinary elementary fermions. For weak bosons they could give string tension assignable to the magnetic flux tube connecting the wormhole contacts. It is not clear whether this contribution is thermalized. This contribution might be present only for the phases with $h_{eff} = n \times h$. This contribution would dominate in hadron masses.
3. Color degrees of freedom contribute to the ground state mass squared since ground state corresponds to an imbedding space spinor mode massless in 8-D sense. The mass squared contribution corresponds to an eigenvalue of CP_2 spinor d'Alembertian. Its eigenvalues correspond to color multiplets and only the covariantly constant right handed neutrino is color singlet. For the other modes the color representation is non-trivial and depends on weak quantum numbers of the fermion. The construction of the massless state from a tachyonic ground state with conformal weight $h_{vac} = -3$ must involve colored super-Kac Moody generators compensating for the anomalous color charge so that one obtains color single for leptons and color triplet for quarks as massless state.
4. Modular degrees of freedom give a contribution depending on the genus g of the partonic 2-surface. This contribution is estimated by considering p-adic variants of elementary particle vacuum functionals Ω_{vac} [34] expressible as products of theta functions with the structure of partition function. Theta functions are expressible as sums of exponent functions $exp(X)$ with X defined as a contraction of the matrix Ω_{ij} defined by Teichmueller parameters between integer valued vectors.

   In ZEO the interpretation of Ω_{vac} is as a complex square root of partition functional (quantum theory as complex square root of thermodynamics in ZEO). The integral of $|\Omega|^2$ over allowed moduli has interpretation as partition function. The exponential $exp(Re(X)) = p^{Re(X)/log(p)}$ has interpretation as an exponential of "Hamiltonian" defined by the vacuum conformal weight defined by moduli. $T = log(p)$ is identified as p-adic temperature as in ordinary p-adic thermodynamics.

   NTU requires that the integration over the moduli parameters reduces to a sum over number theoretically universal moduli parameters. The exponents $exp(X)$ must exist p-adically. PE property alone could guarantee this. The exponentials appearing in theta functions should reduce to products $p^k p^{iy} = exp(k/log(p))p^{iy}$ where k is integer and p^{iy} a root of unity. The vacuum expectation value of $Re(X)$ contributing to the mass squared is obtained from the standard formula as logarithmic temperature derivative of the "integral" $\int |\Omega_{vac}|^2$. The formula is same as for the Super-Virasoro contributions apart from the integration reducing to a sum.

   The considerations of the Section 4.2 [72] suggest that for given p-adic prime p the exponent $k + iy$ corresponds to a linear combinations of poles of fermionic zeta $z_F(s) = \zeta(s)/\zeta(2s)$ in the class $C(p)$ with non-negative integer coefficients. This class corresponds essentially to the conformal weights of a fractal sub-algebra of super-symplectic algebra. It could give rise also to the complex values of action so that Riemann zeta would define the core of TGD.

   The general dependence of the contribution of genus g to mass squared on g follows from the functional form of EPVF as a product theta functions serving as building brick partition functions apart from overall multiplicative constant and gives a nice agreement with the observed charged lepton mass ratios. The basic feature of the formula is exponential dependence on g.

5. The super-symplectic stringy contribution assignable to the magnetic flux tube dominates for weak bosons and is analogous to the stringy contribution to the hadron masses.

p-Adic mass calculations leave open several questions. What is the precise origin of preferred p-adic primes and of p-adic length scale hypothesis? How to understand the preferred number $N = 5$ of Super-Kac-Moody tensor factors? How to calculate the contribution of super-symplectic degrees of freedom—are they thermalized? Why only three lowest genera are light and what are the masses of the predicted bosonic higher genera implying breaking of fermion universality.

3.4 p-Adic Length Scale Hypothesis

p-Adic length scale hypothesis [38, 39] has served as a basic hypothesis of p-adic TGD for several years. This hypothesis states that the scales $L_p = \sqrt{p}l$, $l = 1.376 \cdot 10^4\sqrt{G}$ are fundamental length scale at p-adic condensate level p. The original interpretation of the hypothesis was following:

1. Above the length scale L_p p-adicity sets on and effective course grained space-time or imbedding space topology is p-adic rather than ordinary real topology. Imbedding space topology seems to be more appropriate identification.
2. The length scale L_p serves as a p-adic length scale cutoff for the quantum field theory description of particles. This means that space-time begins to look like Minkowski space so that the QFT $M^4 \to CP_2$ becomes a realistic approximation. Below this length scale string like objects and other particle like 3-surfaces are important.
3. It is un-natural to assume that just single p-adic field would be chosen from the infinite number of possibilities. Rather, there is an infinite number of cutoff length scales. To each prime p there corresponds a cutoff length scale L_p above which p-adic quantum field theory $M^4 \to CP_2$ makes sense and one has a hierarchy of p-adic QFTs. These different p-adic field theories correspond to different hierarchically levels possibly present in the topological condensate. Hierarchical ordering $< p_1 < p_2 < \ldots$ means that only the surface $p_1 < p_2$ can condense on the surface p_2. The condensed surface can in practice be regarded as a point-like particle at level p_2 described by the p-adic conformal field theory below length scale L_{p_2}.

The recent view inspired by adelic physics is that preferred p-adic primes correspond to the so-called ramified primes for the algebraic extension of rationals defining the adele [66]. Weak form of Negentropy Maximization Principle (WNMP) [35] in turn allows to conclude that the length scales corresponding to powers of primes are preferred. Therefore p-adic length scale hypothesis generalizes. There is evidence for 3-adic time scales in biology [16, 17] and 3-adic time scales can be also assigned with Pythagorean scale in geometric theory of harmony [43, 71].

3.5 Mersenne Primes and Gaussian Mersennes Are Special

Mersenne primes and their complex counterparts Gaussian Mersennes pop up in p-adic mass calculations and both elementary particle physics, biology [55], and astrophysics and cosmology [63] provide support for them.

3.5.1 Mersenne Primes

One can also consider the milder requirement that the exponent $\lambda = 2^{\epsilon L_0}$ represents trivial scaling represented by unit in good approximation for some p-adic topology. Not surprisingly, this is the case for $L_0 = mp^k$ since by Fermat's theorem $a^p \ mod \ p = 1$ for any integer a, in particular $a = 2$. This is also the case for $L_0 = mk$ such that $2^k \ mod \ p = 1$ for p prime. This occurs if $2^k - 1$ is Mersenne prime: in this case one has $2^{L_0} = 1$ modulo p so that the sizes of the fractal sub-algebras are exponentially larger than the sizes of $L_0 \propto p^n$ algebras. Note that all scalings a^{L_0} are near to unity for $L_0 = p^n$ whereas now only $a = 2$ gives scalings near unity for Mersenne primes. Perhaps this extended fractality provides the fundamental explanation for the special importance of Mersenne primes.

In this case integrated scalings 2^{L_0} leave the states almost invariant so that even a stronger form of the breaking of the exact conformal invariance would be in question in the super-symplectic case. The representation would be defined by the generators for which conformal weights are odd multiples of n ($M_n = 2^n - 1$) and L_{-kn}, $k > 0$ would generate zero norm states only in order $O(1/M_n)$.

Especially interesting is the hierarchy of primes defined by the so-called Combinatorial Hierarchy resulting from TGD based model for abstraction process. The primes are given by $2, 3, 7 = 2^3 - 1, 127 = 2^7 - 1, 2^{127} - 1, ...$: $L_0 = n \times 127$ would correspond to M_{127}-adicity crucial for the memetic code.

3.5.2 Gaussian Mersennes Are Also Special

If one allows also Gaussian primes, then the notion of Mersenne prime generalizes: Gaussian Mersennes are of form $(1 \pm i)^n - 1$. In this case one could replace the scaling operations by scaling combined with a twist of $\pi/4$ around some symmetry axis: $1 + i = \sqrt{2} exp(i\pi/4)$ and generalized p-adic fractality would mean that for certain values of n the exponentiated operation consisting of n basic operations would be very near to unity.

1. The integers k associated with the lowest Gaussian Mersennes are following: $2, 3, 5, 7, 11, 19, 29, 47, 73, 79, 113$. $k = 113$ corresponds to the p-adic length scale associated with the atomic nucleus and muon. Thus all known charged leptons, rather than only e and τ, as well as nuclear physics length scale, correspond to Mersenne primes in the generalized sense.
2. The primes $k = 151, 157, 163, 167$ define perhaps the most fundamental biological length scales: $k = 151$ corresponds to the thickness of the cell membrane of about ten nanometers and $k = 167$ to cell size about $2.56\,\mu$m. This observation also suggests that cellular organisms have evolved to their present form through four basic evolutionary stages. This also encourages to think that $\sqrt{2}exp(i\pi/4)$ operation giving rise to logarithmic spirals abundant in living matter is fundamental dynamical symmetry in bio-matter.

   Logarithmic spiral provides the simplest model for biological growth as a repetition of the basic operation $\sqrt{2}exp(i\pi/4)$. The naive interpretation would be that growth processes consist of $k = 151, 157, 163, 167$ steps involving scaling by $\sqrt{2}$. This, however, requires the strange looking assumption that growth starts from a structure of size of order CP_2 length. Perhaps this exotic growth process is associated with pair of MEs or magnetic flux tubes of opposite time orientation and energy emerging CP_2 sized region in a mini big bang type process and that the resulting structure serves as a template for the biological growth.
3. $k = 239, 241, 283, 353, 367, 379, 457$ associated with the next Gaussian Mersennes define astronomical length scales. $k = 239$ and $k = 241$ correspond to the p-adic time scales 0.55 ms and 1.1 ms: basic time scales associated with nerve pulse transmission are in question. $k = 283$ corresponds to the time scale of 38.6 min. An interesting question is whether this period could define a fundamental biological rhythm. The length scale $L(353)$ corresponds to about 2.6×10^6 light years, roughly the size scale of galaxies. The length scale $L(367) \simeq \times 3.3 \times 10^8$ light years is of same order of magnitude as the size scale of the large voids containing galaxies on their boundaries (note the analogy with cells). $T(379) \simeq 2.1 \times 10^{10}$ years corresponds to the lower bound for the order of the age of the Universe. $T(457) \sim 10^{22}$ years defines a completely superastronomical time and length scale.

3.6 Questions

The proposed picture leaves open several questions.

1. Could the descriptions by both real and p-adic thermodynamics be possible? Could they be equivalent (possibly in finite measurement resolution) as is suggested by NTU? The consistency of these descriptions would imply temperature quantization and p-adic length scale hypothesis not possible in purely real context.

2. What could the extension of conformal symmetry to super-symplectic symmetry mean? One possible view is that super-symplectic symmetries correspond to dark degrees of freedom and that only the super-symplectic ground states with negative conformal weights affect the p-adic thermodynamics, which applies only to fermionic degrees of freedom at string world sheets. Super-symplectic degrees of freedom would give the dominant contribution to hadron masses and could contribute also to weak gauge boson masses. $N = 5$ for the needed number of tensor factors is, however, a strong constraint and perhaps most naturally obtained when also the super-symplectic Virasoro associated with the trivial zeros of zeta is thermalized.
3. What happens in dark sectors. Preferred extremal property is proposed to mean that the states are annihilated by super-symplectic sub-algebra isomorphic to the original algebra and its commutator with the entire algebra. The conjecture is that this gives rise to Kac-Moody algebras as dynamical symmetries—maybe ADE type algebras, whose Dynkin diagrams characterize the inclusion of HFFs. Does this give an additional tensor factor to Super-Virasoro algebra?
4. Super-conformal symmetry true in the sense that Super-Virasoro conditions hold true. Partition function, however, depends on mass squared only rather than the entire scaling generator L_0 as thought erratically in the first formulation of p-adic calculation. This does not mean breaking of conformal invariance. Super-Virasoro conditions hold true although partition function is for the vibrational part of L_0 determining the mass squared spectrum.

4 p-Adicization and Adelic Physics

This section is devoted to the challenges related to p-adicization and adelization of physics in which the correspondence between real and p-adic numbers via canonical identification serves as the basic building brick. Also the problems associated with p-adic variants of integral, Fourier analysis, Hilbert space, and Riemann geometry should be solved in a manner respecting fundamental symmetries and their p-adic variants must be met. The notion of number theoretical universality (NTU) plays a key role here. One should also answer to questions about the origin of preferred primes and p-adic length scale hypothesis.

4.1 Challenges

The basic challenges encountered are construction of the p-adic variants of real number based physics, understanding their relationship to real physics, and the fusion of various physics to single coherent whole.

The p-adicization of real physics is not just a straightforward formal generalization of scattering amplitudes of existing theories but requires a deeper understanding of the physics involved. The interpretation of p-adic physics as correlate for cognition and imagination is an important guideline and will be discussed in more detail in separate section.

Definite integral and Fourier analysis are basic elements of standard physics and their generalization to the p-adic context defines a highly non-trivial challenge. Also the p-adic variants of Riemann geometry and Hilbert space are suggestive. There are, however, problems.

1. There are problems associated with p-adic definite integral. Riemann sum does not make sense since it approaches zero if the p-adic norm of discretization unit approaches zero. The problems are basically due to the absence of well-orderedness essential for the definition of definite integral and differential forms and their integrals.

   Residue integration might make sense in finite angle resolution. For algebraic extension containing $e^{i\pi/n}$ the number theoretically universal approximation $i\pi = n(e^{i\pi/n} - 1)$ could be used. In twistor approach integrations reduce to multiple residue integrations and since twistor approach generalizes in TGD framework, this approach to integration is very attractive.

   Positivity is a central notion in twistor Grassmannian approach [5]. Since canonical identification maps p-adic numbers to non-negative real numbers, there is a strong temptation to think that positivity relates to NTU [73].
2. There are problems with Fourier analysis. The naive generalization of trigonometric functions by replacing e^{ix} with its p-adic counterpart is not physical. Same applies to e^x. Algebraic extensions are needed to get roots of unity and e as counterparts of the phases and discretization is necessary and has interpretation in terms of finite resolution for angle/phase and its hyperbolic counterpart.
3. The notion of Hilbert space is problematic. The naive generalization of Hilbert space norm square $|x|^2 = \sum x_n\overline{x}_n$ for state $(x_1, x_2, \ldots)$ can vanish p-adically. Also here NTU could help. State would contain as coefficients only roots of e and unity and only the overall factor could be p-adic number. Coefficients could be restricted to the algebraic numbers generating the algebraic extension of rational numbers and would not contain powers of p or even ordinary p-adic numbers expect in the overall normalization factor.

Second challenge relates to the relationship between real and p-adic physics. Canonical identification (CI) $\sum x_n p^n \rightarrow \sum x_n p^{-n}$ or some of its variants should play an important role. CI is expected to map the invariants appearing in scattering amplitudes to their real counterparts.

1. Real and p-adic variants of space-time surfaces should exist and relate to each other somehow. Is this relationship local and involve CI at space-time level or imbedding space level? Or is it only a global and non-local assignment of preferred real extremals to their p-adic counterparts? Or is between these extreme options and involves algebraic discretization of the space-time surface weakening the strong form of SH as already proposed? How do real and p-adic imbedding spaces relate to each other and can this relationship induce local correspondence between preferred extremals (PEs) [25, 67, 70]?

2. NTU in some sense is a highly suggestive approach to these questions and would suggest that canonical identification applies to isometry invariants whereas angles and hyperbolic angles, or rather the corresponding "phases" belonging to an extension of p-adics containing roots of e and roots of unity are mapped to themselves. Note that the roots of e define extensions of rationals, which induce finite-dimensional algebraic extensions of p-adic numbers. This would make possible to define imbedding space in accordance with NTU. Also the Hilbert space could be defined by requiring that its points correspond to number theoretically universal angles expressible in terms of roots of unity.
3. What about real and p-adic variants of WCW? Are they needed at all? Or could their existence be used as a powerful constraint on real physics? The representability of WCW as a union of infinite-dimensional symmetric spaces labelled by zero modes suggests that the same description applies at the level of WCW and imbedding space.

   One cannot circumvent the question about how to generalize functional integral from real WCW to p-adic WCWs. In particular, what is the p-adic variant of the action defining the dynamics of space-time surfaces. In the case of exponent of action the p-adic variant could be defined by assuming algebraic universality: again the roots of e and of unity would be in central role. Also the Kähler structure of WCW implying that Gaussian and metric determinants cancel each other in functional integral would be absolutely crucial.

   One must remember that the exponents of action for scattering amplitudes for the stationary phase extremal cancel from the path integral representation of scattering amplitudes. Also now this mechanism would allow to get rid of the poorly defined exponent for single minimum. If there is sum over scattering amplitudes assignable to different maxima, normalization should give ratios of these exponents for different extrema/maxima and only these ratios should belong to the extension of rationals.

   The zero modes of WCW metric are invariants of super-symplectic group so that canonical identification could relate their real and p-adic variants. Zero modes could break NTU and would be behind p-adic thermodynamics and dependence of mass scale on p-adic prime.

The third challenge relates to the fusion of p-adic physics and real physics to a larger structure. Here a generalization of number concept obtained by glueing reals and various p-adics together along an extension of rational numbers inducing the extensions of p-adic numbers is highly suggestive. Adeles associated with the extension of rationals are highly attractive and closely related notion. Real and various p-adic physics would be correlates for sensory and cognitive aspects of the same universal physics rather than separate physics in this framework. One important implication of this view is that real entropy and p-adic negentropies characterize the same entanglement with coefficients in an extension of rationals.

NTU for hyperbolic and ordinary phases is definitely the central idea. How the invariance of angles under conformal transformations does relate to this? Could one perhaps define a discretized version of conformal symmetry preserving the phases defined by the angles between vectors assignable with the tangent spaces of discretized geometric structures and thus respecting NTU? Of should one apply conformal symmetry at Lie algebra level only?

4.2 NTU and the Correspondence Between Real and p-Adic Physics

p-Adic real correspondence is certainly the basic problem of p-adicization and adelization. One can make several general questions about p-adic real correspondence and canonical identification inspired by p-adic mass calculations.

How generally p-adic real correspondence does apply? Could canonical identification for group invariants combined with direct identification of ordinary and hyperbolic phases identified as roots of unity and e apply at WCW and imbedding space level having maximally symmetric geometries? Could this make sense even at space-time level as a correspondence induced from imbedding space level [77]? Does canonical identification apply locally for the discretizations of space-time surface or only globally for the parameters characterizing PEs (string world sheets and partonic 2-surfaces by SH), which are general coordinate invariant and Poincare invariant quantities?

4.2.1 General Vision About NTU

The following vision seems to be the most feasible one found hitherto.

1. Preservation of symmetries and continuity compete. Lorenz transformations do not commute with canonical identification. This suggests that canonical identification applies only to Lorentz invariants formed from quantum numbers. This is enough in the case of scattering amplitudes. Canonical identification applies only to isometry invariants at the level of WCW and the phases/exponents of ordinary/hyperbolic angles correspond to numbers in the algebraic extension common to extensions of rationals and various p-adics.
2. Canonical identification applies at the level of momentum space and maps p-adic Lorentz invariants of scattering amplitudes to their real counterparts. Phases of angles and their hyperbolic counterparts should correspond to parameters defining extension and should be mapped as such to their p-adic counterparts.
3. The constraints coming from GCI and symmetries do not allow local correspondence but allow to consider its discretized version at space-time level induced by the correspondence at the level of imbedding space.

   This requires the restriction of isometries and other symmetries to algebraic subgroups defined by the extension of rationals. This would imply reduction of symmetry due to finite cognitive/measurement resolution and should be acceptable. If one wants to realize the ideas about imagination, discretization must be applied also for the space-time interior meaning partial breaking of SH and giving rise to dark matter degrees freedom in TGD sense. SH could apply in real sector for realizable imaginations only. Note that the number of algebraic points of space-time surface is expected to be relatively small.

The correspondence must be considered at the level of imbedding space, space-time, and WCW.

1. At the level of imbedding space p-adic–real correspondence is induced by points in extension of rationals and is totally discontinuous. This requires that space-time dimension is smaller than imbedding space dimension.
2. At space-time level the correspondence involves field equations derivable from a local variational principle make sense also p-adically although the action itself is ill-defined as 4-D integral. The notion of p-adic PE makes sense by strong form of holography applied to 2-surfaces in the intersection. p-Adically, however, only the vanishing of Noether currents for a sub-algebra of the super-symplectic algebra might make sense. This condition is stronger than the vanishing of Noether charges defined by 3-D integrals.
3. Correspondence at the level of WCW can make sense and reduces to that for string world sheets and partonic 2-surfaces by SH. Real and p-adic 4-surfaces would be obtained by algebraic continuation as PEs from 2-surfaces by assuming that the space-time surface contains subset of points of imbedding space belonging to the extension of rationals [77]. p-Adic pseudo-constants make p-adic continuation easy. Real continuation need not exist always. p-Adic WCW would be considerably larger than real WCW and make possible a predictive quantum theory of imagination and cognition.

   What I have called intersection of realities and p-adicities can be identified as the set of 2-surfaces plus algebraic discretization of space-time interior. Also the values of induced spinor fields at the points of discretization must be given. The parameters characterizing the extremals (say coefficients of polynomials)—WCW coordinates—would be in extension of rationals inducing a finite-D extension of p-adic number fields.

   The hierarchy of algebraic extensions induces an evolutionary hierarchy of adeles. The interpretation could be as a mathematical correlate for cosmic evolution realized at the level of the core of WCW defined by the intersection? 2-surfaces could be called space-time genes.
4. Also the p-adic variant Kähler action or at least the exponent of Kähler action defining vacuum functional should be obtainable by algebraic continuation. The weakest condition states that the ratios of action exponents for the maxima of Kähler function to the sum of action exponents for maxima belong to the extension. Without this condition the hopes of satisfying NTU seem rather meager.

4.3 NTU at Space-Time Level

What about NTU at space-time level? NTU requires a correspondence between real and p-adic numbers and the details of this corresponds have been a long standing problem.

1. The recent view about the correspondence between real PEs to their p-adic counterparts does not demand discrete local correspondence assumed in the earlier proposal [60]. The most abstract approach would give up the local correspondence at space-time level altogether, and restrict the preferred coordinates of WCW (having maximal group of isometries) to numbers in the extension of rationals considered. WCW would be discretized.

   Intuitively a more realistic view is a correspondence at space-time level in the sense that real and p-adic space-time sheets intersect at points belonging to the extension of rationals and defining "cognitive representations." Only some p-adic space-time surfaces would have real counterpart.
2. The strongest form of NTU would require that the allowed points of imbedding space belonging an extension of rationals are mapped as such to corresponding extensions of p-adic number fields (no canonical identification). At imbedding space level this correspondence would be extremely discontinuous. The "spines" of space-time surfaces would, however, contain only a subset of points of extension, and a natural resolution length scale could emerge and prevent the fluctuation. This could be also seen as a reason for why space-times surfaces must be 4-D. The fact that the curve $x^n + y^n = z^n$ has no rational points for $n > 2$ raises the hope that the resolution scale could emerge spontaneously.
3. The notion of monadic geometry discussed in detail in [77] would realize this idea. Define first a number theoretic discretization of imbedding space in terms of points, whose coordinates in group theoretically preferred coordinate system belong to the extension of rationals considered. One can say that these algebraic points are in the intersection of reality and various p-adicities. Overlapping open sets assigned with this discretization define in the real sector a covering by open sets. In p-adic sector compact-open-topology allows to assign with each point eighth Cartesian power of algebraic extension of p-adic numbers. These compact-open sets define analogs for the monads of Leibniz and p-adic variants of field equations make sense inside them.

   The monadic manifold structure of H is induced to space-time surfaces containing discrete subset of points in the algebraic discretization with field equations defining a continuation to space-time surface in given number field, and unique only in finite measurement resolution. This approach would resolve the tension between continuity and symmetries in p-adic–real correspondence: isometry groups would be replaced by their sub-groups with parameters in extension of rationals considered and acting in the intersection of reality and p-adicities.

   The Galois group of extension acts non-trivially on the "spines" of space-time surfaces. Hence the number theoretical symmetries act as physical symmetries and define the orbit of given space-time surface as a kind of covering space. The coverings assigned to the hierarchy of Planck constants would naturally correspond to Galois coverings and dark matter would represent number theoretical physics.

This would give rise to a kind of algebraic hierarchy of adelic 4-surfaces identifiable as evolutionary hierarchy: the higher the dimension of the extension, the higher the evolutionary level.

4.4 *NTU and WCW*

4.4.1 p-Adic–Real Correspondence at the Level of WCW

It has not been obvious whether one should perform p-adicization and adelization at the level of WCW. Minimalist could argue that scattering amplitudes are all we want and that their p-adicization and adelization by algebraic continuation can be tolerated only if it can give powerful enough constraints on the amplitudes.

1. The anti-commutations for fermionic oscillator operators are number theoretically universal. Supersymmetry suggests that also WCW bosonic degrees of freedom satisfy NTU. This could mean that the coordinates of p-adic WCW consist of super-symplectic invariants mappable by canonical identification to their real counterparts plus phases and their hyperbolic counterparts expressible as genuinely algebraic numbers common to all number fields. This kind of coordinates are naturally assignable to symmetric spaces [77].
2. Kähler structure should be mapped from p-adic to real sector and vice versa. Vacuum functional identified as exponent of action should be NTU. Algebraic continuation defined by SH involves p-adic pseudo-constants. All p-adic continuations by SH should correspond to the same value of exponent of action obtained by algebraic continuation from its real value. The degeneracy associated with p-adic pseudo-constants would be analogous to gauge invariance—imagination in TGD inspired theory of consciousness.
3. Is it possible to have NTU for WCW functional integration? Or is it enough to realize NTU for scattering amplitudes only. What seems clear that functional integral must reduce to a discrete sum. Physical intuition suggests a sum over maxima of Kähler function forming a subset of PEs representing stationary points. One cannot even exclude the possibility that the set of PEs is discrete and that one can sum over all of them.

   Restriction to maximum/stationary phase approximation gives rise to sum over exponents multiplied with Gaussian determinants. The determinant of Kähler metric, however, cancels the Gaussian determinants, and one obtains only a sum over the exponents of action.

The breaking of strong NTU could happen: consider only p-adic mass calculations. This breaking is, however, associated with the parts of quantum states assignable to the boundaries of CD, not with the vacuum functional.

4.4.2 NTU for Functional Integral

Number theoretical vision relies on NTU. In fermionic sector NTU is necessary: one cannot speak about real and p-adic fermions as separate entities and fermionic anti-commutation relations are indeed number theoretically universal.

What about NTU in case of functional integral? There are two opposite views.

1. One can define p-adic variants of field equations without difficulties if preferred extremals are minimal surface extremals of Kähler action so that coupling constants do not appear in the solutions. If the extremal property is determined solely by the analyticity properties as it is for various conjectures, it makes sense independent of number field. Therefore there would be no need to continue the functional integral to p-adic sectors. This in accordance with the philosophy that thought cannot be put in scale. This would be also the option favored by pragmatist.
2. Consciousness theorist might argue that also cognition and imagination allow quantum description. The supersymmetry NTU should apply also to functional integral over WCW (more precisely, its sector defined by CD) involved with the definition of scattering amplitudes.

Key Observations

The general vision involves some crucial observations.

1. Only the expressions for the scattering amplitudes should satisfy NTU. This does not require that the functional integral satisfies NTU.
2. Since the Gaussian and metric determinants cancel in WCW Kähler metric the contributions from maxima are proportional to action exponentials $exp(S_k)$ divided by the $\sum_k exp(S_k)$. Loops vanish by quantum criticality.
3. Scattering amplitudes can be defined as sums over the contributions from the maxima, which would have also stationary phase by the double extremal property made possible by the complex value of α_K. These contributions are normalized by the vacuum amplitude.

   It is enough to require NTU for $X_i = exp(S_i)/\sum_k exp(S_k)$. This requires that $S_k - S_l$ has form $q_1 + q_2 i\pi + q_3 log(n)$. The condition brings in mind homology theory without boundary operation defined by the difference $S_k - S_l$. NTU for both S_k and $exp(S_k)$ would only values of general form $S_k = q_1 + q_2 i\pi + q_3 log(n)$ for S_k and this looks quite too strong a condition.

4. If it is possible to express the 4-D exponentials as single 2-D exponential associated with union of string world sheets, vacuum functional disappears completely from consideration! There is only a sum over discretization with the same effective action and one obtains purely combinatorial expression.

What Does One Mean with Functional Integral?

The definition of functional integral in WCW is one of the key technical problems of quantum TGD [66]. NTU states that the integral should be defined simultaneously in all number fields in the intersection of real and p-adic worlds defined by string world sheets and partonic 2-surfaces with WCW coordinates in algebraic extension of rationals and allowing by strong holography continuation to 4-D space-time surface. NTU is powerful constraint and could help in this respect.

1. Path integral is not in question. Rather, the functional integral is analogous to Wiener integral and perhaps allows identification as a genuine integral in the real sector. In p-adic sectors algebraic continuation should give the integral and here number theoretical universality gives excellent hopes. The integral would have exactly the same form in real and p-adic sector and expressible solely in terms of algebraic numbers characterizing algebraic extension and finite roots of e and roots of unity $U_n = exp(i2\pi/n)$ in algebraic extension of p-adic numbers.

   Since vacuum functional $exp(S)$ is exponential of complex action S, the natural idea is that only rational powers e^q and roots of unity and phases $exp(i2\pi q)$ are involved and there is no dependence on p-adic prime p! This is true in the integer part of q and is smaller than p so that one does not obtain e^{kp}, which is ordinary p-adic number and would spoil the number theoretic universality. This condition is not possible to satisfy for all values of p unless the value of Kähler function is smaller than 2. One might consider the possibility that the allow primes are above some minimum value.

   The minimal solution to NTU conditions is that the ratios of action exponentials for maxima of Kähler function to the sum of these exponentials belong to the extension of rationals considered.
2. What does one mean with functional integral? TGD is expected to be an integrable in some sense. In integrable QFTs functional integral reduces to a sum over stationary points of the action: typically only single point contributes—at least in good approximation.

   For real α_K and Λ vacuum functional decomposes to a product of exponents of real contribution from Euclidian regions ($\sqrt{g_4}$ real) and imaginary contribution Minkowskian regions ($\sqrt{g_4}$ imaginary). There would be no exchange of momentum between Minkowskian and Euclidian regions. For complex values of α_K [30] situation changes and Kähler function as real part of action receives contributions from both Euclidian and Minkowskian regions. The imaginary part of action has interpretation as analog of Morse function and action as it appears in QFTs. Now saddle points must be considered.

PEs satisfy extremely strong conditions [67, 70]. All classical Noether charges for a sub-algebra associated with super-symplectic algebra and isomorphic to the algebra itself vanish at both ends of CD. The conformal weights of this algebra are $n > 0$-ples of those for the entire algebra. What is fascinating that the condition that the preferred extremals are minimal surface extremals of Kähler action could solve these conditions and guarantee also NTU at the level of space-time surfaces. Super-symplectic boundary conditions at the ends of CD would, however, pose number theoretic conditions on the coupling parameters. In p-adic case the conditions should reduce to purely local conditions since p-adic charges are not well-defined as integrals.

3. In TGD framework one is constructing zero energy states rather calculating the matrix elements of S-matrix in terms of path integral. This gives certain liberties but a natural expectation is that functional integral as a formal tool at least is involved.

   Could one *define* the functional integral as a discrete sum of contributions of standard form for the preferred extremals, which correspond to maxima in Euclidian regions and associated stationary phase points in Minkowskian regions? Could one assume that WCW spinor field is concentrated along single maximum/stationary point.

   Quantum classical correspondence suggests that in Cartan algebra isometry charges are equal to the quantal charges for quantum states expressible in number theoretically universal manner in terms of fermionic oscillator operators or WCW gamma matrices? Even stronger condition would be that classical correlation functions are identical with quantal ones for allowed space-time surfaces in the quantum superposition. Could the reduction to a discrete sum be interpreted in terms of a finite measurement resolution?

4. In QFT Gaussian determinants produce problems because they are often poorly defined. In the recent case they could also spoil the NTU based on the exceptional properties of e. In the recent case, however, Gaussian determinant and metric determinant for Kähler metric cancel each other and this problem disappears. One could obtain just a sum over products of roots of e and roots of unity. Thus also Kähler structure seems to be crucial for the dream about NTU.

4.5 Breaking of NTU at the Level of Scattering Amplitudes

NTU in strong sense could be broken at the level of scattering amplitudes. At space-time level the breaking does not look natural in the recent framework. Consider only p-adic mass calculations predicting that mass scale depends on p-adic prime. Also for the action strong form of NTU might fail for small p-adic primes since the value of the real part of action would be larger than p. Should one allow this? What does one actually mean with NTU in the case of action?

Canonical identification is an important element of p-adic mass calculations and might also be needed to map p-adic variants of scattering amplitudes to their real counterparts. The breaking of NTU would take place, when the canonical real valued image of the p-adic scattering amplitude differs from the real scattering amplitude. The interpretation would be in terms of finite measurement resolution. By the finite measurement/cognitive resolution characterized by p one cannot detect the difference.

The simplest form of the canonical identification is $x = \sum_n x_n p^n \rightarrow \sum x_n p^{-n}$. Product xy and sum $x + y$ do not in general map to product and sum in canonical identification. The interpretation would be in terms of a finite measurement resolution: $(xy)_R = x_R y_R$ and $(x + y)_R = x_R + y_R$ only modulo finite measurement resolution. p-Adic scattering amplitudes are obtained by algebraic continuation from the intersection by replacing algebraic number valued parameters (such as momenta) by general p-adic numbers. The real images of these amplitudes under canonical identification are in general not identical with real scattering amplitudes the interpretation being in terms of a finite measurement resolution.

In p-adic thermodynamics NTU in the strong sense fails since thermal masses depend on p-adic mass scale. NTU can be broken by the fermionic matrix elements in the functional integral so that the real scattering amplitudes differ from the canonical images of the p-adic scattering amplitudes. For instance, the elementary particle vacuum functionals in the space of Teichmueller parameters for the partonic 2-surfaces and string world sheets should break NTU [27].

4.6 NTU and the Spectrum of Kähler Coupling Strength

During years I have made several attempts to understand coupling evolution in TGD framework. The most convincing proposal has emerged rather recently and relates the spectrum of $1/\alpha_K$ to that for the zeros of Riemann zeta [30] and to the evolution of the electroweak U(1) couplings strength.

1. The first idea dates back to the discovery of WCW Kähler geometry defined by Kähler function defined by Kähler action (this happened around 1990) [33]. The only free parameter of the theory is Kähler coupling strength α_K analogous to temperature parameter α_K postulated to be is analogous to critical temperature. Whether only single value or entire spectrum of values α_K is possible remained an open question.

   About decade ago I realized that Kähler action is *complex* receiving a real contribution from space-time regions of Euclidian signature of metric and imaginary contribution from the Minkowskian regions. Euclidian region would give Kähler function and Minkowskian regions analog of QFT action of path integral approach defining also Morse function. Zero energy ontology (ZEO) [61] led to the interpretation of quantum TGD as complex square root of thermodynamics so that the vacuum functional as exponent of Kähler action could be identified as a complex square root of the ordinary partition function. Kähler function would correspond to the real contribution Kähler action from Euclidian space-time regions. This led to ask whether also Kähler coupling strength might be complex: in analogy with the complexification of gauge coupling strength in theories allowing magnetic monopoles. Complex α_K could allow to explain CP breaking. I proposed that instanton term also reducing to Chern–Simons term could be behind CP breaking.

The problem is that the dynamics in Minkowskian and Euclidian regions decouple completely and if Euclidian regions serve as space-time correlates for physical objects, there would be no exchanges of classical charges between physical objects. Should one conclude that α_K must be complex?

2. p-Adic mass calculations for two decades ago [34] inspired the idea that length scale evolution is discretized so that the real version of p-adic coupling constant would have discrete set of values labelled by p-adic primes. The simple working hypothesis was that Kähler coupling strength is renormalization group (RG) invariant and only the weak and color coupling strengths depend on the p-adic length scale. The alternative ad hoc hypothesis considered was that gravitational constant is RG invariant. I made several number theoretically motivated ad hoc guesses about coupling constant evolution, in particular a guess for the formula for gravitational coupling in terms of Kähler coupling strength, action for CP_2 type vacuum extremal, p-adic length scale as dimensional quantity [22]. Needless to say these attempts were premature and ad hoc.
3. The vision about hierarchy of Planck constants $h_{eff} = n \times h$ and the connection $h_{eff} = h_{gr} = GMm/v_0$, where $v_0 < c = 1$ has dimensions of velocity [62] forced to consider very seriously the hypothesis that Kähler coupling strength has a spectrum of values in one-one correspondence with p-adic length scales. A separate coupling constant evolution associated with h_{eff} induced by $\alpha_K \propto 1/\hbar_{eff} \propto 1/n$ looks natural and was motivated by the idea that Nature is theoretician friendly: when the situation becomes non-perturbative, Mother Nature comes in rescue and an h_{eff} increasing phase transition makes the situation perturbative again.

Quite recently the number theoretic interpretation of coupling constant evolution [66, 72] in terms of a hierarchy of algebraic extensions of rational numbers inducing those of p-adic number fields encouraged to think that $1/\alpha_K$ has spectrum labelled by primes and values of h_{eff}. Two coupling constant evolutions suggest themselves: they could be assigned to length scales and angles which are in p-adic sectors necessarily discretized and describable using only algebraic extensions involve roots of unity replacing angles with discrete phases.
4. Few years ago the relationship of TGD and GRT was finally understood [50]. GRT space-time is obtained as an approximation as the sheets of the many-sheeted space-time of TGD are replaced with single region of space-time. The gravitational and gauge potential of sheets add together so that linear superposition corresponds to set theoretic union geometrically. This forced to consider the possibility that gauge coupling evolution takes place only at the level of the QFT approximation and α_K has only single value. This is nice but if true, one does not have much to say about the evolution of gauge coupling strengths.

5. The analogy of Riemann zeta function with the partition function of complex square root of thermodynamics suggests that the zeros of zeta have interpretation as inverses of complex temperatures $s = 1/\beta$. Also $1/\alpha_K$ is analogous to temperature. This led to a radical idea to be discussed in detail in the sequel.

   Could the spectrum of $1/\alpha_K$ reduce to that for the zeros of Riemann zeta or—more plausibly—to the spectrum of poles of fermionic zeta $\zeta_F(ks) = \zeta(ks)/\zeta(2ks)$ giving for $k = 1/2$ poles as zeros of zeta and as point $s = 2$? ζ_F is motivated by the fact that fermions are the only fundamental particles in TGD and by the fact that poles of the partition function are naturally associated with quantum criticality whereas the vanishing of ζ and varying sign allow no natural physical interpretation.

   The poles of $\zeta_F(s/2)$ define the spectrum of $1/\alpha_K$ and correspond to zeros of $\zeta(s)$ and to the pole of $\zeta(s/2)$ at $s = 2$. The trivial poles for $s = 2n$, $n = 1, 2, ..$ correspond naturally to the values of $1/\alpha_K$ for different values of $h_{eff} = n \times h$ with n even integer. Complex poles would correspond to ordinary QFT coupling constant evolution. The zeros of zeta in increasing order would correspond to p-adic primes in increasing order and UV limit to smallest value of poles at critical line. One can distinguish the pole $s = 2$ as extreme UV limit at which QFT approximation fails totally. CP_2 length scale indeed corresponds to GUT scale.
6. One can test this hypothesis. $1/\alpha_K$ corresponds to the electroweak U(1) coupling strength so that the identification $1/\alpha_K = 1/\alpha_{U(1)}$ makes sense. One also knows a lot about the evolutions of $1/\alpha_{U(1)}$ and of electromagnetic coupling strength $1/\alpha_{em} = 1/[cos^2(\theta_W)\alpha_{U(1)}$. What does this predict?

   It turns out that at p-adic length scale $k = 131$ ($p \simeq 2^k$ by p-adic length scale hypothesis, which now can be understood number theoretically [66]) fine structure constant is predicted with 0.7% accuracy if Weinberg angle is assumed to have its value at atomic scale! It is difficult to believe that this could be a mere accident because also the prediction evolution of $\alpha_{U(1)}$ is correct qualitatively. Note, however, that for $k = 127$ labelling electron one can reproduce fine structure constant with Weinberg angle deviating about 10% from the measured value of Weinberg angle. Both models will be considered.
7. What about the evolution of weak, color, and gravitational coupling strengths? Quantum criticality suggests that the evolution of these couplings strengths is universal and independent of the details of the dynamics. Since one must be able to compare various evolutions and combine them together, the only possibility seems to be that the spectra of gauge coupling strengths are given by the poles of $\zeta_F(w)$ but with argument $w = w(s)$ obtained by a global conformal transformation of upper half plane—that is Möbius transformation (see http://tinyurl.com/gwjs85b) with real coefficients (element of $GL(2, R)$) so that one as $\zeta_F((as + b)/(cs + d))$. Rather general arguments force it to be and element of $GL(2, Q)$, $GL(2, Z)$ or maybe even $SL(2, Z)$ ($ad - bc = 1$) satisfying additional constraints. Since TGD predicts several scaled variants of weak and color interactions, these copies could be perhaps parameterized by some elements of $SL(2, Z)$ and by a scaling factor K.

Could one understand the general qualitative features of color and weak coupling constant evolutions from the properties of corresponding Möbius transformation? At the critical line there can be no poles or zeros but could asymptotic freedom be assigned with a pole of $cs + d$ and color confinement with the zero of $as + b$ at real axes? Pole makes sense only if Kähler action for the preferred extremal vanishes. Vanishing can occur and does so for massless extremals characterizing conformally invariant phase. For zero of $as + b$ vacuum function would be equal to one unless Kähler action is allowed to be infinite: does this make sense? One can, however, hope that the values of parameters allow to distinguish between weak and color interactions. It is certainly possible to get an idea about the values of the parameters of the transformation and one ends up with a general model predicting the entire electroweak coupling constant evolution successfully.

To sum up, the big idea is the identification of the spectra of coupling constant strengths as poles of $\zeta_F((as + b/)(cs + d))$ identified as a complex square root of partition function with motivation coming from ZEO, quantum criticality, and super-conformal symmetry; the discretization of the RG flow made possible by the p-adic length scale hypothesis $p \simeq k^k$, k prime; and the assignment of complex zeros of ζ with p-adic primes in increasing order. These assumptions reduce the coupling constant evolution to four real rational or integer valued parameters (a, b, c, d). In the sequel this vision is discussed in more detail.

4.7 Other Applications of NTU

NTU in the strongest form says that all numbers involved at "basic level" (whatever this means!) of adelic TGD are products of roots of unity and of power of a root of e. This is extremely powerful physics inspired by conjecture with a wide range of possible mathematical applications.

1. For instance, vacuum functional defined as an exponent of action for preferred externals would be number of this kind. One could define functional integral as adelic operation in all number fields: essentially as sum of exponents of action for stationary preferred extremals since Gaussian and metric determinants potentially spoiling NTU would cancel each other leaving only the exponent.
2. The implications of NTU for the zeros of Riemann zeta [72] will be discussed in more detail in the Appendix. Suffice it to say that the observations about Fourier transform for the distribution of loci of non-trivial zeros of zeta together with NTU lead to explicit proposal for the algebraic form of zeros of zeta. The testable proposal is that zeros decompose to disjoint classes $C(p)$ labelled by primes p and the condition that p^{iy} is root of unity in given class $C(p)$.
3. NTU generalizes to all Lie groups. Exponents $exp(in_iJ_i/n)$ of lie-algebra generators define generalizations of number theoretically universal group elements and generate a discrete subgroup of compact Lie group. Also hyperbolic "phases" based on the roots $e^{m/n}$ are possible and make possible discretized NTU versions of all Lie-groups expected to play a key role in adelization of TGD.

NTU generalizes also to quaternions and octonions and allows to define them as number theoretically universal entities. Note that ordinary p-adic variants of quaternions and octonions do not give rise to a number field: inverse of quaternion can have vanishing p-adic variant of norm squared satisfying $\sum_n x_n^2 = 0$.

NTU allows to define also the notion of Hilbert space as an adelic notion. The exponents of angles characterizing unit vector of Hilbert space would correspond to roots of unity.

4.8 Going to the Roots of p-Adicity

The basic questions raised by the p-adic mass calculations concern the origin of preferred p-adic primes and of p-adic length scale hypothesis. One can also ask whether there might be a natural origin for p-adicity at the level of WCW.

4.8.1 Preferred Primes as Ramified Primes for Extensions of Rationals?

The intuitive feeling is that the notion of preferred prime is something extremely deep and to me the deepest thing I know is number theory. Does one end up with preferred primes in number theory? This question brought to my mind the notion of *ramification of primes* (http://tinyurl.com/hddljlf) (more precisely, of prime ideals of number field in its extension), which happens only for special primes in a given extension of number field, say rationals. Ramification is completely analogous to the degeneracy of some roots of polynomial and corresponds to criticality if the polynomial corresponds to criticality (catastrophe theory of Thom is one application). Could this be the mechanism assigning preferred prime(s) to a given elementary system, such as elementary particle? I have not considered their role earlier also their hierarchy is highly relevant in the number theoretical vision about TGD.

1. Stating it very roughly (I hope that mathematicians tolerate this sloppy language of physicist): as one goes from number field K, say rationals Q, to its algebraic extension L, the original prime ideals in the so-called *integral closure* (http://tinyurl.com/js6fpvr) over integers of K decompose to products of prime ideals of L (prime ideal is a more rigorous manner to express primeness). Note that the general ideal is analog of integer.

   Integral closure for integers of number field K is defined as the set of elements of K, which are roots of some monic polynomial with coefficients, which are integers of K having the form $x^n + a_{n-1}x^{n-1} + \ldots + a_0$. The integral closures of both K and L are considered. For instance, integral closure of algebraic extension of K over K is the extension itself. The integral closure of complex numbers over ordinary integers is the set of algebraic numbers.

Prime ideals of K can be decomposed to products of prime ideals of L: $P = \prod P_i^{e_i}$, where e_i is the ramification index. If $e_i > 1$ is true for some i, *ramification* occurs. P_i:s in question are like co-inciding roots of polynomial, which for in thermodynamics and Thom's catastrophe theory corresponds to criticality. Ramification could therefore be a natural aspect of quantum criticality and ramified primes P are good candidates for preferred primes for a given extension of rationals. Note that the ramification make sense also for extensions of given extension of rationals.

2. A physical analogy for the decomposition of ideals to ideals of extension is provided by decomposition of hadrons to valence quarks. Elementary particles become composite of more elementary particles in the extension. The decomposition to these more elementary primes is of form $P = \prod P_i^{e(i)}$, the physical analog would be the number of elementary particles of type i in the state (http://tinyurl.com/h9528pl). Unramified prime P would be analogous a state with e fermions. Maximally ramified prime would be analogous to Bose–Einstein condensate of e bosons. General ramified prime would be analogous to an e-particle state containing both fermions and condensed bosons. This is of course just a formal analogy.
3. There are two further basic notions related to ramification and characterizing it. *Relative discriminant* is the ideal divided by all ramified ideals in K (integer of K having no ramified prime factors) and relative different for P is the ideal of L divided by all ramified P_i:s (product of prime factors of P in L). These ideals represent the analogs of product of preferred primes P of K and primes P_i of L dividing them. These two integers ideals would characterize the ramification.

In TGD framework the extensions of rationals (http://tinyurl.com/h9528pl) and p-adic number fields (http://tinyurl.com/zq22tvb) are unavoidable and interpreted as an evolutionary hierarchy physically and cosmological evolution would gradually proceed to more and more complex extensions. One can say that string world sheets and partonic 2-surfaces with parameters of defining functions in increasingly complex extensions of prime emerge during evolution. Therefore ramifications and the preferred primes defined by them are unavoidable. For p-adic number fields the number of extensions is much smaller, for instance, for $p > 2$ there are only three quadratic extensions.

How could ramification relate to p-adic and adelic physics and could it explain preferred primes?

1. Ramified p-adic prime $P = P_i^e$ would be replaced with its e:th root P_i in p-adicization. Same would apply to general ramified primes. Each unramified prime of K is replaced with $e = K : L$ primes of L and ramified primes P with $\#\{P_i\} < e$ primes of L: the increase of algebraic dimension is smaller. An interesting question relates to p-adic length scale. What happens to p-adic length scales. Is p-adic prime effectively replaced with e:th root of p-adic prime: $L_p \propto p^{1/2}L_1 \rightarrow p^{1/2e}L_1$? The only physical option is that the p-adic temperature for P would be scaled down $T_p = 1/n \rightarrow 1/ne$ for its e:th root (for fermions

serving as fundamental particles in TGD one actually has $T_p = 1$). Could the lower temperature state be more stable and select the preferred primes as maximally ramified ones? What about general ramified primes?
2. This need not be the whole story. Some algebraic extensions would be more favored than others and p-adic view about realizable imaginations could be involved. p-Adic pseudo-constants are expected to allow p-adic continuations of string world sheets and partonic 2-surfaces to 4-D preferred extremals with number theoretic discretization. For real continuations the situation is more difficult. For preferred extensions—and therefore for corresponding ramified primes—the number of real continuations—realizable imaginations—would be especially large.

   The challenge would be to understand why primes near powers of 2 and possibly also of other small primes would be favored. Why for them the number of realizable imaginations would be especially large so that they would be winners in number theoretical fight for survival?

Can one make this picture more concrete? What kind of algebraic extensions could be considered?

1. In p-adic context a proper definition of counterparts of angle variables as phases allowing definition of the analogs of trigonometric functions requires the introduction of algebraic extension giving rise to some roots of unity. Their number depends on the angular resolution. These roots allow to define the counterparts of ordinary trigonometric functions—the naive generalization based on Taylors series is not periodic—and also allows to defined the counterpart of definite integral in these degrees of freedom as discrete Fourier analysis. For the simplest algebraic extensions defined by $x^n - 1$ for which Galois group is Abelian are unramified so that something else is needed. One has decomposition $P = \prod P_i^{e(i)}$, $e(i) = 1$, analogous to n-fermion state so that simplest cyclic extension does not give rise to a ramification and there are no preferred primes.
2. What kind of polynomials could define preferred algebraic extensions of rationals? Irreducible polynomials are certainly an attractive candidate since any polynomial reduces to a product of them. One can say that they define the elementary particles of number theory. Irreducible polynomials have integer coefficients having the property that they do not decompose to products of polynomials with rational coefficients. It would be wrong to say that only these algebraic extensions can appear but there is a temptation to say that one can reduce the study of extensions to their study. One can even consider the possibility that string world sheets associated with products of irreducible polynomials are unstable against decay to those characterize irreducible polynomials.
3. What can one say about irreducible polynomials? Eisenstein criterion (http://tinyurl.com/47kxjz states following. If $Q(x) = \sum_{k=0,\ldots,n} a_k x^k$ is nth order polynomial with integer coefficients and with the property that there exists at least one prime dividing all coefficients a_i except a_n and that p^2 does not divide a_0, then Q is irreducible. Thus one can assign one or more preferred primes to the

algebraic extension defined by an irreducible polynomial Q of this kind—in fact any polynomial allowing ramification. There are also other kinds of irreducible polynomials since Eisenstein's condition is only sufficient but not necessary.

Furthermore, in the algebraic extension defined by Q, the prime ideals P having the above-mentioned characteristic property decompose to an n:th power of single prime ideal P_i: $P = P_i^n$. The primes are maximally/completely ramified.

A good illustration is provided by equations $x^2 + 1 = 0$ allowing roots $x_\pm = \pm i$ and equation $x^2+2px+p = 0$ allowing roots $x_\pm = -p\pm\sqrt{p}p - 1$. In the first case the ideals associated with $\pm i$ are different. In the second case these ideals are one and the same since $x_+ == -x_- + p$: hence one indeed has ramification. Note that the first example represents also an example of irreducible polynomial, which does not satisfy Eisenstein criterion. In more general case the n conditions defined by symmetric functions of roots imply that the ideals are one and same when Eisenstein conditions are satisfied.

4. What is so nice that one could readily construct polynomials giving rise to given preferred primes. The complex roots of these polynomials could correspond to the points of partonic 2-surfaces carrying fermions and defining the ends of boundaries of string world sheet. It must be, however, emphasized that the form of the polynomial depends on the choices of the complex coordinate. For instance, the shift $x \rightarrow x+1$ transforms $(x^n-1)/(x-1)$ to a polynomial satisfying the Eisenstein criterion. One should be able to fix allowed coordinate changes in such a manner that the extension remains irreducible for all allowed coordinate changes.

   Already the integral shift of the complex coordinate affects the situation. It would seem that only the action of the allowed coordinate changes must reduce to the action of Galois group permuting the roots of polynomials. A natural assumption is that the complex coordinate corresponds to a complex coordinate transforming linearly under subgroup of isometries of the imbedding space.

In the general situation one has $P = \prod P_i^{e(i)}$, $e(i) \geq 1$ so that as of now there are preferred primes so that the appearance of preferred primes is completely general phenomenon.

4.8.2 The Origin of p-Adic Length Scale Hypothesis?

p-Adic length scale hypothesis emerged from p-adic length scale hypothesis. A possible generalization of this hypothesis is that p-adic primes near powers of prime are physically favored. There indeed exists evidence for the realization of 3-adic time scale hierarchies in living matter [17] (http://tinyurl.com/jbh9m27) and in music both 2-adicity and 3-adicity could be present: this is discussed in TGD inspired theory of music harmony and genetic code [43]. See also [75, 78].

One explanation would be that for preferred primes the number of p-adic space-time sheets representable also as real space-time sheets is maximal. Imagined worlds would be maximally realizable. Preferred p-adic primes would correspond to ramified primes for extensions with the property that the number of realizable imaginations is especially large for them. Why primes satisfying p-adic length scale hypothesis or its generalization would appear as ramified primes for extensions, which are winners in number theoretical evolution?

Also the weak form of NMP (WNMP) applying also to the purely number theoretic form of NMP [35] might come in rescue here.

1. Entanglement negentropy for a NE [35] characterized by n-dimensional projection operator is the $log(N_p(n))$ for some p whose power divides n. The maximum negentropy is obtained if the power of p is the largest power of prime divisor of p, and this can be taken as definition of number theoretical entanglement negentropy (NEN). If the largest divisor is p^k, one has $N = k \times log(p)$. The entanglement negentropy per entangled state is $N/n = klog(p)/n$ and is maximal for $n = p^k$. Hence powers of prime are favored, which means that p-adic length scale hierarchies with scales coming as powers of p are negentropically favored and should be generated by NMP. Note that $n = p^k$ would define a hierarchy of $h_{eff}/h = p^k$. During the first years of h_{eff} hypothesis I believe that the preferred values obey $h_{eff} = r^k$, r integer not far from $r = 2^{11}$. It seems that this belief was not totally wrong.
2. If one accepts this argument, the remaining challenge is to explain why primes near powers of two (or more generally p) are favored. $n = 2^k$ gives large entanglement negentropy for the final state. Why primes $p = n_2 = 2^k - r$ would be favored? The reason could be following. $n = 2^k$ corresponds to $p = 2$, which corresponds to the lowest level in p-adic evolution since it is the simplest p-adic topology and farthest from the real topology and therefore gives the poorest cognitive representation of real PE as p-adic PE (Note that $p = 1$ makes formally sense but for it the topology is discrete).
3. WNMP [35, 51] suggests a more feasible explanation. The density matrix of the state to be reduced is a direct sum over contributions proportional to projection operators. Suppose that the projection operator with largest dimension has dimension n. Strong form of NMP would say that final state is characterized by n-dimensional projection operator. WNMP allows "free will" so that all dimensions $n-k, k = 0, 1, \ldots n-1$ for final state projection operator are possible. One-dimensional case corresponds to vanishing entanglement negentropy and ordinary state function reduction isolating the measured system from external world.
4. The negentropy of the final state per state depends on the value of k. It is maximal if $n - k$ is power of prime. For $n = 2^k = M_k + 1$, where M_k is Mersenne prime $n-1$ gives the maximum negentropy and also maximal p-adic prime available so that this reduction is favored by NMP. Mersenne primes would be indeed special. Also the primes $n = 2^k - r$ near 2^k produce large entanglement negentropy and would be favored by NMP.
5. This argument suggests a generalization of p-adic length scale hypothesis so that $p = 2$ can be replaced by any prime.

5 p-Adic Physics and Consciousness

p-Adic physics as physics of cognition and imagination is an important thread in TGD inspired theory of consciousness. In the sequel I describe briefly the basic of TGD inspired theory of consciousness as generalization of quantum measurement theory to ZEO (ZEO), describe the definition of self, consider the question whether NMP is needed as a separate principle or whether it is implied is in statistical sense by the unavoidable statistical increase of $n = h_{eff}/h$ if identified as a factor of the dimension of Galois group extension of rationals defining the adeles, and finally summarize the vision about how p-adic physics serves as a correlate of cognition and imagination.

5.1 From Quantum Measurement Theory to a Theory of Consciousness

The notion of self can be seen as a generalization of the poorly defined definition of the notion of observer in quantum physics. In the following I take the role of skeptic trying to be as critical as possible.

The original definition of self was a subsystem able to remain unentangled under state function reductions associated with subsequent quantum jumps. The density matrix was assumed to define the universal observable. Note that a density matrix, which is power series of a product of matrices representing commuting observables has in the generic case eigenstates, which are simultaneous eigenstates of all observables. Second aspect of self was assumed to be the integration of subsequent quantum jumps to coherent whole giving rise to the experienced flow of time.

The precise identification of self allowing to understand both of these aspects turned out to be difficult problem. I became aware of the solution of the problem in terms of ZEO (ZEO) only rather recently (2014).

1. Self corresponds to a sequence of quantum jumps integrating to single unit as in the original proposal, but these quantum jumps correspond to state function reductions to a fixed boundary of causal diamond CD leaving the corresponding parts of zero energy states invariant—"small" state function reductions. The parts of zero energy states at second boundary of CD change and even the position of the tip of the opposite boundary changes: one actually has wave function over positions of second boundary (CD sizes roughly) and this wave function changes. In positive energy ontology these repeated state function reductions would have no effect on the state (Zeno effect) but in TGD framework there occurs a change for the second boundary and gives rise to the experienced flow of time and its arrow and self: self is generalized Zeno effect.

2. The first quantum jump to the opposite boundary corresponds to the act of "free will" or birth of re-incarnated self. Hence the act of "free will" changes the arrow of psychological time at some level of hierarchy of CDs. The first reduction to the opposite boundary of CD means "death" of self and "re-incarnation" of time-reversed self at opposite boundary at which the temporal distance between the tips of CD increases in opposite direction. The sequence of selves and time-reversed selves is analogous to a cosmic expansion for CD. The repeated birth and death of mental images could correspond to this sequence at the level of sub-selves.
3. This allows to understand the relationship between subjective and geometric time and how the arrow of and flow of clock time (psychological time) emerge. The average distance between the tips of CD increases on the average as along as state function functions occur repeatedly at the fixed boundary: situation is analogous to that in diffusion. The localization of contents of conscious experience to boundary of CD gives rise to the illusion that universe is three-dimensional. The possibility of memories made possibly by hierarchy of CDs demonstrates that this is not the case. Self is simply the sequence of state function reductions at same boundary of CD remaining fixed and the lifetime of self is the total growth of the average temporal distance between the tips of CD.

One can identify several rather abstract state function reductions selecting a sector of WCW.

1. There are quantum measurements inducing localization in the moduli space of CDs with passive boundary and states at it fixed. In particular, a localization in the moduli characterizing the Lorentz transform of the upper tip of CD would be measured. The measured moduli characterize also the analog of symplectic form in M^4 strongly suggested by twistor lift of TGD—that is the rest system (time axis) and spin quantization axes. Of course, also other kinds of reductions are possible.
2. Also a localization to an extension of rationals defining the adeles should occur. Could the value of $n = h_{eff}/h$ be observable? The value of n for given space-time surface at the active boundary of CD could be identified as the order of the smallest Galois group containing all Galois groups assignable to 3-surfaces at the boundary. The superposition of space-time surface would not be eigenstate of n at active boundary unless localization occurs. It is not obvious whether this is consistent with a fixed value of n at passive boundary.

   The measured value of n could be larger or smaller than the value of n at the passive boundary of CD but in statistical sense n would increase by the analogy with diffusion on half line defined by non-negative integers. The distance from the origin unavoidably increases in statistical sense. This would imply evolution as increase of maximal value of negentropy and generation of quantum coherence in increasingly longer scales.
3. A further abstract choice corresponds to the replacement of the roles of active and passive boundary of CD changing the arrow of clock time and correspond to a death of self and re-incarnation as time-reversed self.

Can one assume that these measurements reduce to measurements of a density matrix of either entangled system as assumed in the earlier formulation of NMP, or should one allow both options. This question actually applies to all quantum measurements and leads to a fundamental philosophical questions unavoidable in all consciousness theories.

1. Do all measurements involve entanglement between the moduli or extensions of two CDs reduced in the measurement of the density matrix? Non-diagonal entanglement would allow final states, which are not eigenstates of moduli or of n: this looks strange. This could also lead to an infinite regress since it seems that one must assume endless hierarchy of entangled CDs so that the reduction sequence would proceed from top to bottom. It looks natural to regard single CD as a sub-Universe.

   For instance, if a selection of quantization axis of color hypercharge and isospin (localization in the twistor space of CP_2) is involved, one would have an outcome corresponding to a quantum superposition of measurements with different color quantization axis!

   Going philosophical, one can also argue that the measurement of density matrix is only a reaction to environment and does not allow intentional free will.
2. Can one assume that a mere localization in the moduli space or for the extension of rationals (producing an eigenstate of n) takes place for a fixed CD—a kind of self-measurement possible for even unentangled system? If there is entanglement in these degrees of freedom between two systems (say CDs), it would be reduced in these self-measurements but the outcome would not be an eigenstate of density matrix. An interpretation as a realization of intention would be appropriate.
3. If one allows both options, the interpretation would be that state function reduction as a measurement of density matrix is only a reaction to environment and self-measurement represents a realization of intention.
4. Self-measurements would occur at higher level say as a selection of quantization axis, localization in the moduli space of CD, or selection of extension of rationals. A possible general rule is that measurements at space-time level are reactions as measurements of density matrix whereas a selection of a sector of WCW would be an intentional action. This is because formally the quantum states at the level of WCW are as modes of classical WCW spinor field single particle states.
5. If the selections of sectors of WCW at active boundary of CD commute with observables, whose eigenstates appear at passive boundary (briefly *passive observables*) meaning that time reversal commutes with them—they can occur repeatedly during the reduction sequence and self as a generalized Zeno effect makes sense.

   If the selections of WCW sectors at active boundary do not commute with passive observables, then volition as a choice of sector of WCW must change the arrow of time. Libet's findings show that conscious choice induces neural activity for a fraction of second before the conscious choice. This would imply the correspondences *"big" measurement changing the arrow of time—self-measurement at the level of WCW—intentional action* and *"small" measurement—measurement at space-time level—reaction*.

Self as a generalized Zeno effect makes sense only if there are active commuting with passive observables. If the passive observables form a maximal set, the new active observables commuting with them must emerge. The increase of the size of extension of rationals might generate them by expanding the state space so that self would survive only as long at it evolves.

Otherwise there would be only single unitary time evolution followed by a reduction to opposite boundary. This makes sense only if the sequence of "big" reductions for sub-selves can give rise to the time flow experienced by self: the birth and death of mental images would give rise to flow of time of self.

A hierarchical process starting from given CD and proceeding downwards to shorter scales and stopping when the entanglement is stable is highly suggestive and favors self-measurements. What stability could mean will be discussed in the next section. CDs would be a correlate for self-hierarchy. One can say also something about the anatomy and correlates of self-hierarchy.

1. Self experiences its sub-selves as mental images and even we would represent mental images of some higher level collective self. Everything is conscious but consciousness can be lost or at least it is not possible to have memory about it. The flow of consciousness for a given self could be due to the quantum jump sequences performed by its sub-selves giving rise to mental images.
2. By quantum classical correspondence self has also space-time correlates. One can visualize sub-self as a space-time sheet "glued" by topological sum to the space-time sheet of self. Subsystem is not described as a tensor factor as in the standard description of subsystems. Also sub-selves of selves can entangle negentropically and this gives rise to a sharing of mental images about which stereo vision would be basic example. Quite generally, one could speak of stereo consciousness. Also the experiences of sensed presence [19] could be understood as a sharing of mental images between brain hemispheres, which are not themselves entangled. This is possible also between different brains. In the normal situation brain hemispheres are entangled.
3. At the level of eight-dimensional imbedding space the natural correlate of self would be CD (causal diamond). At the level of space-time the correlate would be space-time sheet or light-like 3-surface. The contents of consciousness of self would be determined by the space-time sheets in the interior of CD. Without further restrictions the experience of self would be essentially four-dimensional. Memories would be like sensory experiences except that they would be about the geometric past and for some reason are not usually colored by sensory qualia. For instance, 0.1 s time scale defining sensory chronon corresponds to the secondary p-adic time scale characterizing the size of electron's CD (Mersenne prime M_{127}), which suggests that Cooper pairs of electrons are essential for the sensory qualia.

5.2 NMP and Self

The view about Negentropy Maximization Principle (NMP) [35] has co-evolved with the notion of self and I have considered many variants of NMP.

1. The original formulation of NMP was in positive energy ontology and made same predictions as standard quantum measurement theory. The new element was that the density matrix of sub-system defines the fundamental observable and the system goes to its eigenstate in state function reduction. As found, the localizations at to WCW sectors define what might be called self-measurements and identifiable as active volitions rather than reactions.
2. In p-adic physics one can assign with rational and even algebraic entanglement probabilities number theoretical entanglement negentropy (NEN) satisfying the same basic axioms as the ordinary Shannon entropy but having negative values and therefore having interpretation as information. The definition of p-adic negentropy (real valued) reads as $S_p = -\sum P_k log(|P_k|_p)$, where $|.|_p$ denotes p-adic norm. The news is that $N_p = -S_p$ can be positive and is positive for rational entanglement probabilities. Real entanglement entropy S is always non-negative.

   NMP would force the generation of negentropic entanglement (NE) and stabilize it. NNE resources of the Universe—one might call them Akashic records—would steadily increase.
3. A decisive step of progress was the realization is that NTU forces all states in adelic physics to have entanglement coefficients in some extension of rationals inducing finite-D extension of p-adic numbers. The same entanglement can be characterized by real entropy S and p-adic negentropies N_p, which can be positive. One can define also total p-adic negentropy: $N = \sum_p N_p$ for all p and total negentropy $N_{tot} = N - S$.

   For rational entanglement probabilities it is easy to demonstrate that the generalization of adelic theorem holds true: $N_{tot} = N - S = 0$. NMP based on N_{tot} rather than N would not say anything about rational entanglement. For extensions of rationals it is easy to find that $N - S > 0$ is possible if entanglement probabilities are of form X_i/n with $|X_i|_p = 1$ and n integer [76]. Should one identify the total negentropy as difference $N_{tot} = N - S$ or as $N_{tot} = N$?

   Irrespective of answer, large p-adic negentropy seems to force large real entropy: this nicely correlates with the paradoxical finding that living systems tend to be entropic although one would expect just the oppositecite [76]: this relates in very interesting manner to the work of biologists Jeremy England [15]. The negentropy would be cognitive negentropy and not visible for ordinary physics.
4. The latest step in the evolution of ideas NMP was the question whether NMP follows from number theory alone just as second law follows form probability theory! This irritates theoretician's ego but is victory for theory. The dimension n of extension is positive integer and cannot grow in statistical sense in evolution! Since one expects that the maximal value of negentropy (define as $N - S$) must increase with n. Negentropy must increase in long run.

5.2.1 Number Theoretic Entanglement Can Be Stable

Number theoretical Shannon entropy can serve as a measure for genuine information assignable to a pair of entanglement systems [35]. Entanglement with coefficients in the extension is always negentropic if entanglement negentropy comes from p-adic sectors only. It can be negentropic if negentropy is defined as the difference of p-adic negentropy and real entropy.

The diagonalized density matrix need not belong to the algebraic extension since the probabilities defining its diagonal elements are eigenvalues of the density matrix as roots of N:th order polynomial, which in the generic case requires n-dimensional algebraic extension of rationals. One can argue that since diagonalization is not possible, also state function reduction selecting one of the eigenstates is impossible unless a phase transition increasing the dimension of algebraic extension used occurs simultaneously. This kind of NE could give rise to cognitive entanglement.

There is also a special kind of NE, which can result if one requires that density matrix serves a universal observable in state function reduction. The outcome of reduction must be an eigenspace of density matrix, which is projector to this sub-space acting as identity matrix inside it. This kind of NE allows all unitarily related basis as eigenstate basis (unitary transformations must belong to the algebraic extension). This kind of NE could serve as a correlate for "enlightened" states of consciousness. Schrödingers cat is in this kind of state stably in superposition of dead and alive and state basis obtained by unitary rotation from this basis is equally good. One can say that there are no discriminations in this state, and this is what is claimed about "enlightened" states too.

The vision about number theoretical evolution suggests that NMP forces the generation of NE resources as NE assignable to the "passive" boundary of CD for which no changes occur during sequence of state function reductions defining self. It would define the unchanging self as negentropy resources, which could be regarded as kind of Akashic records. During the next "re-incarnation" after the first reduction to opposite boundary of CD the NE associated with the reduced state would serve as new Akashic records for the time-reversed self. If NMP reduces to the statistical increase of $h_{eff}/h = n$ the consciousness information contents of the Universe increases in statistical sense. In the best possible world of SNMP it would increase steadily.

5.2.2 Does NMP Reduce to Number Theory?

The heretic question that emerged quite recently is whether NMP is actually needed at all! Is NMP a separate principle or could NMP reduced to mere number theory [35]? Consider first the possibility that NMP is not needed at all as a separate principle.

1. The value of $h_{eff}/h = n$ should increase in the evolution by the phase transitions increasing the dimension of the extension of rationals. $h_{eff}/h = n$ has been

identified as the number of sheets of some kind of covering space. The Galois group of extension acts on number theoretic discretizations of the monadic surface and the orbit defines a covering space. Suppose n is the number of sheets of this covering and thus the dimension of the Galois group for the extension of rationals or factor of it.

2. It has been already noticed that the "big" state function reductions giving rise to death and re-incarnation of self could correspond to a measurement of $n = h_{eff}$ implied by the measurement of the extension of the rationals defining the adeles. The statistical increase of n follows automatically and implies statistical increase of maximal entanglement negentropy. Entanglement negentropy increases in statistical sense.

   The resulting world would not be the best possible one unlike for a strong form of NMP demanding that negentropy does increase in "big" state function reductions. n also decreases temporarily and they seem to be needed. In TGD inspired model of bio-catalysis the phase transition reducing the value of n for the magnetic flux tubes connecting reacting bio-molecules allows them to find each other in the molecular soup. This would be crucial for understanding processes like DNA replication and transcription.
3. State function reduction corresponding to the measurement of density matrix could occur to an eigenstate/eigenspace of density matrix only if the corresponding eigenvalue and eigenstate/eigenspace is expressible using numbers in the extension of rationals defining the adele considered. In the generic case these numbers belong to N-dimensional extension of the original extension. This can make the entanglement stable with respect to state the measurements of density matrix.

   A phase transition to an extension of an extension containing these coefficients would be required to make possible reduction. A step in number theoretic evolution would occur. Also an entanglement of measured state pairs with those of measuring system in containing the extension of extension would make possible the reduction. Negentropy could be reduced but higher-D extension would provide potential for more negentropic entanglement and NMP would hold true in the statistical sense.
4. If one has higher-D eigenspace of density matrix, p-adic negentropy is largest for the entire sub-space and the sum of real and p-adic negentropies vanishes for all of them. For negentropy identified as total p-adic negentropy SNMP would select the entire sub-space and NMP would indeed say something explicit about negentropy.

5.2.3 Or Is NMP Needed as a Separate Principle?

Hitherto I have postulated NMP as a separate principle [35]. Strong form of NMP (SNMP) states that Negentropy does not decrease in "big" state function reductions corresponding to death and re-incarnations of self.

One can, however, argue that SNMP is not realistic. SNMP would force the Universe to be the best possible one, and this does not seem to be the case. Also ethically responsible free will would be very restricted since self would be forced always to do the best deed that is increase maximally the negentropy serving as information resources of the Universe. Giving up separate NMP altogether would allow to have also "Good" and "Evil."

This forces to consider what I christened weak form of NMP (WNMP). Instead of maximal dimension corresponding to N-dimensional projector self can choose also lower-dimensional sub-spaces and 1-D sub-space corresponds to the vanishing entanglement and negentropy assumed in standard quantum measurement theory. As a matter of fact, this can also lead to larger negentropy gain since negentropy depends strongly on what is the large power of p in the dimension of the resulting eigen sub-space of density matrix. This could apply also to the purely number theoretical reduction of NMP.

WNMP suggests how to understand the notions of Good and Evil. Various choices in the state function reduction would correspond to Boolean algebra, which suggests an interpretation in terms of what might be called emotional intelligence [51]. Also it turns out that one can understand how p-adic length scale hypothesis—actually its generalization—emerges from WNMP [66].

1. One can start from ordinary quantum entanglement. It corresponds to a superposition of pairs of states. Second state corresponds to the internal state of the self and second state to a state of external world or biological body of self. In negentropic quantum entanglement each is replaced with a pair of sub-spaces of state spaces of self and external world. The dimension of the sub-space depends on which pair is in question. In state function reduction one of these pairs is selected and deed is done. How to make some of these deeds good and some bad? Recall that WNMP allows only the possibility to generate NNE but does not force it. WNMP would be like God allowing the possibility to do good but not forcing good deeds.

   Self can choose any sub-space of the sub-space defined by $k \leq N$-dimensional projector and 1-D sub-space corresponds to the standard quantum measurement. For $k = 1$ the state function reduction leads to vanishing negentropy, and separation of self and the target of the action. Negentropy does not increase in this action and self is isolated from the target: kind of price for sin.

   For the maximal dimension of this sub-space the negentropy gain is maximal. This deed would be good and by the proposed criterion NE corresponds to conscious experience with positive emotional coloring. Interestingly, there are $2^k - 1$ possible choices, which is almost the dimension of Boolean algebra consisting of k independent bits. The excluded option corresponds to 0-dimensional sub-space—empty set in set theoretic realization of Boolean algebra. This could relate directly to fermionic oscillator operators defining basis of Boolean algebra—here Fock vacuum would be the excluded state. The deed in this sense would be a choice of how loving the attention towards system of external world is.

2. A map of different choices of k-dimensional sub-spaces to k-fermion states is suggestive. The realization of logic in terms of emotions of different degrees of positivity would be mapped to many-fermion states—perhaps zero energy states with vanishing total fermion number. State function reductions to k-dimensional spaces would be mapped to k-fermion states: quantum jumps to quantum states!

   The problem brings in mind quantum classical correspondence in quantum measurement theory. The direction of the pointer of the measurement apparatus (in very metaphorical sense) corresponds to the outcome of state function reduction, which is now 1-D sub-space. For ordinary measurement the pointer has k positions. Now it must have $2^k - 1$ positions. To the discrete space of k pointer positions one must assign fermionic Clifford algebra of second quantized fermionic oscillator operators. The hierarchy of Planck constants and dark matter suggests the realization. Replace the pointer with its space-time k-sheeted covering and consider zero energy states made of pairs of k-fermion states at the sheets of the n-sheeted covering? Dark matter would be therefore necessary for cognition. The role of fermions would be to "mark" the k space-time sheets in the covering.

The cautious conclusion is that NMP as a separate principle is not necessary and follows in statistical sense from the unavoidable increase of $n = h_{eff}/h$ identified as dimension of extension of rationals define the adeles if this extension or at least the dimension of its Galois group is observable.

5.3 p-Adic Physics as Correlate of Cognition and Imagination

The items in the following list give motivations for the proposal that p-adic physics could serve as a correlate for cognition and imagination.

1. By the total disconnectedness of the p-adic topology, p-adic world decomposes naturally into blobs, objects. This happens also in sensory perception. The pinary digits of p-adic number can be assigned to a p-tree. Parisi proposed in the model of spin glass [12] that p-adic numbers could relate to the mathematical description of cognition and also Khrennikov [18] has developed this idea. In TGD framework that idea is taken to space-time level: p-adic space-time sheets represent thought bubbles and they correlate with the real ones since they form cognitive representations of the real world. SH allows a concrete realization of this.
2. p-Adic non-determinism due to p-adic pseudo-constants suggests interpretation in terms of imagination. Given 2-surfaces could allow completion to p-adic preferred extremal but not to a real one so that pure "non-realizable" imagination is in question.
3. Number theoretic negentropy has interpretation as negentropy characterizing information content of entanglement. The superposition of state pairs could be interpreted as a quantum representation for a rule or abstracted association

containing its instances as state pairs. Number theoretical negentropy characterizes the relationship of two systems and should not be confused with thermodynamical entropy, which characterizes the uncertainty about the state of single system.

The original vision was that p-adic non-determinism could serve as a correlate for cognition, imagination, and intention. The recent view is much more cautious. Imagination need not completely reduce to p-adic non-determinism since it has also real physics correlates—maybe as partial realizations of SH as in nerve pulse pattern, which does not propagate down to muscles.

A possible interpretation for the solutions of the p-adic field equations would be as geometric correlates of cognition, imagination, and perhaps even intentionality. Plans, intentions, expectations, dreams, and possibly also cognition as imagination in general could have p-adic cognitive space-time sheets as their geometric correlates. A deep principle seems to be involved: incompleteness is the characteristic feature of p-adic physics but the flexibility made possible by this incompleteness is absolutely essential for imagination and cognitive consciousness in general.

The most feasible view is that the intersections of p-adic and real space-time surfaces define cognitive representations of real space-time surfaces (PEs, [25, 67, 70]). One could also say that real space-time surface represents sensory aspects of conscious experience and p-adic space-time surfaces its cognitive aspects. Both real and p-adics rather than real or p-adics.

The identification of p-adic pseudo-constants as correlates of imagination at space-time level is indeed a further natural idea.

1. The construction of PEs by SH from the data at 2-surfaces is like boundary value problem with number theoretic discretization of space-time surface as additional data. PE property in real context implies strong correlations between string world sheets and partonic 2-surfaces by boundary conditions a them. One cannot choose these 2-surfaces completely independently in real context.
2. In p-adic sectors the integration constants are replaced with pseudo-constants depending on finite number of pinary digits of variables depending on coordinates normal to string world sheets and partonic 2-surfaces. The fixing of the discretization of space-time surface would allow to fix the p-adic pseudo-constants. Once the number theoretic discretization of space-time surface is fixed, the p-adic pseudo-constants can be fixed. Pseudo-constant could allow a large number of p-adic configurations involving string world sheets, partonic 2-surfaces, and number theoretic discretization but not allowed in real context.

   Could these p-adic PEs correspond to imaginations, which in general are not realizable? Could the realizable intentional actions belong to the intersection of real and p-adic WCWs? Could one identify non-realistic imaginations as the modes of WCW spinor fields for which 2-surfaces are not extendable to real space-time surfaces and are localized to 2-surfaces? Could they allow only a partial continuation to real space-time surface. Could nerve pulse pattern representing imagined motor action and not proceeding to the level of muscles correspond to a partially real PE?

Could imagination and problem solving be search for those collections of string world sheets and partonic 2-surfaces, which allow extension to (realization as) real PEs? If so, p-adic physics would be there as an independent aspect of existence and this is just the original idea. Imagination could be realized in state function reduction, which always selects only those 2-surfaces, which allow continuation to real space-time surfaces. The distinction between only imaginable and also realizable would be the extendability by using strong form of holography.

3. An interesting question is why elementary particles are characterized by preferred p-adic primes (primes near powers of 2, in particular Mersenne primes). Could the number of realizable imaginations for these primes be especially large?

I have the feeling that this view allows respectable mathematical realization of imagination in terms of adelic quantum physics. It is remarkable that SH derivable from—you can guess, SGCI (the Big E again!), plays an absolutely central role in it.

Appendix: Super-Symplectic Conformal Weights and Zeros of Riemann Zeta

Since fermions are the only fundamental particles in TGD one could argue that the conformal weight of the generating elements of supersymplectic algebra could be negatives for the poles of fermionic zeta ζ_F. This demands $n > 0$ as does also the fractal hierarchy of super-symplectic symmetry breakings. NTU of Riemann zeta in some sense is strongly suggested if adelic physics is to make sense.

For ordinary conformal algebras there are only finite number of generating elements ($-2 \leq n \leq 2$). If the radial conformal weights for the generators of g consist of poles of ζ_F, the situation changes. ζ_F is suggested by the observation that fermions are the only fundamental particles in TGD.

1. Riemann Zeta $\zeta(s) = \prod_p (1/(1-p^{-s})$ identifiable formally as a partition function $\zeta_B(s)$ of arithmetic boson gas with bosons with energy $log(p)$ and temperature $1/s = 1/(1/2 + iy)$ should be replaced with that of arithmetic fermionic gas given in the product representation by $\zeta_F(s) = \prod_p (1 + p^{-s})$ so that the identity $\zeta_B(s))/\zeta_F(s) = \zeta_B(2s)$ follows. This gives

$$\frac{\zeta_B(s)}{\zeta_B(2s)} \ .$$

$\zeta_F(s)$ has zeros at zeros s_n of $\zeta(s)$ and at the pole $s = 1/2$ of $zeta(2s)$. $\zeta_F(s)$ has poles at zeros $s_n/2$ of $\zeta(2s)$ and at pole $s = 1$ of $\zeta(s)$.

The spectrum of $1/T$ would be for the generators of algebra $\{(-1/2 + iy)/2, n > 0, -1\}$. In p-adic thermodynamics the p-adic temperature is $1/T = 1/n$ and corresponds to "trivial" poles of ζ_F. Complex values of temperature do not make sense in ordinary thermodynamics. In ZEO quantum theory can be regarded as a square root of thermodynamics and complex temperature parameter makes sense.

2. If the spectrum of conformal weights of the generating elements of the algebra corresponds to poles serving as analogs of propagator poles, it consists of the "trivial" conformal $h = n > 0$- the standard spectrum with $h = 0$ assignable to massless particles excluded—and "non-trivial" $h = -1/4 + iy/2$. There is also a pole at $h = -1$.

   Both the non-trivial pole with real part $h_R = -1/4$ and the pole $h = -1$ correspond to tachyons. I have earlier proposed conformal confinement meaning that the total conformal weight for the state is real. If so, one obtains for a conformally confined two-particle states corresponding to conjugate non-trivial zeros in minimal situation $h_R = -1/2$ assignable to N-S representation.

   In p-adic mass calculations ground state conformal weight must be $-5/2$ [34]. The negative fermion ground state weight could explain why the ground state conformal weight must be tachyonic $-5/2$. With the required five tensor factors one would indeed obtain this with minimal conformal confinement. In fact, arbitrarily large tachyonic conformal weight is possible but physical state should always have conformal weights $h > 0$.
3. $h = 0$ is not possible for generators, which reminds of Higgs mechanism for which the naive ground states correspond to tachyonic Higgs. $h = 0$ conformally confined massless states are necessarily composites obtained by applying the generators of Kac-Moody algebra or super-symplectic algebra to the ground state. This is the case according to p-adic mass calculations [34], and would suggest that the negative ground state conformal weight can be associated with super-symplectic algebra and the remaining contribution comes from ordinary super-conformal generators. Hadronic masses, whose origin is poorly understood, could come from super-symplectic degrees of freedom. There is no need for p-adic thermodynamics in super-symplectic degrees of freedom.

A General Formula for the Zeros of Zeta from NTU

Dyson's comment about Fourier transform of Riemann Zeta [1] (http://tinyurl.com/hjbfsuv) is interesting from the point of NTU for Riemann zeta.

1. The numerical calculation of Fourier transform for the imaginary parts iy of zeros $s = 1/2 + iy$ of zeta shows that it is concentrated at discrete set of frequencies coming as $log(p^n)$, p prime. This translates to the statement that the zeros of zeta form a 1-dimensional quasicrystal, a discrete structure Fourier spectrum by definition is also discrete (this of course holds for ordinary crystals as a special case). Also the logarithms of powers of primes would form a quasicrystal, which is very interesting from the point of view of p-adic length scale hypothesis. Primes label the "energies" of elementary fermions and bosons in arithmetic number theory, whose repeated second quantization gives rise to the hierarchy of infinite primes [46]. The energies for general states are logarithms of integers.

2. Powers p^n label the points of quasicrystal defined by points $log(p^n)$ and Riemann zeta has interpretation as partition function for boson case with this spectrum. Could p^n label also the points of the dual lattice defined by iy.
3. The existence of Fourier transform for points $log(p_i^n)$ for any vector y_a in class $C(p)$ of zeros labelled by p requires $p_i^{iy_a}$ to be a root of unity inside $C(p)$. This could define the sense in which zeros of zeta are universal. This condition also guarantees that the factor $n^{-1/2-iy}$ appearing in zeta at critical line are number theoretically universal ($p^{1/2}$ is problematic for Q_p: the problem might be solved by eliminating from p-adic analog of zeta the factor $1-p^{-s}$).
   (a) One obtains for the pair (p_i, s_a) the condition $log(p_i)y_a = q_{ia}2\pi$, where q_{ia} is a rational number. Dividing the conditions for (i, a) and (j, a) gives

   $$p_i = p_j^{q_{ia}/q_{ja}}$$

   for every zero s_a so that the ratios q_{ia}/q_{ja} do not depend on s_a. From this one easily deduce $p_i^M = p_j^N$, where M and N are integers so that one ends up with a contradiction.
   (b) Dividing the conditions for (i, a) and (i, b) one obtains

   $$\frac{y_a}{y_b} = \frac{q_{ia}}{q_{ib}}$$

   so that the ratios q_{ia}/q_{ib} do not depend on p_i. The ratios of the imaginary parts of zeta would be therefore rational number which is very strong prediction and zeros could be mapped by scaling y_a/y_1 where y_1 is the zero which smallest imaginary part to rationals.
   (c) The impossible consistency conditions for (i, a) and (j, a) can be avoided if each prime and its powers correspond to its own subset of zeros and these subsets of zeros are disjoint: one would have infinite union of sub-quasicrystals labelled by primes and each p-adic number field would correspond to its own subset of zeros: this might be seen as an abstract analog for the decomposition of rational to powers of primes. This decomposition would be natural if for ordinary complex numbers the contribution in the complement of this set to the Fourier transform vanishes. The conditions $(i, a)$and (i, b) require now that the ratios of zeros are rationals only in the subset associated with p_i.

For the general option the Fourier transform can be delta function for $x = log(p^k)$ and the set $\{y_a(p)\}$ contains N_p zeros. The following argument inspires the conjecture that for each p there is an infinite number N_p of zeros $y_a(p)$ in class $C(p)$ satisfying

$$p^{iy_a(p)} = u(p) = e^{\frac{r(p)}{m(p)}i2\pi} ,$$

where $u(p)$ is a root of unity that is $y_a(p) = 2\pi(m(a) + r(p))/log(p)$ and forming a subset of a lattice with a lattice constant $y_0 = 2\pi/log(p)$, which itself need not be a zero.

In terms of stationary phase approximation the zeros $y_a(p)$ associated with p would have constant stationary phase whereas for $y_a(p_i \neq p)$ the phase $p^{iy_a(p_i)}$ would fail to be stationary. The phase e^{ixy} would be non-stationary also for $x \neq log(p^k)$ as function of y.

1. Assume that for $x = qlog(p)$, where q not a rational, the phases e^{ixy} fail to be roots of unity and are random implying the vanishing/smallness of $F(x)$.
2. Assume that for a given p all powers p^{iy} for $y \notin \{y_a(p)\}$ fail to be roots of unity and are also random so that the contribution of the set $y \notin \{y_a(p)\}$ to $F(p)$ vanishes/is small.
3. For $x = log(p^{k/m})$ the Fourier transform should vanish or be small for $m \neq 1$ (rational roots of primes) and give a non-vanishing contribution for $m = 1$. One has
$$\begin{aligned} F(x = log(p^{k/m}) &= \sum_{1 \le a \le N(p)} e^{k\frac{M(a,p)}{mN(p)} i2\pi} u(p) \ , \\ u(p) &= e^{\frac{r(p)}{m(p)} i2\pi} \ . \end{aligned}$$
Obviously one can always choose $N(a, p) = N(p)$.
4. For the simplest option $N(p) = 1$ one would obtain delta function distribution for $x = log(p^k)$. The sum of the phases associated with $y_a(p)$ and $-y_a(p)$ from the half axes of the critical line would give
$$F(x = log(p^n)) \propto X(p^n) \equiv 2cos(n\frac{r(p)}{m(p)}2\pi) \ .$$
The sign of F would vary.
5. For $x = log(p^{k/m})$ the value of Fourier transform is expected to be small by interference effects if $M(a, p)$ is random integer, and negligible as compared with the value at $x = log(p^k)$. This option is highly attractive. For $N(p) > 1$ and $M(a, p)$ a random integer also $F(x = log(p^k)$ is small by interference effects. Hence it seems that this option is the most natural one.
6. The rational $r(p)/m(p)$ would characterize given prime (one can require that $r(p)$ and $m(p)$ have no common divisors). $F(x)$ is non-vanishing for all powers $x = log(p^n)$ for $m(p)$ odd. For $p = 2$, also $m(2) = 2$ allows to have $|X(2^n)| = 2$. An interesting ad hoc ansatz is $m(p) = p$ or $p^{s(p)}$. One has periodicity in n with period $m(p)$ that is logarithmic wave. This periodicity serves as a test and in principle allows to deduce the value of $r(p)/m(p)$ from the Fourier transform.

What could one conclude from the data (http://tinyurl.com/hjbfsuv)?

1. The first graph gives $|F(x = log(p^k)|$ and second graph displays a zoomed up part of $|F(x = log(p^k)|$ for small powers of primes in the range $[2, 19]$. For the first graph the eighth peak ($p = 11$) is the largest one but in the zoomed graphs this is not the case. Hence something is wrong or the graphs correspond to different approximations suggesting that one should not take them too seriously.

In any case, the modulus is not constant as function of p^k. For small values of p^k the envelope of the curve decreases and seems to approach constant for large values of p^k (one has $x < 15$ ($e^{15} \simeq 3.3 \times 10^6$).

2. According to the first graph $|F(x)|$ decreases for $x = klog(p) < 8$, is largest for small primes, and remains below a fixed maximum for $8 < x < 15$. According to the second graph the amplitude decreases for powers of a given prime (say $p = 2$). Clearly, the small primes and their powers have much larger $|F(x)|$ than large primes.

There are many possible reasons for this behavior. Most plausible reason is that the sums involved converge slowly and the approximation used is not good. The inclusion of only 10^4 zeros would show the positions of peaks but would not allow reliable estimate for their intensities.

1. The distribution of zeros could be such that for small primes and their powers the number of zeros is large in the set of 10^4 zeros considered. This would be the case if the distribution of zeros $y_a(p)$ is fractal and gets "thinner" with p so that the number of contributing zeros scales down with p as a power of p, say $1/p$, as suggested by the envelope in the first figure.
2. The infinite sum, which should vanish, converges only very slowly to zero. Consider the contribution $\Delta F(p^k, p_1)$ of zeros not belonging to the class $p_1 \neq p$ to $F(x = log(p^k)) = \sum_{p_i} \Delta F(p^k, p_i)$, which includes also $p_i = p$. $\Delta F(p^k, p_i)$, $p \neq p_1$ should vanish in exact calculation.

   (a) By the proposed hypothesis this contribution reads as

$$\Delta F(p, p_1) = \sum_a cos \left[X(p^k, p_1)(M(a, p_1) + \tfrac{r(p_1)}{m(p_1)})2\pi) \right] \ .$$
$$X(p^k, p_1) = \frac{log(p^k)}{log(p_1)} \ .$$

   Here a labels the zeros associated with p_1. If p^k is "approximately divisible" by p^1 in other words, $p^k \simeq np_1$, the sum over finite number of terms gives a large contribution since interference effects are small, and a large number of terms are needed to give a nearly vanishing contribution suggested by the non-stationarity of the phase. This happens in several situations.

   (b) The number $\pi(x)$ of primes smaller than x goes asymptotically like $\pi(x) \simeq x/log(x)$ and prime density approximately like $1/log(x) - 1/log(x)^2$ so that the problem is worst for the small primes. The problematic situation is encountered most often for powers p^k of small primes p near larger prime and primes p (also large) near a power of small prime (the envelope of $|F(x)|$ seems to become constant above $x \sim 10^3$).

   (c) The worst situation is encountered for $p = 2$ and $p_1 = 2^k - 1$—a Mersenne prime and $p_1 = 2^{2^k} + 1$, $k \leq 4$—Fermat prime. For $(p, p_1) = (2^k, M_k)$ one encounters $X(2^k, M_k) = (log(2^k)/log(2^k - 1)$ factor very near to unity for large Mersennes primes. For $(p, p_1) = (M_k, 2)$ one encounters $X(M_k, 2) = (log(2^k - 1)/log(2) \simeq k$. Examples of Mersennes and Fermats are $(3, 2), (5, 2), (7, 2), (17, 2), (31, 2), (127, 2), (257, 2), \ldots$ Powers 2^k, $k = 2, 3, 4, 5, 7, 8, \ldots$ are also problematic.

(d) Also twin primes are problematic since in this case one has factor $X(p = p_1 + 2, p_1) = \frac{log(p_1+2)}{log(p_1)}$. The region of small primes contains many twin prime pairs: (3,5), (5,7), (11,13), (17,19), (29,31),....

These observations suggest that the problems might be understood as resulting from including too small number of zeros.

3. The predicted periodicity of the distribution with respect to the exponent k of p^k is not consistent with the graph for small values of prime unless the periodic $m(p)$ for small primes is large enough. The above-mentioned effects can quite well mask the periodicity. If the first graph is taken at face value for small primes, $r(p)/m(p)$ is near zero, and $m(p)$ is so large that the periodicity does not become manifest for small primes. For $p = 2$ this would require $m(2) > 21$ since the largest power $2^n \simeq e^{15}$ corresponds to $n \sim 21$.

To summarize, the prediction is that for zeros of zeta should divide into disjoint classes $\{y_a(p)\}$ labelled by primes such that within the class labelled by p one has $p^{iy_a(p)} = e^{(r(p)/m(p))i2\pi}$ so that has $y_a(p) = [M(a,p) + r(p)/m(p))]2\pi/log(p)$.

More Precise View About Zeros of Zeta

There is a very interesting blog post by Mumford (http://tinyurl.com/zemw27o), which leads to much more precise formulation of the idea and improved view about the Fourier transform hypothesis: the Fourier transform or its generalization must be defined for all zeros, not only the non-trivial ones and trivial zeros give a background term allowing to understand better the properties of the Fourier transform.

Mumford essentially begins from Riemann's "explicit formula" in von Mangoldt's form.

$$\sum_p \sum_{n\geq 1} log(p)\delta_{p^n}(x) = 1 - \sum_k x^{s_k - 1} - \frac{1}{x(x^2-1)} ,$$

where p denotes prime and s_k a non-trivial zero of zeta. The left-hand side represents the distribution associated with powers of primes. The right-hand side contains sum over cosines

$$\sum_k x^{s_k-1} = 2\frac{\sum_k cos(log(x)y_k)}{x^{1/2}} ,$$

where y_k is the imaginary part of non-trivial zero. Apart from the factor $x^{-1/2}$ this is just the Fourier transform over the distribution of zeros.

There is also a slowly varying term $1 - \frac{1}{x(x^2-1)}$, which has interpretation as the analog of the Fourier transform term but sum over trivial zeros of zeta at $s = -2n$, $n > 0$. The entire expression is analogous to a "Fourier transform" over the distribution of all zeros. Quasicrystal is replaced with union on 1-D quasicrystals.

Therefore the distribution for powers of primes is expressible as "Fourier transform" over the distribution of both trivial and non-trivial zeros rather than only non-trivial zeros as suggested by numerical data to which Dyson [1] referred to (http://tinyurl.com/hjbfsuv). Trivial zeros give a slowly varying background term large for small values of argument x (poles at $x = 0$ and $x = 1$—note that also $p = 0$ and $p = 1$ appear effectively as primes) so that the peaks of the distribution are higher for small primes.

The question was how can one obtain this kind of delta function distribution concentrated on powers of primes from a sum over terms $cos(log(x)y_k)$ appearing in the Fourier transform of the distribution of zeros.

Consider $x = p^n$. One must get a constructive interference. Stationary phase approximation is in terms of which physicist thinks. The argument was that a destructive interference occurs for given $x = p^n$ for those zeros for which the cosine does not correspond to a real part of root of unity as one sums over such y_k: random phase approximation gives more or less zero. To get something non-trivial y_k must be proportional to $2\pi \times n(y_k)/log(p)$ in class $C(p)$ to which y_k belongs. If the number of these y_k:s in $C(p)$ is infinite, one obtains delta function in good approximation by destructive interference for other values of argument x.

The guess that the number of zeros in $C(p)$ is infinite is encouraged by the behaviors of the densities of primes one hand and zeros of zeta on the other hand. The number of primes smaller than real number x goes like

$$\pi(x) = \#(primes < x) \sim \frac{x}{log(x)}$$

in the sense of distribution. The number of zeros along critical line goes like

$$\#(zeros < t) = (t/2\pi) \times log(\frac{t}{2\pi})$$

in the same sense. If the real axis and critical line have same metric measure, then one can say that the number of zeros in interval T per number of primes in interval T behaves roughly like

$$\frac{\#(zeros < T)}{\#(primes < T)} = log(\frac{T}{2\pi}) \times \frac{log(T)}{2\pi}$$

so that at the limit of $T \to \infty$ the number of zeros associated with given prime is infinite. This assumption of course makes the argument a poor man's argument only.

Possible Relevance for TGD

What this speculative picture from the point of view of TGD?

1. A possible formulation for NTU for the poles of fermionic Riemann zeta $\zeta_F = \zeta(s)/\zeta(2s)$ could be as a condition that is that the exponents $p^{ks_a(p)/2} = p^{k/4}p^{iky_a(p)/2}$ exist in a number theoretically universal manner for the zeros $s_a(p)$ for given p-adic prime p and for some subset of integers k. If the proposed conditions hold true, exponent reduces $p^{k/4}e^{k(r(p/m(p)i2\pi}$ requiring that k is a multiple of 4. The number of the non-trivial generating elements of super-symplectic algebra in the monomial creating physical state would be a multiple of 4. These monomials would have real part of conformal weight -1. Conformal confinement suggests that these monomials are products of pairs of generators for which imaginary parts cancel.
2. Quasicrystal property might have an application to TGD. The functions of light-like radial coordinate appearing in the generators of super-symplectic algebra could be of form r^s, s zero of zeta or rather, its imaginary part. The eigenstate property with respect to the radial scaling rd/dr is natural by radial conformal invariance.

   The idea that arithmetic QFT assignable to infinite primes is behind the scenes in turn suggests light-like momenta assignable to the radial coordinate have energies with the dual spectrum $log(p^n)$. This is also suggested by the interpretation of ζ as square root of thermodynamical partition function for boson gas with momentum $log(p)$ and analogous interpretation of ζ_F.

   The two spectra would be associated with radial scalings and with light-like translations of light-cone boundary respecting the direction and light-likeness of the light-like radial vector. $log(p^n)$ spectrum would be associated with light-like momenta whereas p-adic mass scales would characterize states with thermal mass. Note that generalization of p-adic length scale hypothesis raises the scales defined by p^n to a special physical position: this might relate to ideal structure of adeles.
3. Finite measurement resolution suggests that the approximations of Fourier transforms over the distribution of zeros taking into account only a finite number of zeros might have a physical meaning. This might provide additional understand about the origins of generalized p-adic length scale hypothesis stating that primes $p \simeq p_1^k$, p_1 small prime—say Mersenne primes—have a special physical role.

References

Mathematics

1. Baez, J.: Quasicrystals and the Riemann Hypothesis. The n-Category Cafe. Available at: https://golem.ph.utexas.edu/category/2013/06/quasicrystals_and_the_riemann.html (2013)
2. Freed, D.S.: The Geometry of Loop Groups (1985)
3. Hitchin, N.: Kählerian twistor spaces. Proc. Lond. Math. Soc. **8**(43), 133–151 (1981). Available at: http://tinyurl.com/pb8zpqo
4. Reid, M.: McKay correspondence. Available at: http://arxiv.org/abs/alg-geom/9702016

Physics

5. Arkani-Hamed, N., et al.: Scattering amplitudes and the positive Grassmannian. Available at: http://arxiv.org/pdf/1212.5605v1.pdf
6. Golovko, V.A.: Why does superfluid helium leak out of an open container? Available at: http://arxiv.org/pdf/1103.0517.pdf (2012)
7. Huang, Y.-T., Elvang, H.: Scattering amplitudes. Available at: http://arxiv.org/pdf/1308.1697v1.pdf (2013)
8. Klebanov, I.R.: TASI Lectures: Introduction to the AdS/CFT Correspondence. Available at: http://arxiv.org/abs/hep-th/0009139 (2000)
9. Lamb Shift: Available at: http://en.wikipedia.org/wiki/Lamb_shift.
10. Matos de, C.J., Tajmar, M.: Local Photon and Graviton Mass and Its Consequences. Available at: http://arxiv.org/abs/gr-gc0603032 (2006)
11. Nottale, L., Da Rocha, D.: Gravitational Structure Formation in Scale Relativity. Available at: http://arxiv.org/abs/astro-ph/0310036 (2003)
12. Parisi, G.: Field Theory, Disorder and Simulations. World Scientific, Singapore (1992)
13. Penrose, R.: The central programme of twistor theory. Chaos Solitons Fractals **10**, 581–611 (1999)
14. Tajmar, M., et al.: Experimental Detection of Gravimagnetic London Moment. Available at: http://arxiv.org/abs/gr-gc0603033 (2006)

Biology and Neuroscience

15. England, J., Perunov, N., Marsland, R.: Statistical Physics of Adaptation. Available at: http://arxiv.org/pdf/1412.1875v1.pdf (2014)
16. Fiaxat, J.D.: A hypothesis on the rhythm of becoming. World Futur. **36**, 31–36 (1993)
17. Fiaxat, J.D.: The hidden rhythm of evolution. Available at: http://byebyedarwin.blogspot.fi/p/english-version_01.html (2014)
18. Khrennikov, A.Y.: Classical and quantum mechanics on information spaces with applications to cognitive, psychological, social, and anomalous phenomena. Found. Phys. **29**, 1065–2098 (1999)
19. Lavallee, F.C., Persinger, M.A.: Theoretical and experimental evidence of macroscopic entanglement between human brain activity and photon emissions: implications for quantum consciousness and future applications. J. Conscious. Explor. Res. **1**(7), 785–807 (2010)

Books Related to TGD

20. Pitkänen, M.: Topological Geometrodynamics (1983)
21. Pitkänen, M.: Basic properties of CP_2 and elementary facts about p-adic numbers. In: Towards M-matrix. In online book. Available at: http://tgdtheory.fi/public_html/pdfpool/append.pdf (2006)
22. Pitkänen, M.: Is it possible to understand coupling constant evolution at space-time level? In: Towards M-Matrix. In online book. Available at: http://tgdtheory.fi/public_html/tgdquantum/tgdquantum.html#rgflow (2006)
23. Pitkänen, M.: About nature of time. In: TGD Inspired Theory of Consciousness. In online book. Available at: http://tgdtheory.fi/public_html/tgdconsc/tgdconsc.html#timenature (2006)

24. Pitkänen, M.: About the new physics behind qualia. In: Quantum Hardware of Living Matter. In online book. Available at: http://tgdtheory.fi/public_html/bioware/bioware.html#newphys (2006)
25. Pitkänen, M.: Basic extremals of Kähler action. In: Physics in Many-Sheeted Space-Time. In online book. Available at: http://tgdtheory.fi/public_html/tgdclass/tgdclass.html#class (2006)
26. Pitkänen, M.: Conscious information and intelligence. In: TGD Inspired Theory of Consciousness. In online book. Available at: http://tgdtheory.fi/public_html/tgdconsc/tgdconsc.html#intsysc (2006)
27. Pitkänen, M.: Construction of elementary particle vacuum functionals. In: p-Adic Physics. In online book. Available at: http://tgdtheory.fi/public_html/padphys/padphys.html#elvafu (2006)
28. Pitkänen, M.: Construction of WCW Kähler geometry from symmetry principles. In: Quantum Physics as Infinite-Dimensional Geometry. In online book. Available at: http://tgdtheory.fi/public_html/tgdgeom/tgdgeom.html#compl1 (2006)
29. Pitkänen, M.: Cosmic strings. In: Physics in Many-Sheeted Space-Time. In online book. Available at: http://tgdtheory.fi/public_html/tgdclass/tgdclass.html#cstrings (2006)
30. Pitkänen, M.: Does Riemann zeta code for generic coupling constant evolution? In: Towards M-Matrix. In online book. Available at: http://tgdtheory.fi/public_html/tgdquantum/tgdquantum.html#fermizeta (2006)
31. Pitkänen, M.: Does TGD predict the spectrum of Planck constants? In: Hyper-Finite Factors and Dark Matter Hierarchy. In online book. Available at: http://tgdtheory.fi/public_html/neuplanck/neuplanck.html#Planck (2006)
32. Pitkänen, M.: Fusion of p-adic and real variants of quantum TGD to a more general theory. In: TGD as a Generalized Number Theory. In online book. Available at: http://tgdtheory.fi/public_html/tgdnumber/tgdnumber.html#mblocks (2006)
33. Pitkänen, M.: Identification of the WCW Kähler function. In: Quantum Physics as Infinite-Dimensional Geometry. In online book. Available at: http://tgdtheory.fi/public_html/tgdgeom/tgdgeom.html#kahler (2006)
34. Pitkänen, M.: Massless states and particle massivation. In: p-Adic Physics. In online book. Available at: http://tgdtheory.fi/public_html/padphys/padphys.html#mless (2006)
35. Pitkänen, M.: Negentropy maximization principle. In: TGD Inspired Theory of Consciousness. In online book. Available at: http://tgdtheory.fi/public_html/tgdconsc/tgdconsc.html#nmpc (2006)
36. Pitkänen, M.: New particle physics predicted by TGD: part I. In: p-Adic Physics. In online book. Available at: http://tgdtheory.fi/public_html/padphys/padphys.html#mass4 (2006)
37. Pitkänen, M.: New particle physics predicted by TGD: part II. In: p-Adic Physics. In online book. Available at: http://tgdtheory.fi/public_html/padphys/padphys.html#mass5 (2006)
38. Pitkänen, M.: Overall view about TGD from particle physics perspective. In: p-Adic Physics. In online book. Available at: http://tgdtheory.fi/public_html/padphys/padphys.html#TGDoverall (2006)
39. Pitkänen, M.: p-Adic numbers and generalization of number concept. In: TGD as a Generalized Number Theory. In online book. Available at: http://tgdtheory.fi/public_html/tgdnumber/tgdnumber.html#padmat (2006)
40. Pitkänen, M.: p-Adic physics as physics of cognition and intention. In: TGD Inspired Theory of Consciousness. In online book. Available at: http://tgdtheory.fi/public_html/tgdconsc/tgdconsc.html#cognic (2006)
41. Pitkänen, M.: Quantum antenna hypothesis. In: Quantum Hardware of Living Matter. In online book. Available at: http://tgdtheory.fi/public_html/bioware/bioware.html#tubuc (2006)
42. Pitkänen, M.: Quantum astrophysics. In: Physics in Many-Sheeted Space-Time. In online book. Available at: http://tgdtheory.fi/public_html/tgdclass/tgdclass.html#qastro (2006)
43. Pitkänen, M.: Quantum model for hearing. In: TGD and EEG. In online book. Available at: http://tgdtheory.fi/public_html/tgdeeg/tgdeeg.html#hearing (2006)
44. Pitkänen, M.: TGD and astrophysics. In: Physics in Many-Sheeted Space-Time. In online book. Available at: http://tgdtheory.fi/public_html/tgdclass/tgdclass.html#astro (2006)

45. Pitkänen, M.: TGD and cosmology. In: Physics in Many-Sheeted Space-Time. In online book. Available at: http://tgdtheory.fi/public_html/tgdclass/tgdclass.html#cosmo (2006)
46. Pitkänen, M.: TGD as a generalized number theory: infinite primes. In: TGD as a Generalized Number Theory. In online book. Available at: http://tgdtheory.fi/public_html/tgdnumber/tgdnumber.html#visionc (2006)
47. Pitkänen, M.: TGD as a generalized number theory: p-adicization program. In: TGD as a Generalized Number Theory. In online book. Available at: http://tgdtheory.fi/public_html/tgdnumber/tgdnumber.html#visiona (2006)
48. Pitkänen, M.: TGD as a generalized number theory: quaternions, octonions, and their hyper counterparts. In: TGD as a Generalized Number Theory. In online book. Available at: http://tgdtheory.fi/public_html/tgdnumber/tgdnumber.html#visionb (2006)
49. Pitkänen, M.: The classical part of the twistor story. In: Towards M-Matrix. In online book. Available at: http://tgdtheory.fi/public_html/tgdquantum/tgdquantum.html#twistorstory (2006)
50. Pitkänen, M.: The relationship between TGD and GRT. In: Physics in Many-Sheeted Space-Time. In online book. Available at: http://tgdtheory.fi/public_html/tgdclass/tgdclass.html#tgdgrt (2006)
51. Pitkänen, M.: Time and consciousness. In: TGD Inspired Theory of Consciousness. In online book. Available at: http://tgdtheory.fi/public_html/tgdconsc/tgdconsc.html#timesc (2006)
52. Pitkänen, M.: Was von Neumann right after all? In: Hyper-Finite Factors and Dark Matter Hierarchy. In online book. Available at: http://tgdtheory.fi/public_html/neuplanck/neuplanck.html#vNeumann (2006)
53. Pitkänen, M.: WCW spinor structure. In: Quantum Physics as Infinite-Dimensional Geometry. In online book. Available at: http://tgdtheory.fi/public_html/tgdgeom/tgdgeom.html#cspin (2006)
54. Pitkänen, M.: Construction of quantum theory: more about matrices. In: Towards M-Matrix. In online book. Available at: http://tgdtheory.fi/public_html/tgdquantum/tgdquantum.html#UandM (2012)
55. Pitkänen, M.: Quantum mind, magnetic body, and biological body. In: TGD Based View About Living Matter and Remote Mental Interactions. In online book. Available at: http://tgdtheory.fi/public_html/tgdlian/tgdlian.html#lianPB (2012)
56. Pitkänen, M.: SUSY in TGD Universe. In: p-Adic Physics. In online book. Available at: http://tgdtheory.fi/public_html/padphys/padphys.html#susychap (2012)
57. Pitkänen, M.: TGD based view about classical fields in relation to consciousness theory and quantum biology. In: TGD and EEG. In online book. Available at: http://tgdtheory.fi/public_html/tgdeeg/tgdeeg.html#maxtgdc (2012)
58. Pitkänen, M.: Are dark photons behind biophotons. In: TGD Based View About Living Matter and Remote Mental Interactions. In online book. Available at: http://tgdtheory.fi/public_html/tgdlian/tgdlian.html#biophotonslian (2013)
59. Pitkänen, M.: Comments on the recent experiments by the group of Michael Persinger. In: TGD Based View About Living Matter and Remote Mental Interactions. In online book. Available at: http://tgdtheory.fi/public_html/tgdlian/tgdlian.html#persconsc (2013)
60. Pitkänen, M.: What p-adic icosahedron could mean? And what about p-adic manifold? In: TGD as a Generalized Number Theory. In online book. Available at: http://tgdtheory.fi/public_html/tgdnumber/tgdnumber.html#picosahedron (2013)
61. Pitkänen, M.: Why TGD and what TGD is? In: Topological Geometrodynamics: An Overview. In online book. Available at: http://tgdtheory.fi/public_html/tgdview/tgdview.html#WhyTGD (2013)
62. Pitkänen, M.: Criticality and dark matter. In: Hyper-Finite Factors and Dark Matter Hierarchy. In online book. Available at: http://tgdtheory.fi/public_html/neuplanck/neuplanck.html#qcritdark (2014)
63. Pitkänen, M.: More about TGD inspired cosmology. In: Physics in Many-Sheeted Space-Time. In online book. Available at: http://tgdtheory.fi/public_html/tgdclass/tgdclass.html#cosmomore (2014)

64. Pitkänen, M.: Quantum gravity, dark matter, and prebiotic evolution. In: Genes and Memes. In online book. Available at: http://tgdtheory.fi/public_html/genememe/genememe.html#hgrprebio (2014)
65. Pitkänen, M.: Recent view about Kähler geometry and spin structure of WCW . In: Quantum Physics as Infinite-Dimensional Geometry. In online book. Available at: http://tgdtheory.fi/public_html/tgdgeom/tgdgeom.html#wcwnew (2014)
66. Pitkänen, M.: Unified number theoretical vision. In: TGD as a Generalized Number Theory. In online book. Available at: http://tgdtheory.fi/public_html/tgdnumber/tgdnumber.html#numbervision (2014)
67. Pitkänen, M.: About preferred extremals of Kähler action. In: Physics in Many-Sheeted Space-Time. In online book. Available at: http://tgdtheory.fi/public_html/tgdclass/tgdclass.html#prext (2015)
68. Pitkänen, M.: Can One Apply Occam's Razor as a General Purpose Debunking Argument to TGD? In online book.Available at: http://tgdtheory.fi/public_html/tgdquantum/tgdquantum.html#simplicity (2016)
69. Pitkänen, M.: From Principles to Diagrams. In online book.Available at: http://tgdtheory.fi/public_html/tgdquantum/tgdquantum.html#diagrams (2016)
70. Pitkänen, M.: How the Hierarchy of Planck Constants Might Relate to the Almost Vacuum Degeneracy for Twistor Lift of TGD? Available at: http://tgdtheory.fi/public_html/tgdquantum/tgdquantum.html#hgrtwistor (2016)

Articles Related to TGD

71. Pitkänen, M.: Geometric Theory of Harmony. Available at: http://tgdtheory.fi/public_html/articles/harmonytheory.pdf (2014)
72. Pitkänen, M.: Could One Realize Number Theoretical Universality for Functional Integral? Available at: http://tgdtheory.fi/public_html/articles/ntu.pdf (2015)
73. Pitkänen, M.: Positivity of N = 4 Scattering Amplitudes from Number Theoretical Universality. Available at: http://tgdtheory.fi/public_html/articles/positivity.pdf (2015)
74. Pitkänen, M.: About Minimal Surface Extremals of Kähler Action. Available at: http://tgdtheory.fi/public_html/articles/minimalkahler.pdf (2016)
75. Pitkänen, M.: Combinatorial Hierarchy: Two Decades Later. Available at: http://tgdtheory.fi/public_html/articles/CH.pdf (2016)
76. Pitkänen, M.: Is the Sum of p-Adic Negentropies Equal to Real Entropy? Available at: http://tgdtheory.fi/public_html/articles/adelicinfo.pdf (2016)
77. Pitkänen, M.: p-Adicizable Discrete Variants of Classical Lie Groups and Coset Spaces in TGD Framework. Available at: http://tgdtheory.fi/public_html/articles/padicgeom.pdf (2016)
78. Pitkänen, M.: Why Mersenne Primes Are So Special? Available at: http://tgdtheory.fi/public_html/articles/whymersennes.pdf (2016)
79. Pitkänen, M.: Some Questions Related to the Twistor Lift of TGD. Available at: http://tgdtheory.fi/public_html/articles/graviconst.pdf (2017)

Nash Limit Cycles: A Game-Theoretical Analysis of Cultural Integration in America

Bourama Toni

Abstract To be and to become an *American*: past, present, and future. We analyze the dynamics of the socio-cultural evolution of America as a player in a 1-agent game, integrating multiple immigrant-based cultural identities: American, in continuous random cross-cultural interactions with one another, assign a probability to each of the American ten core values which we recall, thereby defining at every instant a population state as a vector of probabilities. Adapting a methodology from Evolutionary Game Theory (replicator-like in human context), we uncover all possible dynamical game scenarios, to include *Nash Equilibria, Eventual Nash Equilibria*, *Nash Limit Cycles*, and *Isochrons*, a state of self-sustained oscillations in decision-making and where individual preferences evolve with the same constant phase. (Social Isochrons). We conjecture a game scenario for America stability and prosperity as the co-existence of asymptotically stable community Nash Equilibria based on the respective cultural identities, around a Nash Limit Cycle inside-attractive and outside-repulsive, together with a national Nash Equilibrium, asymptotically stable, around a Nash Limit Cycle inside-repulsive while outside-attractive, all in the interior of a 9-simplex, convex compact subset of the $\mathbf{R}^{10}$ Euclidean space. However, socio-cultural structures being highly *hierarchical*, such an analysis should be extended to the much richer *non-Archimedean/p-adic* simplex.

Keywords Game theory • Evolution matrix game • Nash Equilibrium • Stability • Limit cycles • American core values • Culture integration

---

B. Toni (✉)
Department of Mathematics & Economics, Virginia State University, 1 Hayden Drive, Petersburg, VA 23806, USA
e-mail: btoni@vsu.edu

B. Toni (ed.), *New Trends and Advanced Methods in Interdisciplinary Mathematical Sciences*, STEAM-H: Science, Technology, Engineering, Agriculture, Mathematics & Health, DOI 10.1007/978-3-319-55612-3_12

1 Overview

Game theory,[1] a branch of mathematics, is concerned with the modeling and analysis of strategic rational interaction between multiple decision-making entities called *players* whose actions should result in the *best possible* consequences or outcomes according to their preferences [8, 19, 25]. *Game* refers to any interactive situation where benefits and costs are shared and depend on all players with a fundamental premise that all players are rational decision-makers. A *player* could be any decision-making entity such as an automaton, a machine, a program, a person or an animal, a living cell or a molecule. Players have possibly conflicting or common objectives, with a clearly quantifiable *utility or payoff*. The strategies or objectives may be tightly coupled, or may allow a probability distribution for a random selection. The *solution concept or equilibrium* refers to a balance of players' strategies in such a way that no player has a motive to change unilaterally. The equilibrium is *strong* if it does not allow any group deviation, and it is *Pareto efficient*[2] if there is no other outcome to make all players better off. The rationality assumption indeed forces the appearance of an equilibrium solution.

An intuitive explanation of the concept of *Nash equilibrium* could be traced back to Augustin Antoine Cournot in his *Recherches into the Principles of the Theory of Wealth in 1838* in which one could find as well an evolutionary or dynamic idea of the best response correspondence. Picking on Cournot works Francis Ysidro Edgeworth in his *Mathematical Psychics* derived the concept of competitive equilibria in a two-players economy. Then Emile Borel, in *Algèbre et calcul des probabilités* published in Comptes Rendus de l'Académie des Sciences, vol. 185 (1927), dealt with mixed strategies, probability distributions leading to stable game.

However, the modern analysis is accepted to have been initiated by von Neumann and Morgenstern [25] in their book *Theory of Games and Economic Behavior*, drawing from which Nash [19, 20] provides us with the modern methodological framework.

Games are classified into *noncooperative* and *cooperative* game, distinguished along whether decisions are made jointly or not for strategies. The study of multiple equilibria, notably the Nash equilibrium, is part of noncooperative game theory. In a cooperative game, one analyzes how presumably strictly rational/selfish players could benefit from alliance by reaching bargaining agreements. The game could also be termed *static* or *dynamic* in reference to its evolution over time.

[1]Game theory is the theme of the 2001 Best Picture and Best Director, *Beautiful Mind*, featuring Russell Crowe playing John Nash, the nerdy Princeton student who's life and dating experience led him to the Nash Equilibrium solution concept that ultimately ended in a 1994 Noble Price Award in Economics. The underlying ideas of game theory transpired throughout history from the Bible Talmud, the works of Descartes and Sun Tzu to the writings of Darwin. A very hot topic today, game theory is used for Stock Markets, Portfolios, Military War Games, Bridge games, Hunting Party, Marketing, and Politics.

[2]Vilfredo Pareto, Italian mathematician, defines the efficiency as "A situation where there is no way to rearrange things to make at least one person better off without making anyone else worse off."

Some instances of games may be described by admissible mechanisms for coordination to ensure emergence of desirable collective behavior with respect to a given objective. For example, the *commuting time to get home from work* depends not only on the chosen route but also on decisions taken by other drivers. A solution concept or a *Nash Equilibrium (NE)*: *in traffic everyone is driving on the right. No single driver (rational) has an interest in driving on the left.*

Classical applications of game theory were in Economics, Behavioral Sciences and Biology. Most recent ones are related to networked systems as reflected in online advertisement on the internet, information evolution, belief propagation in social networks; deployment of distributed passive and active sonars for underwater sensor field design. Game can be used indeed to model the interactions taking place in a network, with the network nodes, largely interdependent, acting as players competing or forming coalition to maximize their *quality of service*. For instance, in the field of Signal Processing for communication networks one may design a game to address issues such as data security, spectrum sensing in cognitive radio, multimedia resource management, and image segmentation.

As the science of strategy, game theory has all the features of a mathematical construct. That is, it has a precise set of concepts and assumptions, many fundamental theorems, and many real-life applications allowing a mathematical analysis of human and social behavior. At its core is the rational choice assumption; but here rationality seems concerned with pursuit of happiness by decision-makers not necessarily altruist. A game has typically three dominant mathematical forms (1) Normal Form or Strategic Form Game (SFG), (2) Extensive Form Game (EFG), (3) Coalition Form Game (CFG) dealing with options for subsets of players. The SFG consists of players making the relevant decisions with the strategies available, the payoffs are the rewards contingent upon the actions of all the players in the game. The EFG places the emphasis on the timing of the decisions to be made, as well as the information available, and is representable by a decision or game tree. See more details in [8, 12, 13, 26]. In passing the so-called *zero-sum game* not of interest here is a game in which the gain of one player is balanced by the loss of the other with gains and losses summing up to zero.

The notion of *Rationality* is a fundamental assumption in game theory. That is, every player's motivation is to maximize the payoff selfishly. Here our game is at the interfaces of Strategic Form Games and Evolutionary Social Games or Population Games to model strategic interactions between a large number of individual players/agents, with a finite number of strategies/actions/values, randomly pairwise interacting with continuous payoffs/utilities. Our players adapt strategies by trial-and-error processes, by learning over time that some strategies work better than others: successful strategies reflected in "successful individuals" in the society at large spread by various means including imitation, "natural selection," learning. The population is therefore stratified in communities delineated by class, ethnicity, religion, national origin, cultural identities, etc. in strategic cross-cultural interactions with a finite number n of "survival strategies" based upon some clearly defined core values. Payoffs/utilities depend on own choices, selfish or altruist, a personal degree of rationality, as well as on the distribution of others' strategies and also the distribution of information pertaining to the core values.

Fig. 1 Towards Nash Limit Cycles, July 2013

Most importantly our current focus in this work is on America as a single player and the American social dynamics with respect to cultural integration and assimilation of immigrants in particular. We propose a game-theoretical analysis of the evolution dynamics of the American society as a whole to uncover and understanding the underlying social currents and to predict its future stability and prosperity, and also to offset the potential of radicalism or social instabilities challenging a smooth complete successful cultural integration.

Early on in July 2013 we were fortunate to discuss with the late John Nash himself, (see picture enclosed), this game-theoretical approach to America cultural evolution. His inputs were valuable to us then, as they are today, and we dedicate this work and what it is worth to his memory and his everlasting contributions to the understanding of human economic, social, and behavior evolution. It has been a privilege to get to know John Nash. His genuine kindness and keen show of interest in my game-theoretical work and the STEAM-H series is fondly remembered: he kindly autographed my first two early books (Fig. 1).

In the next section, Section 2, we present the ten American core values as defined and proposed by behavioral scientists. (See [5].) These are values a foreign visitor could observe in America as they fit most American and stand in sharp contrast to the values commonly held in other countries. Section 3 introduces the basic tenets of game theory along with some Dynamical Systems preliminaries. In Section 4 we discuss and describe evolutionary game as it applied to America as single player game, presenting as well subgames in relation to the subpopulations in America; The section exposes also various game scenarios for America with a conjectured scenario for stability and progress for the society. Section 5 concludes with perspectives for future research, in particular to extend the mathematical analysis into the much richer p-adic and non-Archimedean simplex to fully account for all the hierarchical "nuances" in human preferences and behaviors.

2 America Core Values and Culture

Our premise is that culture[3] consists of *preference distribution* along with *equilibrium behavior distribution*, but also individuals in any given culture are multidimensional. Immigration has been a major driving and defining factor of American culture. The behavior and choices of immigrants are influenced by many factors, e.g., home country traditions, in interaction with other American, while the preferences and equilibrium behaviors are shaped by *socialization and self-persuasion* and vary continuously, impacted by cross-cultural contacts, leading as a result, to some type of cultural hybridization. Discrepancies between the ideal and actual choices increase with time and fuel discontent. One would like to make choices in agreement with personal preferences, but the gains and payoffs are dependent upon the degree of choice coordination. The interactions, while seemingly random, are enhanced by factors such as common language, shared symbols and meanings, communication rules, down to culinary habits, family size.

America, as a social entity, has been experiencing rapidly changing demographics, with varying cultural norms and values. An American culture distinctly emerged as a compromise, a *modus vivendi*, and *modus operandi* out of multiple early immigrant cultures, behaviors, and preferences. These cross-cultural interactions generate a single hybrid culture, with its own peculiar language and rules of laws. The basic beliefs, assumptions, and values by which most American live could be comprehensively described by some ten core values deeply ingrained in most American [15].

1. **Individualism and Privacy**. (A_1) : Each person is seen as unique, special, and independent. Such belief put a premium on individual initiative, expression, orientation, and values privacy. Other cultures emphasize group orientation, conformity as a way to societal harmony. American believe they control their own destiny. Other cultures' focus is on extended family, and its corollary of loyalty and responsibility to family, with age given status and respect.
2. **Equality/Egalitarianism**. (A_2) : Each person expects to be treat equally, with a minimum sense of hierarchy, leading to a directness in relations with others, informal sense of self and space. One by-product is to allow the challenging of authority as opposed to respect for authority and social order at all cost

[3]Some view culture as a vehicle for providing generally accepted solutions to problems. Others define it as part of the knowledge owned by a substantial segment of a group but not necessarily by the general population.

prevalent in other cultures. Gender equity rather than different roles for men and women. In other countries people seem to draw a sense of security and certainty from Class and Authority, Rank and Status, considering it reassuring to know, from birth, who you are and where you fit in the complex social system.

3. **Materialism**. (A_3) : Each person enjoys the right to be "well off" and to "pursuit material happiness." People are often judged by their possessions. An undesirable by-product is the well-documented *American Greed*. It is a higher priority to obtain, maintain, and protect material objects rather than developing and enjoying interpersonal relationships, for instance.
4. **Science and Technology**. (A_4) : This value is seen as the driving force for changes and the primary source of goods. Scientific and technological approaches are highly valued, and lead to a problem-solving focus, with a mental process and learning style that are linear, logical, and sequential.
5. **Progress and Change**. (A_5) : This leads to a great national optimism and the so-called "Manifest Destiny": nothing is impossible and greatness is ours, as opposed to just accepting life's difficulties. Other societies value stability, tradition, continuity, and rich and ancient heritage, seeing change as disruptive and potentially destructive.
6. **Work and Leisure**. (A_6) : This refers to a strong work ethic, and work for its intrinsic values and as the basis of recognition and power. Idleness is seen as a threat to society while leisure should be a reward for hard work and individual achievement. In other cultures, work is seen as a necessity of life, and rewards are based on seniority and relationships.
7. **Competition**. (A_7) : That is, the "Be always First" mentality, encouraging an aggressive and competitive nature. It has led to *Free Enterprise* as an economic system, in the belief that competition fosters rapid progress. Cooperation is instead promoted in other cultures.
8. **Mobility**. (A_8) : This refers to a vertical social, economic, and physical mobility, in a society with people on the move to better self.
9. **Volunteerism**. (A_9) : Philanthropy is highly valued as a personal choice not a communal expectation, involving associations or denominations rather than kin-groups as in other cultures.
10. **Action and Achievement Oriented**. (A_{10}) : That is, a practical mind-set with emphasis on getting things done, a focus of function and pragmatism, and a tendency to be brief and business-like through an explicit and direct communication style, with emphasis on content and meaning found in the choice of words. In other cultures the emphasis is on context, meaning is found around the words, and communication is implicit and indirect. Formal handshakes rather than hugs and bows! *Dress for success* rather than as a sign of position, wealth, prestige, or religious rules.

These core values are perceived sometimes by individuals from other cultural identities, at least during an initial period, with a negative and derogatory connotation. However, a requirement for a complete cultural integration in America is for newcomers to be familiarized as quickly and efficient as possible with these values.

We consider the cultural evolution strategies to be associated with these ten core values, and define a probability distribution for each strategy A_i, $i = 1, \ldots, 10$. We are interested in determining the conditions for social stability and progress in providing a mathematical framework to understand the American social dynamics.

Our analysis could also help clarify the challenges of the two most common social objectives of *multiculturalism* and *social integration*. The former, aiming at preserving the multiplicity of existing cultures, has been consistently set in policies in countries such as Canada, while the latter appears to be the preference for many others countries including America and seeks to favor interactions across boundaries of class, ethnicity, religion, and national origin. Social conflicts (friction and even violence) result in both cases, in particular when immigrant lifestyles are perceived to threaten these core American values.

Our approach will help understand how groups differ in their resistance to cultural integration, as well as the necessity of cultural adaptations/adjustments in America and the consequences of immigration. Indeed in reference to the ten values, individuals optimal contributions to a so-called public good in modern economics are tightly dependent on their interactions with others, and require oftentimes behavioral adjustments, e.g., speaking one language at home and another at work. Cultural and social integration is accelerated by the need to interact efficiently with individuals belonging to other cultures or holding different core values with varied probabilities. It is believed that an integrated society results in economic efficiency, increased productivity, and a fair distribution of income. Such a belief partially justifies *Civil Rights Laws* and *Anti-discrimination Statutes* in the United States.

Our work amounts indeed to the study of the dynamics of cultural integration in America with respect to ten American core values defining who we are as American, proposing a game-theoretical approach both as a modeling and a design tool. We first next describe the basics of the theory.

3 Basics of Game Theory

3.1 Strategic Form Games

When all players in a game act simultaneously and without knowledge of others' actions, the game model is called a *strategic form game* we denote *SFG* [8, 12, 13].

Definition 3.1. A *strategic form game* is described by a triplet $\mathcal{G} =< \mathbf{P}, (\mathbf{S}_i)_{i\in\mathbf{I}_n}, (u_i)_{i\in\mathbf{I}_n} >$, where $\mathbf{I}_n = \{1, \ldots, n\}$, and with

1. $\mathbf{P}$ is a finite set of the n players $P_{i=1,\ldots,n}$
2. $\mathbf{S}_i$ is the non-empty set of *pure strategies* or available actions for player P_i
3. $u_i : \mathbf{S} = \prod \mathbf{S}_{i\in\mathbf{P}} \to \mathbf{R}$ is the *utility* or *payoff* function for player P_i
4. The vector $s = (s_1, \ldots, s_n)$ denotes the *strategy/ action profile* or *outcome*

5. We denote by $s_{-i} = (s_1, \ldots, s_{i-1}, s_{i+1}, \ldots, s_n)$ the vector of strategies taken by all other players except player P_i. That is, $s = (s_i, s_{-i})$. And accordingly $\mathbf{S}_{-i} = \prod_{j \neq i} \mathbf{S}_j$ describes all action profiles for all players except for P_i.

Note that we use the terms *action* and *strategy* interchangeably, as well as *utility* and *payoff.*

Remark 3.2. 1. In SFG each player is interested only in action profiles s that maximize its own utility function $u_i(s)$.
2. Every player P_i can control only its strategy S_i.

Definition 3.3. A pure strategy/action or value $\alpha \in S_i$ is said to be *(strictly) dominated* by a strategy $\beta \in S_i$, $\forall i = 1, \ldots, n$ denoted by $\alpha < \beta$ when $u_i(\alpha, s_{-i}) < u_i(\beta, s_{-i})$.

Remark 3.4. Dominated strategies/actions/profiles are rationally unjustifiable. However, removing them in a given game may lead to the emergence of other dominated strategies in the subgame, which results inductively in the so-called *iteratively dominated strategies. Iteratively undominated strategies* survive all rounds of elimination by dominance, resulting sometimes in instances of games called *dominance solvable* where the undominated strategies form a singleton [7, 8].

Most of the time games cannot be solved by removing dominated strategies. We then refer to the fundamental solution concept that is needed to capture the steady state of play with each player acting optimally given a "conjecture" about the behavior of other players. It is the *Nash Equilibrium* of the game to characterize strategy profiles resilient to unilateral deviations. It is akin to a compromise in a game where every player endeavors to maximize its own selfish payoff. Formally

Definition 3.5. A strategy profile $s^* = (s_1^*, \ldots, s_n^*) = (s_i^*, s_{-i}^*)$ is a Nash Equilibrium (NE) if

$$u_i(s^*) = u_i(s_i^*, s_{-i}^*) \geq u_i(s_i, s*_{-i}), \quad \forall s_i \in S_i. \tag{1}$$

Remark 3.6. 1. The *Nash Equilibrium* must be *strategically stable* to unilateral deviations: No player can unilaterally profitably deviate. It is *strong* if it is stable to group deviations, as to mitigate mutation and the impact of mutant players.
2. The *Nash Equilibrium* exists for a broad class of games as presented below.
3. A NE may result from repeated interactions, even in a game with partial information and primitive decision-making as in the *Harper's 33 Ducks Experiment* [10].

In the botanic garden of Cambridge University, UK, the biologist David Harper experimented with a flock of 33 ducks and 2 bread tossers, 20 m apart feeding the ducks at regular intervals. Tosser 1 (the least profitable site) has a frequency of supply of 12 pieces of bread per minute whereas Tosser 2 (the most profitable site) has a frequency of 24 pieces per minute. After about a minute, the number of ducks at the least profitable site stabilizes around 11, with thus 22 ducks at the most profitable site, which indicates a Nash equilibrium has been reached: Switching unilaterally to another site for a duck would result in getting less food. Indeed the amount of food a duck gets does not depend on its choice but also on others' choice. During a transient period, behaving as a typical optimizer most of the ducks rushed to the most profitable site, for some to realize they could get more food at the least profitable site, and therefore switched site. After switching a few times the ducks stick to a given choice. Repeated interactions have led to a Nash Equilibrium in a game where the entities, the duck, could hardly qualify as rational players. Such a process has been termed *tâtonnement.*

3.2 *Computing Nash Equilibrium*

The *Nash Equilibrium* may computed through a so-called *Best Response correspondence* as follows.

Definition 3.7. For every $s_{-i} \in \mathbf{S}_{-i}$ the *best response* of player P_i is given

$$B_i(s_{-i}) = \{s_i \in S_i \quad u_i(s_i, s_{-i}) \geq u_i(s_i', s_{-i}), \quad \forall s_i' \in S_i\} \tag{2}$$

Therefore a strategy profile s^* is a *Nash Equilibrium* if its components satisfy $s_i^* \in B_i(s_{-i}^*)$, or if and only if $s^* \in B(s^*) = \prod_i B_i(s_{-i}^*)$. That is, Nash equilibria are intersection points of the best responses or fixed points of the best response correspondence.

Remark 3.8. Nash in [19, 20] characterizes his equilibrium concept using *Brouwer–Kakutani* fixed-points theorems, based on topological-geometrical assumptions: compactness of the sets of strategies and the continuity of the payoff functions. That is, every compact continuous game has at least one mixed Nash equilibrium. Moreover

Theorem 3.9 (Nash [19]). *Every finite game has at least a Nash equilibrium*

This result certainly yields a very attractive solution concept, but with no presumption of efficiency, as seen below in the classic Prisoner's Dilemma. The claim follows from the Kakutani's fixed point theorem, a generalization of the Brouwer's fixed point theorem. We here recall both. The proofs are readily available in the fixed point theorems literature [3, 16, 17].

Theorem 3.10 (Brouwer's Fixed Point Theorem).[4] *Every continuous self-mapping of a convex and compact subset of an n-dimensional Euclidean space has at least one fixed point.*

---

[4]First studied by the French mathematician Poincaré and Picard in relation to differential equations, the general case has been proved by Jacques Hadamard and Jan Brouwer who allegedly derived it from the observation of motion in a cup of coffee upon stirring to dissolve a lump of sugar.

That is, let f be a continuous mapping: $U \subset \mathbf{R}^n \to U \subset \mathbf{R}^n$ with U convex and compact. Therefore there exists $x^* \in U$ such that $f(x^*) = x^*$. This is an existence, non-constructive result, with no indication as to how to find the fixed point. An example of a convex and compact subset could be the closed unit balls[5] in $\mathbf{R}^n$, i.e., set of points $x = (x_1, \ldots, x_n) \in \mathbf{R}^n$ that satisfy the inequality $\sum_{i=1}^{n} x_i^2 \leq 1$. Note also that the theorem is the Schauder fixed point theorem replacing Euclidean space by Banach space in the above statement. Proof of the theorem could be achieved by various methods, to include: Proof using the notion of *degree of a continuous mapping*; proof using Homology; proof using oriented area; proof using Stokes' theorem; proof using the Hex game; as well as the Hirsch's proof; The combinatorial proof is closer to our study.

Remark 3.11 (Combinatorial Proof). Let f be a continuous self-mapping on the unit simplex Δ^n, i.e., the set $\{x \in \mathbf{R}^{n+1}, x_i \geq 0 | \sum_{i=0}^{n} x_i = 1\}$. Then $\sum_i x_i = 1 = \sum_i (f(x))_i$, $\forall x \in \Delta^n$. By the Pigeonhole principle, for every such x, there is an index $j = 0, \ldots, x$ such that $(f(x))_j \leq x_j$. By Sperner's lemma, there must a point $x^* \in \Delta^n$ such that $\forall j$, $(f(x^*))_j \leq (x^*)_j$. Therefore $f(x^*) = x^*$ from the equality of the sums of the coordinates.

Brouwer's fixed point theorem was generalized further by Kakutani. Consider a non-empty subset $K \subset \mathbf{R}^n$ with the Euclidean norm, and the associated power set 2^K. Define a set-valued mapping $f : K \to 2^K$, such that $f(x)$ is a non-empty subset $f(x) \subset K$. Now a point $x^* \in K$ is called a fixed point if $x^* \in f(x^*)$.

Definition 3.12. The set-valued mapping f is upper semi-continuous in $x \in K$, if for every sequence $(x_k)_{k\in\mathbf{N}}$ in K with $x_k \to x$ and every sequence $(y_k)_{k\in\mathbf{N}}$ in K with $y_k \in f(x_k)$ for all $k \in \mathbf{N}$ and $y_k \to y$ we have $y \in f(x)$.

Note that upper semi-continuity amounts to continuity for a point-map. The function f is also said to have a closed graph.

Theorem 3.13 (Kakutani's Fixed Point Theorem). *For $K \subset \mathbf{R}^n$ compact convex subset and f upper semi-continuous on K,, the mapping f then has a fixed point in the above sense, i.e., there is some $x^* \in K$ such that $x^* \in f(x^*)$.*

Obviously Kakutani's theorem implies the Brouwer's fixed point theorem. It was used by John Nash for his equilibrium concept in game theory, and subsequently it has many applications in game theory and economics. It is also used to prove the minimax theorem in zero-sum games as in the original work by Kakutani.[6]

---

[5]The shape of the unit ball depends on the chosen norm in the metric space: it is the usual round ball with the usual Euclidean norm; but it has "corners" with the taxicab norm (squared shape) and with max norm (diamond shape) similar to $[-1, 1]^n$.

[6] It is said as an anecdote that a puzzled Kakutani once asked an attendee to his talk what is Kakutani's fixed point theorem.

3.3 Classic Examples

Example 3.14 (Prisoner's Dilemma [9, 13, 20, 25]). Following the story discussed by Melcin Dresher, Meriil Flood, and Albert Tucker, two criminal suspects in separate Police custody are offered a deal: If both testify (strategy D of *Defection*) they get a sentence reduction of 1 year; if both do not testify (strategy C of *Cooperation*), their sentence is reduced by 3 years for lack of hard evidence. If one testifies (strategy D) against the other who does not (strategy C), the sentence is reduced by 5 years. The game has therefore the payoff matrix

$$A = \begin{pmatrix} * & C & D \\ C & 3 & 0 \\ D & 5 & 1 \end{pmatrix} \tag{3}$$

With a lower payoff, C (cooperation) is a strategy *dominated* by strategy D (defection). The unique Nash Equilibrium is at (D, D) but a higher payoff is obtained at (C, C). That is, the players are better off at playing C; Such a Nash equilibrium is said to be *Pareto inefficient*.

Another example is the game with cyclic behavior.

Example 3.15 (Rock-Scissors-Paper [8, 13, 21, 23, 25]). Simultaneously each player displays a sign of Rock (R: simple fist), Scissors (S: fingers in V sign) or Paper (P: flat hand) with the rule: rock crushes scissors, scissors cut paper, and paper wraps rock. Such a game does not have a typical NE, but in repeated plays one ends up selecting as the best strategy randomly mixed strategies (mixed NE) with probability $1/3$. This game was often used to experiment human social evolutionary behaviors [13, 18, 19].

Remark 3.16 (Prize of Anarchy). A *Pareto efficient* point s^p is such that there exists no other strategy profile s satisfying $u_i(s) \geq u_i(s^p)$ and $u_i(s) > u_i(s^p)$ for some $i = 1, \ldots, n$.

A *Pareto efficient* Nash Equilibrium is highly desirable in decision-making. A measure of efficiency of the so-called *Prize of Anarchy* (PoA) defined by

$$PoA = \frac{max_{\mathbf{S}} \sum_{i \in N} u_i(s)}{min_{\mathbf{NE}} \sum_{i \in N} u_i(s)} \tag{4}$$

So the closer the PoA is to 1, the higher the efficiency of the Nash Equilibrium. One way of improving the efficiency of Nash Equilibrium in a game design is to keep the Nash equilibrium and transform the game into a game with different structure such as supermodular game; potential game; repeated game; stochastic game, and Bayesian game. See more details in [1, 6–8].

3.4 Dynamical Preliminaries

From the work of Poincaré the mathematical concept of a dynamical system is an outgrowth of the qualitative theory of differential equations: Given a metric space $< \mathbf{X}, d >$ and an additive semigroup[7] $I \subset \mathbf{R}$ containing 0, a flow or dynamical system on $\mathbf{X}$ is determined by a continuous mapping

$$\Phi : \mathbf{X} \times I \rightarrow \mathbf{X} \tag{5}$$

satisfying the *identity property*

$$\Phi(x, 0) = s, \tag{6}$$

as well as the *semigroup property*

$$\Phi(\Phi(s, t), s) = \Phi(x, t + s). \tag{7}$$

That is, the states of the system, when the time t evolves in the real semigroup I, are elements of the metric space $\mathbf{X}$. Here we consider an Archimedean metric with the triangle inequality, i.e.,

$$d(x, z) \leq d(x, y) + d(y, z), \quad \forall x, , y, z \in \mathbf{X}. \tag{8}$$

In the case where the space $\mathbf{X}$ is endowed with an ultrametric d with a strong or ultrametric inequality replacing the triangle inequality, i.e.,

$$(x, z) \leq max(d(x, y), d(y, z)), \tag{9}$$

the dynamical system is non-Archimedean, as in the class of the so-called *p-adic dynamics.*

Letting $I = \mathbf{N}_0 = \{0, 1, 2, \ldots\}$ leads to *time-discrete dynamical system*, whereas the *time-continuous dynamical system* requires $I = \mathbf{R}$ or $\mathbf{R}_{\geq 0} = \{t \in \mathbf{R} | t \geq 0\}$.

The curve $\gamma_x(.) : \mathbf{R} \rightarrow \mathbf{R}$ such that $\gamma_x(t) = \phi(x, t)$ defines the orbit or trajectory of x under the flow ϕ, simply denoted sometimes $x(t)$. Note that *orbit* or *trajectory* through $x \in \mathbf{X}$ is therefore described by the point set

$$\gamma_{\mathbf{R}}(x) = \{y = \Phi(x, t) | t \in \mathbf{R}\} = \cup_{t \in \mathbf{R}} \{\Phi(x, t)\}. \tag{10}$$

The flow thus induces a vector field $\mathscr{X}$ on $\mathbf{X}$ given by

$$\mathscr{X}(x) := \dot{x}(0) \in T_x\mathbf{X} \tag{11}$$

---

[7]A semigroup is an algebraic structure endowed with an associative closed binary operation.

where $\dot{x}(0)$ is the initial velocity of $x(t)$ with $T_x\mathbf{X}$ denoting the *tangent space* to $\mathbf{X}$ at x. By the fundamental theorem on flows, $x(t)$ is the unique solution to the *first order dynamical system*

$$\dot{x} = \mathscr{X}(x(t)), \quad t \in \mathbf{R}. \tag{12}$$

By a recursive formula the nth order dynamical systems takes the form

$$x^{(n)} = \mathscr{X}^n. \tag{13}$$

corresponding to a flow on the *phase space* $\Omega = \cup_{x,disjoint}(T_x\mathbf{X})^{n-1}$ representing all possible states of the system, with $\mathbf{X}$ called the *configuration space* of the system.

The asymptotic behavior of the first order system is determined by defining the $\omega(resp.\alpha)$*-limit set* for every $x \in \mathbf{X}$ as

$$\begin{aligned}\omega(x) = \lim_{\omega} \gamma(x) =& \{y \in \mathbf{X} | y = \lim_{n\to\infty} \Phi(x, t_n), \quad \text{for a sequence } (t_n)_{n\geq 0} \subset \mathbf{R}_{\geq 0} \text{ with } t_n \to \infty\} \\ =& \cap_{s\in\mathbf{R}} \{\Phi(x,t))^{-} : t > s\}\end{aligned} \tag{14}$$

Note that the omega (alpha) limit set is closed and invariant; for $\mathbf{X}$ compact, it is non-empty, compact, and connected.

Definition 3.17. 1. A point $x_0 \in \mathbf{X}$ is a *rest point or an equilibrium point* of the flow Φ if $\Phi(x_0, t) = x_0$, $\forall t \in \mathbf{R}$.
2. Define the orbit mapping or *movement* through $x \in \mathbf{X}$ as the mapping

$$\gamma(.) = \Phi_x(.) : t \to \Phi(x, t). \tag{15}$$

Φ_x is *periodic* if there is some $p > 0$ with $\Phi(x, t + T) = \Phi(x, t)$, $\forall t \in \mathbf{R}$. The number T is the period of the movement.
3. The equilibrium point x_0 is *stable* if

$$\forall \varepsilon > 0, \quad \exists \delta = \delta(\varepsilon) | \quad d(x, y) \leq \delta \Longrightarrow d(x, \Phi(y, t)) \leq \varepsilon, \quad \forall t \geq 0. \tag{16}$$

4. The point x_0 is *asymptotically stable* if it is stable and $\exists \delta_0 > 0$, such that $\lim_{t\to\infty} \Phi(y, t) = x$, for all $y \in \mathbf{X}$ with $d(x, y) \leq \delta_0$.

Stability is also investigated using the classic method of Lyapunov. Given an autonomous system

$$\dot{x} = f(x) \tag{17}$$

where $f \in \mathscr{C}^1(U, \mathbf{R}^n)$, $U \subset \mathbf{R}^n$ open and connected. The *direct Lyapunov method* gives sufficient conditions for stability and asymptotic stability as follows [3, 16].

Definition 3.18. A *Lyapunov function* for the system is any function $V \in \mathscr{C}^1(X)$ such that

$$\dot{V}(x) = \nabla V(x).f(x) \leq 0 \quad x \in X, \tag{18}$$

where ∇ denotes the gradient vector.

Proposition 3.19. *A rest point x_0 is stable if there exists a Lyapunov function V that is positive definite. That is,*

$$V(x_0) = 0, \quad V(x) > 0, \quad \forall x \in X, x \neq x_0. \tag{19}$$

If in addition

$$\begin{aligned} \dot{V}(x_0) &= \frac{dV}{dt}(x_0) = 0 \\ \dot{V}(x) &< 0, \quad x \neq x_0 \end{aligned} \tag{20}$$

then x_0 is asymptotically stable.

Example 3.20 (Linear System). Consider a linear system

$$\dot{x} = Ax, \quad x \in \mathbf{R}^n, \tag{21}$$

where A is a real $n \times n$ matrix. Assume that A is nonsingular with $x = 0$ the only rest point, and that all the eigenvalues of A have negative real parts. Take C an arbitrary symmetric and positive definite matrix to define the positive definite symmetric matrix

$$B = \int_0^\infty e^{tA^T} C e^{tA} dt. \tag{22}$$

The function

$$V(x) = x^T Bx \tag{23}$$

is a positive definite Lyapunov with respect to $x = 0$, which is therefore an asymptotically stable rest point.

In general, (see also in [3, 16, 17])

Proposition 3.21. *If the autonomous system above has a rest point x_0 such that the corresponding Jacobian matrix $(\frac{\partial f_i}{\partial x_j}(x_0))_{1 \leq i,j \leq n}$ has only eigenvalues with negative real parts, then x_0 is asymptotically stable.*

3.4.1 Special Dynamics

Limit Cycles are isolated periodic orbits, forward or backward limit cycles of neighboring trajectories; they model self-sustained oscillations in the phase space of the dynamical system. They appear in an attempt to detect stable cyclic behavior. The main bifurcation route towards periodicity remains the *Hopf Bifurcation*, and for a one-parameter family of continuous-time systems the non-degenerate Hopf bifurcation is the only generic bifurcation in which a limit cycle emerges and disappears. See [3, 17]. Given the planar parameter-dependent system

$$\dot{x} = f(x, \alpha), \quad x \in \mathbf{R}^2, \alpha \in \mathbf{R}, f \text{ smooth and analytic}, \tag{24}$$

with the Jacobian eigenvalues at the fixed point $x^* = 0$,

$$\lambda_{1,2} = \mu(\alpha) \pm i\omega(\alpha), \tag{25}$$

with

$$\mu(\alpha) < 0, \quad \text{for } \alpha < 0, \quad \mu(0) = 0, \quad \mu(\alpha) > 0, \text{for } \alpha > 0, \tag{26}$$

we have

Theorem 3.22 ([17]). *Assume the genericity conditions*[8] *hold, i.e.,*

1.

$$[\frac{\partial \mu(\alpha)}{\partial \alpha}]_{\alpha=0} \neq 0, \quad \text{(Transversality)} \tag{27}$$

2.

$$l_1(0) \neq 0, \quad \text{(Nondegeneracy)} \tag{28}$$

where $l_1(0)$ is the first Lyapunov coefficient.

then the system undergoes an Andronov–Hopf bifurcation at $\alpha = 0$.

That is, either the stable focus equilibrium x^* at $\mu(\alpha) < 0$, for $\alpha < 0$, and $l_1(0) < 0$ becomes unstable for $\alpha > 0$ and surrounded by a stable limit cycle (*Supercritical, soft or non-catastrophic Hopf*), or for $l_1(0) > 0$, surrounded, when $\alpha < 0$, by an unstable limit cycle, shrinking and disappearing as α crosses the critical value 0 (*subcritical, sharp, or catastrophic Hopf.*)

[8]Genericity conditions ensure the existence of smooth invertible coordinate transformation depending smoothly on parameters together with parameter changes and (possibly) time reparametrizations reducing the system to the *normal* form.

The existence of limit cycles can be also detected using the *Poincaré–Bendixon* theorem together with the Dulac criterion, based on trajectories trapping regions. This has been used by Hofbauer and Sigmund for some 3-strategy games such as the classical game of cyclical competition, *Rock-Scissor-Paper*. All Hopf bifurcations were found to be degenerate, i.e., $l_1(0) = 0$ [11–14, 23].

4 Evolutionary Games

Evolutionary game theory [2, 12–14, 18, 23, 24, 26] is concerned with strategic interactions in large populations of players, where changes over time in behavior are driven by factors such as natural selection, "pursuit of happiness," and economical success of the individuals. Maynard Smith [18] in a series of pioneering papers adapted methods of traditional game theory to the context of biological natural selection, to explain, for instance, the existence of ritualized animal conflicts.

4.1 Matrix Games

Consider a two-player game with mixed strategies defined on two simplices as

$$
\begin{aligned}
X &= \{x = (x_1, \ldots, x_m) \in \mathbf{R}^m | \quad x_i \geq 0, \quad \sum_{i=1}^{m} x_i = 1\} \\
Y &= \{y = (y_1, \ldots, y_n) \in \mathbf{R}^n | \quad x_i \geq 0, \quad \sum_{i=1}^{n} y_i = 1\}
\end{aligned}
\tag{29}
$$

Define the payoff functions on $X \times Y$ by

$$
\begin{aligned}
\Phi_1(x, y) &= x^T A y \\
\Phi_2(x, y) &= x^T B y,
\end{aligned}
\tag{30}
$$

where A and B are $m \times n$-matrices.

Definition 4.1. The point $p^* = (x^*, y^*) \in X \times Y$ is called a Nash equilibrium if

$$
\begin{aligned}
x^{*T} A y^* &\geq x^T A y^* \quad \forall x \in X \\
x^{*T} B y^* &\geq x^{*T} B y \quad \forall y \in Y
\end{aligned}
\tag{31}
$$

Remark 4.2. The game is *symmetric* for $X = Y$ and $A = B^T$. The Nash equilibrium is determined accordingly. If $x^* = y^*$, then the conditions for equilibrium are equivalent to

$$x^{*T}Ax^{*} \geq x^{T}Ax^{*}, \forall x \in X. \tag{32}$$

The two players are thus identical, with only symmetric pairs (x^*, y^*) being of interests [16].

4.2 Evolution Matrix Game

Consider a game where the single player is represented by a population whose individuals have n mixed strategies for interaction with each other, leading to a probability distribution.

Let $s_i \in [0, 1] \subset \mathbf{R}$, be the probability for strategic value A_i to be chosen. The vector of probabilities or frequencies $s = (s_1, \ldots, s_n) \in [0, 1]^n \subset \mathbf{R}^n$ describes a state of the population as well as a behavioral strategy profile, with the components s_i satisfying the probability constraint $\sum_i s_i = 1$. The state vector S lies on the (n-1) simplex space of strategies

$$\Delta = \{s = (s_1, \ldots, s_n) \in \mathbf{R}^n \quad | \quad 0 \leq s_i \leq 1, \sum_i s_i = 1\} \tag{33}$$

compact and convex spanned by the set of vectors e_i of the standard basis

$$e_i = (0, \ldots, 0, 1_i, 0, \ldots, 0), \quad i = 1, \ldots, n. \tag{34}$$

However, our evolution game here is played by the American society as a single player whose individuals dispose of a finite number of strategies, ten core values, in their "pursuit of happiness." Immigrants strive to integrate such a population with actions and strategies based on these values. The ultimate outcome is the stability and prosperity of America. We study the evolution dynamics over time in terms of time-continuous dynamical models. Similar models have been introduced earlier by Taylor and Jonker [23] as the *Replicator Dynamics* with application in biological, genetic, and chemical systems where organisms, genes, and molecules evolve over time through replication. In [12–14, 23, 26–28] Hofbauer et al. and Zeeman studied 3-strategy games under the replicator dynamics to conclude on the occurrence of only "simple" dynamical behaviors such as sinks, sources, centers, and saddles. Many interesting results were obtained using analytical and numerical tools from non-linear dynamical system theory with respect to changes in the payoff and behavioral parameters.

The strategies are given by the ten American core values described above. Let $p_i \in [0, 1]$, $i = 1, \ldots, 10$ be the probability for the core value A_i to be chosen by American at any given moment. Therefore the state of population is determined by the vector of probabilities

$$p = (p_1, \ldots, p_{10}), \quad \sum_{i=1}^{10} p_i = 1. \tag{35}$$

This forms the space of population states in the form of a 9-simplex[9] Δ, that is

$$\Delta = \{p = (p_1, \ldots, p_{10}) \in [0,1]^{10} \subset \mathbf{R}^{10} | \quad \sum_{i=1}^{10} p_i = 1\}. \tag{36}$$

Remark 4.3. 1. Pure populations states are given by the corners vectors of the standard Euclidean basis $e_i = (0, \ldots, 0, 1_i, 0, \ldots, 0), i = 1, \ldots, 10$. The interpretation is that in such a state all American choose the single core value A_i with a probability p_i giving any other value a zero probability. These states are also called *homogeneous*. For instance, American choose to value only *Individualism* disregarding the other nine core values. No doubt such a state would be of concern.

2. Note that the probability s_i for the core strategic value A_i changes with respect to payoffs by learning, copying, inheriting, etc. And in turn payoffs are function of the probabilities yielding a *feedback loop dynamic*. The difference $h_i(s)$ has been identified in certain "game dynamics" with a relative increase of the frequency [23, 24].
3. Strategies with higher payoffs reproduce faster. Consequently individuals have a tendency to vary the probability assigned to a core value based on some perceived relative advantage.
4. Subscribing to competing strategic values lead to the emergence of mutant subgroups, with the by-products of the appearance of a variety of life forms, beliefs, cultures, languages, practices, or techniques.
5. The standard population is a *mixed state* of a probability vector p. That is, each core value has been given a probability $p_i \in [0, 1]$ by each American in interaction with others. The value of the probability p_i depends upon personal preferences and behaviors, and certainly influenced by cultural identities, social class, upbringing, etc. The resulting behavior is therefore *stochastic*: playing the core value A_i with probability p_i.
6. In contrast to other types of dynamics such as Replicator, Replicator-Mutator, or Logit Dynamics, [12, 13] any kind of revision protocol is described by changes in the values of the probability assigned to each core value, not by switching of strategy. Successful strategies are represented by "successful individuals" and could be learned (imitated), possibly inherited.

[9]Simplex generalizes the concept of triangle in arbitrary dimension, with an n-simplex being the convex hull of its (n + 1) vertices.

7. We consider the American society has stratified in communities or clusters based on cultural identities and ethnicity, which greatly impact the probabilities assigned to the core values.

The interaction between an A_i value individual and an A_j value individual result in a *payoff* also called *reward* or *utility* denoted by $a_{ij} \in \mathbf{R}$. Therefore all the payoffs form the *evolution payoff matrix*

$$A = (a_{ij})_{1 \leq i,j \leq 10} \tag{37}$$

defining the game as an evolution matrix game. According to the context, the real entries of this 10×10 matrix could be assumed to satisfy various properties of matrices. That is, the matrix A could be *symmetric*, i.e., $A = A^T$; *skew-symmetric*, $A = -A^T$; *cyclic symmetric* (Toeplitz circulant with indices counted cyclically modulo n); or a *banded matrix* (sparse with nonzero entries confined to a diagonal band).

Accordingly we define a payoff or utility function $u = (u_1, \ldots, u_{10}) : \Delta \subset [0,1]^{10} \to \mathbf{R}^{10}$, also called the payoff vector field.

Definition 4.4. 1. Given a population state determined by the probability vector $p = (p_1, \ldots, p_{10}) \in \Delta$, the *expected payoff* of an A_i American is defined as

$$u_i(p) = e_i A p^T = \sum_{j=1}^{10} a_{ij} p_j = (Ap)_i. \tag{38}$$

The payoff functions u_i are continuous linear functions.

2. The *average payoff/utility* in the population state p is given by

$$\bar{u}(p) = pAp^T = \sum_{1 \leq i,j \leq 10} p_i A p_j. \tag{39}$$

3. We denote by

$$h_i(p) = u_i(p) - \bar{u}(p), \tag{40}$$

that is, the excess payoff between the individual expected payoff and the average payoff in the population state actually impacts the frequency and the probability value of a core value.

Note that the sign of excess payoff $h_i(p)$ for the core value A_i in the population state p determines its variation in frequency as well as its corresponding probability. The *standard or regular* population state p is a mixed state in which each core value receives a nonzero probability, that is, $p_i \in (0,1)$. We denote for a general n the support of a state p by

$$supp(p) = \{i \in \{1, \ldots, n\} | p_i > 0\} \tag{41}$$

We now define the equilibrium of such a population game, which amounts to a national cultural and norm compromise in America, an agreed upon modus vivendi/operandi.

Definition 4.5. A population state $p^* = (p_1^*, \ldots, p_n^*) \in \Delta$ is called a *Nash Equilibrium,* denoted NE, if

$$\bar{u}(p^*) = p^* A p^{*T} \geq p A p^{*T}, \quad \forall p \in \Delta. \tag{42}$$

Equivalently we have

$$\begin{aligned} \bar{u}(p^*) &= p^* A p^* = max_{i=1,\ldots,n} u_i(p^*) \\ u_i(p^*) &= u_i(p_i^*, p_{-i}^*) \geq u(p_i, p_{-i}^*) \end{aligned} \tag{43}$$

where again

$$u_i(p_i, p_{-i}^*) = u_i(p_1^*, \ldots, p_{i-1}^*, p_i, p_{i+1}^*, \ldots, p_n^*), \forall p_i \in \Delta_i = [0, 1].$$

Remark 4.6 (Interpretation). In the Nash Equilibrium state p^* the unilateral change of the probability of a single core value does not lead to a higher payoff. Under the rationality assumption of the game one would expect such a state to be maintained. That is not in general the case because individuals do not always behave rationally. Deviations are expected dictated by changes in preferences, and the stability of the Nash Equilibrium state p^* is not guaranteed. We are therefore interested in dynamic conditions that could ensure the *evolutionary stability* of that equilibrium as defined next.

Definition 4.7 (Evolutionary Stability). A Nash Equilibrium state $p^* \in \Delta$ is *evolutionarily stable* if

$$\bar{u}(p^*) = p^* A p^* = p_0 A p^{*T} \quad \text{for some } p_0 \in \Delta,\ p_0 \neq p^* \tag{44}$$

implies

$$\bar{u}_{(p_0)} = p_0 A p_0 < p^* A p_0^T. \tag{45}$$

That is, if the same payoff of the NE state p^* could be achieved in some other state p_0, then this other state cannot be an NE state. Conditions are needed to characterize dynamically both states, necessary and/or sufficient conditions. First we note

Proposition 4.8. *Let $p^* \in \Delta$ be an NE state with support supp(p^*). Then we get*

$$u_i(p^*) = \bar{u}(p^*) \quad \forall i \in supp(p^*) \tag{46}$$

In an NE state, the individual expected payoff is the same as the average in the state for the nonzero probabilities making the NE state.

Proof. From the definition of NE, for $i \in supp(p^*)$, and

$$u_i(p^*) \leq \bar{u}(p^*) = max_i u_i(p^*).$$

Hence the claim. □

As indicated above the payoff functions u_i are continuous linear functions on the 9-simplex Δ, a convex compact subset of $\mathbf{R}^{10}$. This ensures

Lemma 4.9. $\forall p^* \in \Delta, \exists \tilde{p}_i \in \Delta_i = [0, 1]$, *such that*

$$u_i(\tilde{p}_i, p^*_{-i}) \geq u_i(p_i, p^*_{-i}) \quad | \quad \forall p_i \in \Delta_i \tag{47}$$

This in turn leads to a constructive proof of the well-known Nash's theorem we recall below. First we define a continuous self-mapping f on the simplex.

$$f = (f_1, \ldots, f_n) : \Delta \to \Delta \quad f(p^*) = \tilde{p} = (f_i(p^*)) = (\tilde{p}_i), \tag{48}$$

with $\tilde{p}_i$ determined as below. The mapping f is continuous on the compact convex set Δ. Therefore by Brouwer–Kakutani's fixed point theorem, there is at least a fixed point p^*, which is proved to be the Nash Equilibrium state of the population game. Note first that for such a mapping one can construct the following sequence $(p^k)_{k\geq 0} \subset \Delta$, by performing an iteration: take a value $p^0 \in \Delta$ and construct

$$p^{k+1} = f(p^k) = p^{k+1}(p^0), \quad k \in \mathbf{N}_0. \tag{49}$$

Such a sequence converges to the fixed point p^* given by $p^* = f(p^*)$. Along the same lines as Nash's original proof, the fixed point $p^* \in \Delta$ is shown to be a Nash Equilibrium as defined above. That is,

Theorem 4.10 (Existence of an NE State). *Under the above conditions, the population game admits a Nash Equilibrium state $p^* \in \Delta$.*

Proof. We here outline the proof as an adaptation of the Nash's proof by now readily available in the game theory literature [18–20]. The NE state p^* is indeed the fixed point of the above function f whose existence is ensured by the fixed point theorems. We recall, denoting $I_n = \{1, \ldots, n\}$,

$$\bar{u}(p^*) = p^*Ap^{*T} = max_{i=1,\ldots,n} u_i(p^*), \tag{50}$$

where $u_i(p^*) = e_iAp^{*T}$. Now define for $i = 1, \ldots, n$ and $p \in \Delta$

$$\tilde{h}_i = max_{i\in I_n}(0, h_i(s))$$
$$f_i(p) = \tilde{p}_i = \frac{p_i + \tilde{h}_i}{1 + \sum_{j=1}^n \tilde{h}_j(p)}. \tag{51}$$

This leads to a continuous self-mapping on Δ given by $f = (f_1, \ldots, f_n)$, which has at least a Brouwer–Kakutani fixed point $p^* \in \Delta$, i.e., $f(p^*) = p^*$. This implies

$$\tilde{h}_i(p^*) = 0, \quad \forall i = 1, \ldots, n.$$

Therefore

$$\bar{u}(p^*) = max_{i \in I_n} u_i(p^*). \tag{52}$$

Thus p^* is an NE. Hence the claim. □

Remark 4.11. One could interpret this result as though the rationality assumption is forcing the game into an NE state, raising the question of determining conditions under which the NE state is evolutionarily stable.

If there exists an integer r such that for a point $\tilde{p} \in \Delta$ we have $f^r(\tilde{p}) = p^*$ (fixed point), where f is the constructed continuous mapping on the simplex. We say that the point $\tilde{p}$ is an *Eventual Nash Equilibrium.* The co-existence of *Nash Equilibria* alongside *Eventual Nash Equilibria* is indeed an interesting feature for some dynamical systems, and should be investigated. That is, a non-equilibrium state may go to a Nash Equilibrium state in a finite time.

4.2.1 America Evolution Dynamics

The population game as described above is so far in a *static form.* Individuals as rational decision-makers to reach greater success in America adapt their strategies over time by varying the probabilities assigned to each core value for a more efficient distribution of their preferences and behaviors. Indeed strategies with higher payoff/rewards spread quickly in the society leading to cultural norms through, e.g., learning, copying/imitating. Payoffs depend on the frequencies/probabilities of the core value in the society, which themselves change according to the payoffs/rewards, resulting in a feedback loop dynamic characterizing the evolution of the American society [21, 23]. We therefore introduce a dynamic in the game to account for the changes over time of the probability distribution.

The per capita growth rate of the time-continuous dependent probability $p_i(t)$ is the logarithm derivative

$$\frac{d(logp_i(t))}{dt} = \frac{\dot{p}_i}{p_i} \tag{53}$$

determined by the difference between the expected payoff to the core value A_i and the average payoff to the population state, leading to the time-continuous dynamical system ($\mathscr{A}$)

$$\frac{\dot{p}_i(t)}{p_i} = h_i(p) = u_i(p) - \bar{u}(p), \tag{54}$$

or equivalently

$$\dot{p}_i(t) = p_i(u_i(p) - \bar{u}(p)) = p_i((Ap)_i - pAP^T) = g_i(p), \qquad (\mathscr{A})$$

Remark 4.12. 1. This is a system of differential equations for $i = 1, \ldots, n$ with $n = 10$ in our context. It appears in various areas such as population genetics, chemical networks, and is famously known as the *replicator dynamics*[10] introduced by Taylor and Jonker [24] and coined as such by Schuster and Sigmund [12–14, 23].

2. Under the constraint $\sum_{i=1}^{10} = 1$ the system of equations actually reduces to 9 differential equations we analyze on a 9-simplex.
3. Moreover adding a constant to a column entries of the payoff matrix A does not change the equations and its dynamic properties. One may then set the diagonal entries to zeros, or set the last row of A to zero in analyzing the dynamics. Such simplifications are indeed undertaken in practice.
4. The well-defined power product $V(p) = \prod_i p_i^{\alpha_i}$ satisfies

$$\dot{V} = V \sum \alpha_i[(Ap)_i - pAp^T]. \qquad (55)$$

5. The frequency of a core value in the society as given by the probability increases when it has above average utility/payoff. Individuals have limited and localized knowledge of the whole system, according to the distribution of information. Some core values could become extinct as time goes to infinity. Indeed whenever a core value A_i is recursively strongly dominated as defined above, it will not survive the America Evolution equation ($\mathscr{A}$). That is, for the assigned probability p_i, we have $\lim_{t\to\infty} p_i(t) = 0$. The corresponding core value therefore goes extinct. This process might explain the steady variation of cultural values and norms [1, 7, 9, 22].
6. Another outcome is that, if for a core value A_i, there is no A_i player in the society at one point in time, then it can never appear in the future. Therefore the population states given by $p_i = 1$, and $p_j = 0$, for $j \neq i$, are fixed points of the dynamic.
7. Finally stochastic effects are not considered here in our evolution equation; they can lead to the elimination of core values in subpopulations.

Fundamentally we have

Proposition 4.13. *The mean average payoff or utility of a population state increases along any trajectory of the evolution equation* ($\mathscr{A}$), *giving by the equation*

$$\dot{\bar{u}} = 2 \sum_{i=1}^{n} p_i(u_i(p) - \bar{u})^2. \qquad (56)$$

[10]A replicator is an entity with the means to make accurate copies of itself. It can be a gene, an organism, a belief, a technique, a convention, a cultural norm.

It is similar to a result by the biologist Fisher with respect to Natural Selection [8].

Proof. Recall

$$u_i(p)\sum_{j=1}^{n} a_{ij}p_j, \quad \bar{u} = \sum_{i=1}^{n} p_i u_i = pAp^T. \tag{57}$$

Therefore

$$\dot{\bar{u}} = 2\sum p_j a_{ij}\dot{p}_i = 2\sum_{i=1}^{n} p_i(u_i - \bar{u})^2. \tag{58}$$

□

We analyze the system on the unit 9-simplex subset of $\mathbf{R}^{10}$, and the state space of the vectors of probabilities. It is actually a cubic polynomial dynamic, whose class has been extensively investigated. We denoted $\dot{\Delta} = int(\Delta)$ the *interior* of the simplex given by the set

$$\dot{\Delta} := int(\Delta) = \{p = (p_1, \ldots, p_n)| \quad p_i > 0 \quad \forall i = 1, \ldots, n\}. \tag{59}$$

The *boundary faces* are denoted by $\partial\Delta$ and defined as

$$\partial\Delta(J) = \{p \in \Delta : p_i = 0, \quad \forall i \in J\}, \tag{60}$$

where J is any nontrivial subset of $\{1, \ldots, n\}$.

Proposition 4.14. *The hyperplanes* $\sum p_i = 1$ *and* $x_i = 0$ *are invariant*

Proof. Differentiating the constraint $\sum p_i = 1$ yields

$$\frac{d(\sum p_i)}{dt} = \sum \dot{p}_i = pap^T - (\sum p_i)pAp^T = 0. \tag{61}$$

Thus the invariance of the hyperplane. Likewise we obtain the invariance of the simplex faces. □

Our purpose is to describe the long term behavior of such a dynamical system.

Proposition 4.15. *The rest points of the evolution equation are the Nash Equilibria of the game.*

Proof. The rest points are the zeros of $g(p) = (g_i(p)) = 0$. That is, points $p \in \Delta$ such that $(Ap)_i = pAp^T$, for all $i \in supp(x)$. Thus an interior rest point is a solution of the system of linear equations

$$(Ap)_1 = (Ap)_2 = \ldots = (Ap)_n. \tag{62}$$

By definition a Nash Equilibrium p^* is an interior solution of such system. Thus an NE is an interior rest point. □

Rest points of the evolution equation ($\mathscr{A}$) or *replicator dynamics* in biological context and Nash equilibria are closely related as in the so-called *Folk Theorem of Evolutionary Game Theory* [12–14, 23, 24, 26].

Theorem 4.16. *1. If p^* is a Nash Equilibrium, then it is a rest point for ($\mathscr{A}$)*
2. If p^* is an evolutionarily stable NE, then it is asymptotically stable for ($\mathscr{A}$), globally if p^* is interior.
3. If the rest point $\hat{p}$ is stable, then it is a Nash Equilibrium
4. Every interior rest point $\hat{p} \in \dot{\Delta}$ is a Nash Equilibrium
5. If a rest point $\hat{p}$ for ($\mathscr{A}$) is also the forward limit point of an interior orbit $x(t)$ of ($\mathscr{A}$), i.e., $x(t) \in \dot{\Delta} = int(\Delta)$, then $\hat{p}$ is a Nash Equilibrium.

Remark 4.17. The converse is not true in the *Folk Theorem of Evolutionary Game Theory.*

For a boundary rest point $\hat{p}$ the difference $h_i(\hat{p}) = (A\hat{p})_i - \hat{p}A\hat{p}^T$ is also an eigenvalue $\lambda_{\hat{p}}$ for the Jacobian $J(\hat{p})$ with an eigenvector transversal to the face $\hat{p}_i = 0$. That entails

Lemma 4.18. *A rest point $\hat{p}$ for the evolution equation ($\mathscr{A}$) is a Nash Equilibrium if and only if its transversal eigenvalues are nonpositive.*

Consequently one derives the existence of Nash Equilibria for the evolution dynamics ($\mathscr{A}$). That is,

Theorem 4.19. *Every population game or subgame has at least one Nash Equilibrium*

Following [13] consider indeed the perturbation $\mathscr{A}_\varepsilon$ of the evolution equation ($\mathscr{A}$) following (Hofbauer)

$$\begin{aligned} \dot{p}_i(t) &= g_i(p) + \varepsilon(1 - np_i) \\ &= p_i((Ap)_i - pAp^T) + \varepsilon(1 - np_i) \end{aligned} \tag{63}$$

for small $\varepsilon > 0$. The perturbed equation $\mathscr{A}_\varepsilon$ again satisfies $\sum_i \dot{p}_i = 0$ on Δ with a flow on the boundary $\partial\Delta$ directed towards the interior $\dot{\Delta}$. Thus there is at least one rest point $\hat{p}(\varepsilon)$in the interior leading to

$$h_i(\hat{p}(\varepsilon)) - n\varepsilon = \frac{\varepsilon}{\hat{p}_i(\varepsilon)} < 0 \tag{64}$$

Therefore any accumulation point p^* of $\hat{p}(\varepsilon)$ as $\varepsilon \to 0$ is an NE. Consequently there is only odd number of regular Nash Equilibria, i.e., with nonsingular Jacobian (see also [12, 13, 23].

Remark 4.20 (Interpretation). For the evolution of America with respect to the ten core values, there is always at least one Nash Equilibrium, Pareto efficient or not. It is therefore important to study the conditions of its stability and efficiency. Moreover the American society been stratified according to communities preferences and behaviors there are several communities subgames reflected in cultural differences. These subgames are determined by the number n_c of core values receiving nonzero probabilities. For instance, a community might settle on the three core values of *Individualism, Freedom, and Mobility* which receive a nonzero probability while giving a zero probability to the remaining core values. The subgame played by such a community has at least a Nash Equilibrium. It is therefore interesting to analyze the relationship between the local community Nash Equilibrium NE_c and the National Nash Equilibrium NE_A.

We are therefore interesting in the overall dynamics of the national population game for the entire American society as well as in the dynamics of the community subgame as it relates to the overall evolution of America. This indeed amounts to uncover all possible dynamics on a 9-*simplex* embedded into the ten-dimensional Euclidean/Archimedean[11] metric space $\mathbf{R}^{10}$. Such a metric space has its own intrinsic limitations to handle efficiently the inherent complexity of the America social evolution. This prompts the need and search of a better and richer space, possibly a non-Archimedean space such as the p-adic spaces and their extension to Berkovich spaces.

Another important question is the following: These ten American core values are currently present in the America society ensuring the status of the country as a powerful and rich society. Are these values *permanent and resistant* to shocks of any nature and size, random or otherwise? It amounts to the notion of *permanence* for the *replicator dynamics* which we adapt to our context preferring the term *self-sustained.*

Definition 4.21. The evolution equation ($\mathscr{A}$) is *self-sustained* if $\exists$ a compact set $K \subset int(\Delta)$ such that $\forall x \in int(\Delta)$ $\exists$ a real time T such that $p(t) \in K \quad \forall t > T$.

In other words all the core values ultimately probabilistically settle in a compact set in the interior of the 9-simplex where all the probabilities are nonzero; they become in a sense the "laws of the land" or ingrained social norms further defining what is meant to be an American. Following [13, 23] we have

Theorem 4.22. *If the evolution equation ($\mathscr{A}$) is self-sustained, then it has a unique rest p^* in the interior $\dot{\Delta}$ which is therefore the unique national Nash Equilibrium NE_A.*

[11] Archimedes of Syracuse is an ancient Greek geometer and physicist. An algebraic structure is Archimedean when, given any two nonzero elements, neither is *infinitesimal* with respect to the other. That is, for x and y positive elements for a linearly ordered group, x is *infinitesimal* with respect to y if $\forall n \in \mathbf{N}, nx = x + \cdots + x < y$.

Moreover along each interior orbit $p(t)$ the time averages converge to p^*

$$\bar{a}_i := \frac{1}{T}\int_0^T p_i(t)dt \longrightarrow p_i^*, \tag{65}$$

as $T \to \infty$, and $i = 1, \ldots, n$.

The system is said to be *ergodic.* Consequently

Remark 4.23. If an interior orbit $p(t)$ has a boundary ω-limit point the time average need not converge. And every orbit converges to the boundary $\partial\Delta$ in the absence of an interior rest point.

Consequently we also have

Theorem 4.24. *Assume that $\exists$ an interior point p such that*

$$pAq^T > qAq^T, \quad \text{for all rest points } q \in int(\Delta).$$

As for the *self-sustainability* of the system, one is also interested in conditions for the existing NE to be an *evolutionarily stable state* ESS, resistant to deviations individual or groups as well as resistant to “mutation.” Recall an NE $p^* = (p_1^*, \ldots, p_n^*) \in \Delta$ is an ESS if

$$p^*Ap^{*T} > pAp^T \tag{66}$$

for all $p \neq p^*$ in some neighborhood of p^*, with the inequality holding for all $p \in \Delta$ if p^* is an interior point. Now define the function $V(p) = \prod_i p_i^{\alpha_i}$ as some power product. Such a function acts as a local strict Lyapunov function for the evolution equation for $\alpha_i = p_i^*$. That is,

$$V(p^*) = \prod_i p_i^{p_i^*} \tag{67}$$

satisfies

$$\dot{V}(p^*) > 0. \tag{68}$$

In particular an ESS is asymptotically stable for the evolution dynamic, globally stable when it is an interior point.

Now recall our evolution equation applied to individuals with the cognitive capacities to imitate and to learn, not with only inherited programmed behaviors as for animals [1, 4, 21]. Therefore one should expect the emergence of subpopulations in America with selective core values forming communities and playing a subgame with a local Nash equilibrium NE_c which could be *strategically stable* with a well-defined *basin of attraction* possibly bounded by a limit cycle we term *Nash Limit Cycle.* We next analyze the dynamics of these subgames.

4.3 Dynamics of Subgames

Immigration has always been a driving force for the American cultural dynamics generating a diversity of well-defined cultural identities, resulting sometimes in cultural enclaves in contrast to the cultural integration efforts, e.g., New York, Chinatown, or the many Little Italys.

4.3.1 Communities Subgames in Two Dimensions

We consider here communities in America playing a subgame consisting of selecting only two American core values with nonzero probabilities and assigning a zero probability to the remaining values. For instance, take a community which values the most *Individualism* and *Materialism*, in reaction to an original cultural background that emphasizes *uniformism* and discourages the *pursuit of material happiness*. We analyze both the internal dynamics of such a community and its contribution to the overall national social equilibrium.

We take therefore the number $2 = n_c < n = 10$, with the 2-component vectors of probability $p = (p_1, p_2) \in \Delta_1$, the 1-*simplex* $= [0, 1]$. Then one may set $p_2 = 1-p_1$. The resulting evolution subequation is rewritten with $q = p_1$ for simplicity

$$\dot{q} = q(1-q)((Aq)_1 - (Aq)_2), \quad \text{on } [0,1], \tag{69}$$

which can also be rewritten under the invariance modulo addition of constants to column entries

$$\dot{q} = q(1-q)(a+bq), \quad \text{on } [0,1], \tag{70}$$

The rest points are given by

1. $q = 0$, i.e., $p = (0, 1)$,
2. $q = 1$, i.e., $p = (1, 0)$,
3. p such that $(Ap)_1 = (Ap)_2$, or $a + bp = 0$.

The linear stability analysis leads to the following:

1. In addition to the fixed points $q = 0$ and $q = 1$ there is an interior equilibrium point p^*, $0 < p^* < 1$, if and only if $0 < -a < b$ or $0 < a < -b$.
2. If we have such an interior equilibrium p^*, it is asymptotically stable if and only if $b < 0$, i.e., $0 < a < -b$. Then both the other equilibria are unstable. This interior equilibrium, globally stable, is the unique NE; this could be interpreted as the stable co-existence of the two core values.
3. On the other hand, if there exists an interior equilibrium that is unstable the two values given by $x = 0$, and $x = 1$ are also NE and both attracting. That is a *bistability* case of the two values.

4. In the absence of an interior equilibrium one of the probabilities converges to 0, and the corresponding core value is *dominated* by the other.

Therefore we note that the NE $p^* = (p_1^*, p_2^*)$ is either unstable or an evolutionary focus where all trajectories $p(t)$ close around the equilibrium, and the time average of the probability $p(t)$ around an orbit is the equilibrium position, making the system into an *ergodic* system.

4.3.2 Communities Subgames in Three Dimensions

Here we look at a three-dimensional subgame corresponding to a community in America selecting only three core values with a nonzero probability. One particular interesting scenario is when such a community is playing a game similar to the 3-D *rock-scissors-paper* (RSP) ordering the selected core values in a dominated structure [13]. For instance, the community takes a first core value dominated by a second one in the absence of the third one, and likewise second one dominated by the third one, which is, in turn, dominated by the first one [13]. The payoff matrix could be normalized to

$$A = \begin{pmatrix} 0 & -a_2 & a_3 \\ a_4 & 0 & a_5 \\ -a_6 & a_7 & 0 \end{pmatrix} \tag{71}$$

yielding

1. There is a unique NE and it is located in the interior. It has been proved in [26, 27] that for $det(A) > 0$ or equivalently $p^*Ap^{*T} > 0$, p^* is globally asymptotically stable.
2. For $det(A) = 0$ all orbits are interior closed centered at p^*. That is the classic equilibrium of *center type.*
3. For $det(A) < 0$ all interior orbits converge to the boundary $\partial\Delta$. That is, the ω-limit set of an interior point $p \in int(\Delta)$ is the *heteroclinic* cycle of the three corners *saddle* points e_i.

Remark 4.25. Zeeman has classified all possible phase portraits of the $n = 3$ evolution dynamics to 33 generic phase portraits (up to flow reversal), proving as well the absence of isolated periodic orbits, i.e., no *limit cycles* [26–28]. One could interpret this as follows: in the subpopulation selecting three core values, there is no self-sustained oscillations in the community evolution.

4.3.3 Communities Subgames in Higher Dimensions

Consider communities in America defined by at least $n = 4$ core values with nonzero probability. The probability constraint leads to a three-dimensional system of evolution equations. Some particular examples display periodic and chaotic attractors. One such case corresponds to the well-known three-dimensional *Lotka–Volterra* equations [5, 11, 13]

$$\dot{x} = x_i(r_i + \sum_j c_{ij}x_j), \quad i = 1, 2, 3. \tag{72}$$

There exists the possibility of a continua of periodic orbits evolving into many coexisting limit cycles. Indeed Hofbauer proved in [13] that the n-dimensional *Lotka–Volterra* equation admits stable limit cycles if and only if $n \geq 3$.

4.4 Game-Theoretic Evolution Scenarios for America

America evolution dynamics is determined by the time evolution of the system of nine differential equations we studied on a 9-simplex. The game has a national Nash Equilibrium akin to a national *modus vivendi* and *modus operandi* informed by the ten American core values to which all immigrants have to subscribe. The immigration flow has generated a diversity of cultural identities with cross-cultural interactions converging to what an American is. The game-theoretical model ensures the emergence of a Nash equilibrium, in the interior of the simplex. Conditions could be realized for a globally asymptotically stable equilibrium, evolutionarily stable, possibly with every American assigning a nonzero probability to each of the ten core values.

The existence of multiple communities stratifies the American society. They are delineated by the number of core values they select with a nonzero probability. This leads to the occurrence of subgames within the national game, with the emergence of local Nash Equilibria. The co-existence of these local Nash Equilibria NE_c along the national Nash Equilibrium NE_A is by now a fact of the American social dynamics. The dynamical relationships between these equilibria create conditions of stability/instability, progress/stagnation, or worse radicalism and unrest. Various social scenarios are possible based on the mathematical game-theoretical models.

4.4.1 Absence of Local Nash Equilibria

In principle America does not seek to promote *multiculturalism* understood as the co-existence of multiple cultural identities along the core culture that has emerged from the early immigrants that settle America as a distinct cultural entity. Cultural integration has been a continuing challenging process even though cultural diversity has been perceived by some as a basic source of prosperity. Following a complete cultural integration, there exists a unique national Nash Equilibrium NE_A, evolutionarily stable, i.e., a globally asymptotically stable equilibrium point of the evolution dynamics in the interior on the 9-simplex. That is, the corresponding

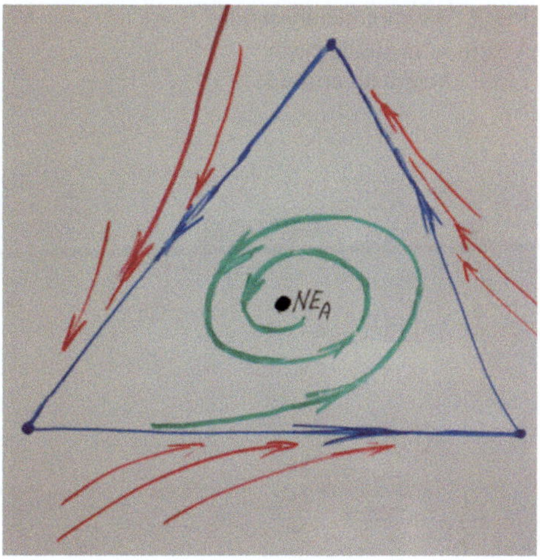

Fig. 2 NE_A asymptotically stable with inside-repulsive and outside-attractive heteroclinic boundary

population state features nonzero probabilities. Social stability is thus ensured and the core values remain ingrained in the social fabric for the greater good of the American society. This process generates the national *Nash Limit Cycle*, semi-stable and connecting the boundary pure population state equilibria described by the nine corners of the 9-simplex. It accounts for the sustained oscillations in decision-making. Such a scenario is represented by Fig. 2.

The national equilibrium NE_A is asymptotically stable (ESS) and is the ω-limit point of all trajectories inside the simplex. Of course one may assume circumstances in which the NE_A is unstable becoming the α-limit point of trajectories converging to the boundary. This amounts to culturally uncertain times, with the potential of generating attractive local community equilibria NE_c.

4.4.2 Co-existence of Multiple Local Nash Equilibria

In reality a complete cultural integration is an extremely long process, in particular with a continuous flow of immigrants from diverse cultural identities. Therefore communities continue to exist within America subscribing to the ten core values with varied probabilities oftentimes assigning zero probability to some of these values, not necessarily due to the so-called *bounded rationality* but due to the distribution of information and the interpretation of these values.

Consequently there exist multiple subgames with their own Nash Equilibria and Eventual Nash Equilibria, possibly locally evolutionarily stable, locally asymptotically stable equilibrium points denoted (AS), with a basin of attraction in the interior of the 9-simplex. The local equilibria could be also unstable, or surrounded by limit cycles stable or unstable we call *Nash Limit Cycles* which also account for sustained

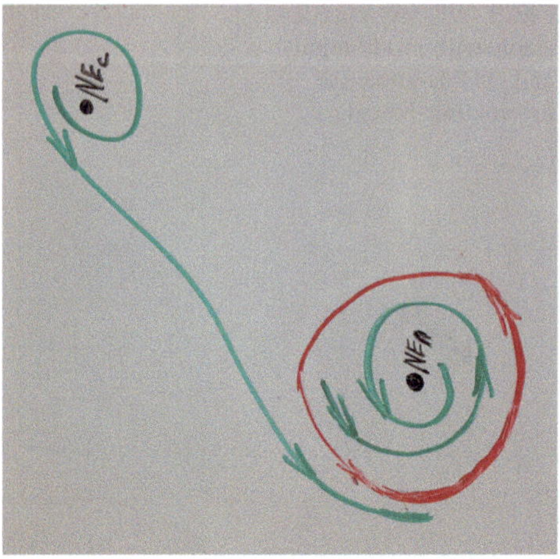

Fig. 3 Asymptotically stable NE_A in semi-stable Nash Limit Cycle with unstable NE_c

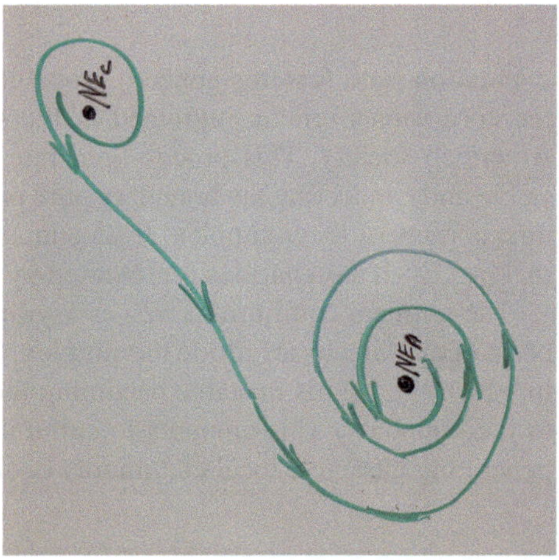

Fig. 4 Unstable NE_c with asymptotically stable NE_A : conducive to cultural integration

oscillations in decision-making with respect to the core values. This scenario is represented in various forms using the notations: NLC for Nash Limit Cycle, NE_A for national Nash Equilibrium, NE_c for local community Nash Equilibrium (Figs. 3 and 4).

Remark 4.26. One should envision the possibility of a strong local Nash Equilibrium, asymptotically stable with a forward extending basin of attraction, possibly stemming from a distorted interpretation of the American core values, or purposely selecting only a small number of these values with nonzero probability. Such a game scenario, as in Fig. 5, could breed indeed *radicalism* understood as a negative or steadily decreasing contribution to the national Nash Equilibrium. It could pose a serious threat to the society (Fig. 6).

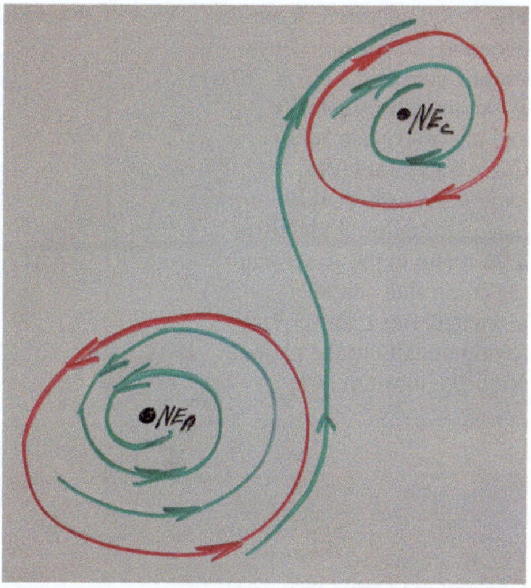

Fig. 5 Unstable Nash Limit Cycle around asymptotically stable NE_A with semi-stable Nash Limit Cycle around asymptotically stable NE_c : challenging scenario for cultural integration with potential for social instability

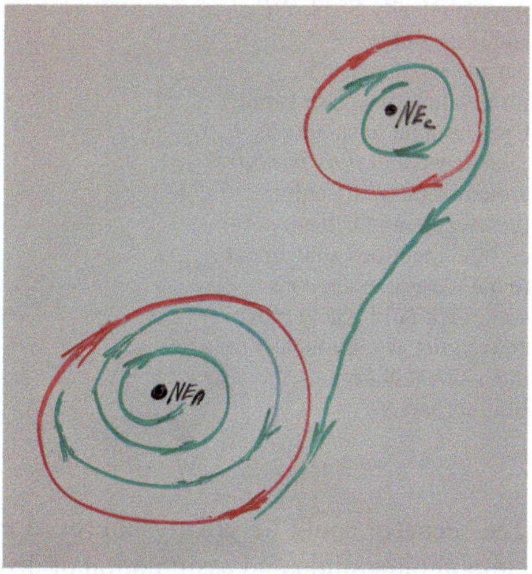

Fig. 6 Unstable Nash Limit Cycle around asymptotically stable NE_c with stable Nash Limit Cycle around unstable NE_A : conducive to cultural integration with possibly transient period of sustained oscillations in decision-making with respect to the ten core values

Remark 4.27 (Conjecture). The interesting question is to investigate conditions for a stable, prosperous, and forward looking American society. We conjecture the following game scenario: the co-existence of an asymptotically semi-stable (ESS) national Nash Equilibrium NE_A surrounding by a stable Nash Limit Cycle (population state with sustained oscillations in decision-making towards equilibrium, repulsive on the inside towards the NE_A but attractive on the outside) and multiple communities Nash Equilibria NE_c unstable (individuals in process of adjusting preferences and behaviors) surrounded by a semi-unstable Nash Limit Cycle (directing neighboring trajectories to converge to the national Nash Limit Cycle while attractive on the inside). See Fig. 7.

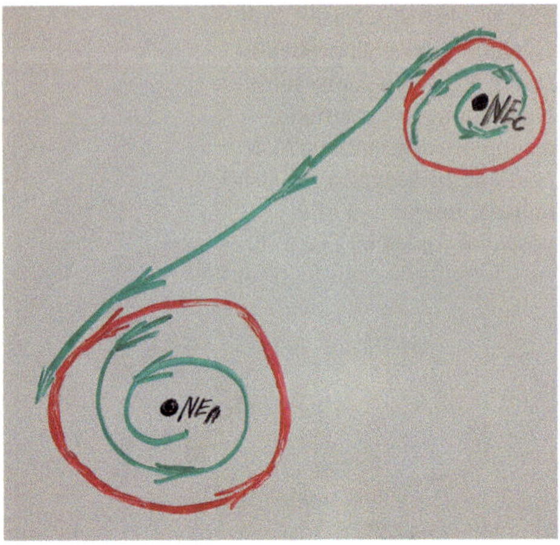

Fig. 7 Conjectured Game Scenario: semi-stable Nash Limit Cycle on asymptotically stable NE_A and unstable Nash Limit Cycle on unstable NE_c : *Double Transient oscillatory period* for cultural identities adjustment to the American ten core values with ultimately most converging asymptotically to the national Nash Equilibrium NE_A

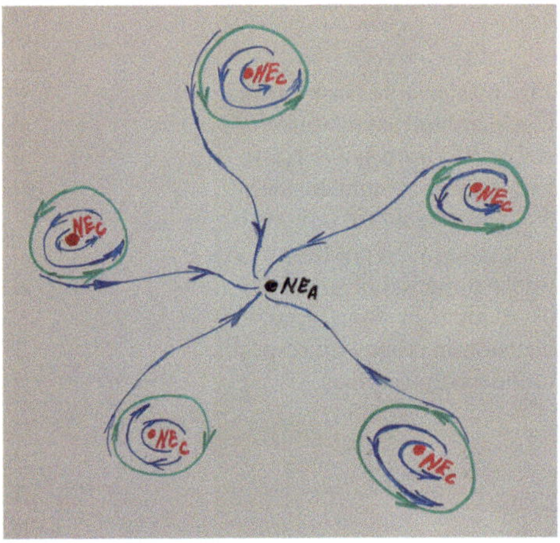

Fig. 8 Improved Conjectured Game Scenario: Asymptotically Stable NE_A and semi-stable outside-repulsive and inside-attractive Nash Limit Cycle around unstable NE_c : *Transient oscillatory period* around community cultural preferences and behaviors for cultural identities adjustment to the American ten core values with ultimately most converging asymptotically to the national Nash Equilibrium NE_A

The scenario could be greatly improved into the more realistic and conducive for cultural integration and assimilation of diverse cultural identities into the American social fabric. See Fig. 8.

We also note the existence of what we call *Nash Isochrons* to describe these groups of individuals associated with trajectories evolving with the same phase as the Nash Limit Cycles.

The challenge is twofold: first a mathematical framework in terms of designing a game with these desirable dynamical features. Second a social and cultural design through possibly social laws and incentives to practically realize such a scenario.

5 Concluding Remarks

Game theory, a modeling tool, could also be seen as a design tool, in particular in the context of our study. Policies could help design admissible utility or payoffs in the continuous interactions of the individuals so that the resulting game or subgame possesses the desirable properties to enforce and sustain the evolution of the ten core values. Adaptation rules could also be designed to lead to the desirable nation-wide behavior. The Nash Limit Cycles around the local Nash Equilibria should be all unstable surrounding asymptotically stable local equilibria. Trajectories outside the narrow basin of attraction converge to the national Nash Equilibrium. Individuals in a given community hold a given set of preferences and behaviors but subscribe positively to the national set of core values. Of course these preferences mostly driven by behavioral compromises produce inter and intra-personal stresses. Most individuals maintain two types of equilibrium behavior, one for their communities where there are expected to *fit in*, and the other for the greater society in which they do not want to experience their original cultural lifestyles being negatively stigmatized. It is expected that such a process of socialization could be distorted whenever publicly expressed preferences with respect to the ten American core values come at odd with privately held preferences from one's cultural background. However through these cross-cultural interactions, an American culture emerges hybrid in its components. (See also [1, 4, 9, 21, 22].)

We have proposed a dynamic theoretical framework based on individual choice to understand and predict the evolution of America. We intend to uncover the full spectrum of dynamical possibilities in a 9-simplex, in particular for dimensions $4 \leq n \leq 10$. The results could be extended to other societies holding less than ten core values. Future work will also involve a type of *revision protocol* for the core values. Though our evolution equation appears *replicator*-like dynamics in biological context, one could also design an evolution game with different dynamics such as *best reply*-like dynamics, *the logit*-like dynamics, or *Brown–Nash–von Neumann*-like dynamics. See more details in [8, 12–14, 23, 24, 26]. Note that in our model we recognize neither replication nor mutation nor "switching" of values. The dynamical possibilities are driven only by the variation of the probabilities each individual assigns to the core values in their random interactions with others.

Now in order to account rigorously for all the "nuances" in decisions and selections of the core values, such as the distinction between negligible and zero probability one should consider *p-adic probability.* Individuals, recent immigrants or otherwise, do not actually assign zero probability, rather negligible probability. Moreover the selection of these ten core values presumes some *hierarchy structure* among these values, best described in the *non-Archimedean/p-adic* context.

References

1. Axelrod, R.: The Evolution of Cooperation. Basic Books, New York (1984)
2. Balkenborg, D., Schlag, K.: Evolutionarily stable sets. Int. J. Game Theory **29**, 571–595 (2000)
3. Benaim, M., Hirsch, M.W.: Mixed equilibria and dynamical systems arising from repeated games. Games Econ. Behav. **29**, 36–72 (1999)
4. Bisin, A., Verdier, T.: The economics of cultural transmission and the dynamics of preferences. J. Econ. Theory **97**, 298–319 (2001)
5. Bomze, I.M.: Lotka-Volterra equation and replicator dynamics: a two-dimensional classification. Biol. Cybern. **48**, 201–211 (1983)
6. Farell, J., Ware, R.: Evolutionary stability in the repeated prisoner's dilemma. Theor. Popul. Biol. **36**, 161–166 (1989)
7. Fisher, R.A.: On the dominance ratio. Proc. R. Soc. Edinb. **42**, 321–341 (1922)
8. Fudenberg, D., Tirole, J.: Game Theory. MIT Press, Cambridge (1991)
9. Gilboa, I., Matsui, A.: Social stability and equilibrium. Econometrica **59**, 859–867 (1991)
10. Harper, D.G.: Competitive foraging in mallards: "Ideal free" ducks. Anim. Behav. **30**(2), 575–584 (1982)
11. Hofbauer, J.: On the occurrence of limit cycles in the Lotka-Volterra differential equation. J. Nonlinear Anal. **5**, 1003–1007 (1981)
12. Hofbauer, J., Sigmund, K.: Evolutionary Games and Replicator Dynamics. Cambridge University Press, Cambridge (1998)
13. Hofbauer, J., Sigmund, K.: Evolutionary game dynamics. Bull. Am. Math. Soc. **40**, 479–519 (2003)
14. Hofbauer, J., Schuster, P., Sigmund, K.: A note on evolutionary stable strategies and game dynamics. J. Theor. Biol. **81**, 609–612 (1979)
15. IBR: American Ten Core Values. https://www.andrews.edu/~tidwell/bsad560/USValues.html
16. Krabs, W., Pickl, S.: Dynamical Systems, Stability, Controllability and Chaotic Behavior. Springer, Berlin (2010)
17. Kuznetsov, Y.: Elements of Applied Bifurcation Theory. Springer, Berlin (1998)
18. Maynard Smith, J.: Evolution and the Theory of Games. Cambridge University Press, Cambridge (1982)
19. Nash, J.: Equilibrium points in n-person games. Proc. Natl. Acad. Sci. U. S. A. **36**, 48–49 (1950)
20. Nash, J.: Non-cooperative games. Ann. Math. **54**, 287–295 (1951)
21. Nowak, M.A.: Evolutionary Dynamics. Harvard University Press, Cambridge (2006)
22. Sandholm, W.H.: Preference evolution, two-speed dynamics, and rapid social change. Rev. Econ. Dyn. **4**, 637–679 (2001)
23. Schuster, P., Sigmund, K.: Replicator dynamics. J. Theor. Biol. **100**, 533–538 (1983)
24. Taylor, P.D., Jonker, L.B.: Evolutionary stable strategies and game dynamics. Math. Biosci. **40**, 145–156 (1978)
25. von Neumann, J., Morgenstern, O.: Theory of Games and Economic Behaviour. Princeton University Press, Princeton (1947)
26. Weibull, J.: Evolutionary Game Theory. MIT Press, Cambridge (1995)
27. Zeeman, E.: Population dynamics from game theory. In: Global Theory of Dynamical Systems. Lecture Notes in Mathematics, vol. 819. Springer, New York (1980)
28. Zeeman, E.: Dynamics of the evolution of animal conflicts. J. Theor. Biol. **89**, 248–270 (1981)

Index

B. Toni (ed.), *New Trends and Advanced Methods in Interdisciplinary Mathematical Sciences*, STEAM-H: Science, Technology, Engineering, Agriculture, Mathematics & Health, DOI 10.1007/978-3-319-55612-3

MIX
Papier aus verantwortungsvollen Quellen
Paper from responsible sources
FSC® C105338

FSC
www.fsc.org

If you have any concerns about our products,
you can contact us on
ProductSafety@springernature.com

In case Publisher is established outside the EU,
the EU authorized representative is:
Springer Nature Customer Service Center GmbH
Europaplatz 3, 69115 Heidelberg, Germany

Printed by Libri Plureos GmbH
in Hamburg, Germany